Perturbation of the Boundary in Boundary-Value Problems of Partial Differential Equations

LONDON MATHEMATICAL SOCIETY LECTURE NOTE SERIES

Managing Editor: Professor N. J. Hitchin, Mathematical Institute,
University of Oxford, 24-29 St Giles, Oxford OX1 3LB, United Kingdom

The titles below are available from booksellers, or from Cambridge University Press at www.cambridge.org

100 Stopping time techniques for analysts and probabilists, L. EGGHE
105 A local spectral theory for closed operators, I. ERDELYI & WANG SHENGWANG
107 Compactification of Siegel moduli schemes, C.-L. CHAI
109 Diophantine analysis, J. LOXTON & A. VAN DER POORTEN (eds)
113 Lectures on the asymptotic theory of ideals, D. REES
116 Representations of algebras, P.J. WEBB (ed)
119 Triangulated categories in the representation theory of finite-dimensional algebras, D. HAPPEL
121 Proceedings of *Groups - St Andrews 1985*, E. ROBERTSON & C. CAMPBELL (eds)
128 Descriptive set theory and the structure of sets of uniqueness, A.S. KECHRIS & A. LOUVEAU
130 Model theory and modules, M. PREST
131 Algebraic, extremal & metric combinatorics, M.-M. DEZA, P. FRANKL & I.G. ROSENBERG (eds)
141 Surveys in combinatorics 1989, J. SIEMONS (ed)
144 Introduction to uniform spaces, I.M. JAMES
146 Cohen-Macaulay modules over Cohen-Macaulay rings, Y. YOSHINO
148 Helices and vector bundles, A.N. RUDAKOV *et al*
149 Solitons, nonlinear evolution equations and inverse scattering, M. ABLOWITZ & P. CLARKSON
150 Geometry of low-dimensional manifolds 1, S. DONALDSON & C.B. THOMAS (eds)
151 Geometry of low-dimensional manifolds 2, S. DONALDSON & C.B. THOMAS (eds)
152 Oligomorphic permutation groups, P. CAMERON
153 L-functions and arithmetic, J. COATES & M.J. TAYLOR (eds)
155 Classification theories of polarized varieties, TAKAO FUJITA
158 Geometry of Banach spaces, P.F.X. MÜLLER & W. SCHACHERMAYER (eds)
159 Groups St Andrews 1989 volume 1, C.M. CAMPBELL & E.F. ROBERTSON (eds)
160 Groups St Andrews 1989 volume 2, C.M. CAMPBELL & E.F. ROBERTSON (eds)
161 Lectures on block theory, BURKHARD KÜLSHAMMER
163 Topics in varieties of group representations, S.M. VOVSI
164 Quasi-symmetric designs, M.S. SHRIKANDE & S.S. SANE
166 Surveys in combinatorics, 1991, A.D. KEEDWELL (ed)
168 Representations of algebras, H. TACHIKAWA & S. BRENNER (eds)
169 Boolean function complexity, M.S. PATERSON (ed)
170 Manifolds with singularities and the Adams-Novikov spectral sequence, B. BOTVINNIK
171 Squares, A.R. RAJWADE
172 Algebraic varieties, GEORGE R. KEMPF
173 Discrete groups and geometry, W.J. HARVEY & C. MACLACHLAN (eds)
174 Lectures on mechanics, J.E. MARSDEN
175 Adams memorial symposium on algebraic topology 1, N. RAY & G. WALKER (eds)
176 Adams memorial symposium on algebraic topology 2, N. RAY & G. WALKER (eds)
177 Applications of categories in computer science, M. FOURMAN, P. JOHNSTONE & A. PITTS (eds)
178 Lower K- and L-theory, A. RANICKI
179 Complex projective geometry, G. ELLINGSRUD *et al*
180 Lectures on ergodic theory and Pesin theory on compact manifolds, M. POLLICOTT
181 Geometric group theory I, G.A. NIBLO & M.A. ROLLER (eds)
182 Geometric group theory II, G.A. NIBLO & M.A. ROLLER (eds)
183 Shintani zeta functions, A. YUKIE
184 Arithmetical functions, W. SCHWARZ & J. SPILKER
185 Representations of solvable groups, O. MANZ & T.R. WOLF
186 Complexity: knots, colourings and counting, D.J.A. WELSH
187 Surveys in combinatorics, 1993, K. WALKER (ed)
188 Local analysis for the odd order theorem, H. BENDER & G. GLAUBERMAN
189 Locally presentable and accessible categories, J. ADAMEK & J. ROSICKY
190 Polynomial invariants of finite groups, D.J. BENSON
191 Finite geometry and combinatorics, F. DE CLERCK *et al*
192 Symplectic geometry, D. SALAMON (ed)
194 Independent random variables and rearrangement invariant spaces, M. BRAVERMAN
195 Arithmetic of blowup algebras, WOLMER VASCONCELOS
196 Microlocal analysis for differential operators, A. GRIGIS & J. SJÖSTRAND
197 Two-dimensional homotopy and combinatorial group theory, C. HOG-ANGELONI *et al*
198 The algebraic characterization of geometric 4-manifolds, J.A. HILLMAN
199 Invariant potential theory in the unit ball of $\mathbf{C}^n$, MANFRED STOLL
200 The Grothendieck theory of dessins d'enfant, L. SCHNEPS (ed)
201 Singularities, JEAN-PAUL BRASSELET (ed)
202 The technique of pseudodifferential operators, H.O. CORDES
203 Hochschild cohomology of von Neumann algebras, A. SINCLAIR & R. SMITH
204 Combinatorial and geometric group theory, A.J. DUNCAN, N.D. GILBERT & J. HOWIE (eds)
205 Ergodic theory and its connections with harmonic analysis, K. PETERSEN & I. SALAMA (eds)
207 Groups of Lie type and their geometries, W.M. KANTOR & L. DI MARTINO (eds)
208 Vector bundles in algebraic geometry, N.J. HITCHIN, P. NEWSTEAD & W.M. OXBURY (eds)
209 Arithmetic of diagonal hypersurfaces over finite fields, F.Q. GOUVÉA & N. YUI
210 Hilbert C^*-modules, E.C. LANCE
211 Groups 93 Galway / St Andrews I, C.M. CAMPBELL *et al* (eds)
212 Groups 93 Galway / St Andrews II, C.M. CAMPBELL *et al* (eds)
214 Generalised Euler-Jacobi inversion formula and asymptotics beyond all orders, V. KOWALENKO *et al*
215 Number theory 1992–93, S. DAVID (ed)
216 Stochastic partial differential equations, A. ETHERIDGE (ed)
217 Quadratic forms with applications to algebraic geometry and topology, A. PFISTER
218 Surveys in combinatorics, 1995, PETER ROWLINSON (ed)
220 Algebraic set theory, A. JOYAL & I. MOERDIJK
221 Harmonic approximation, S.J. GARDINER

222 Advances in linear logic, J.-Y. GIRARD, Y. LAFONT & L. REGNIER (eds)
223 Analytic semigroups and semilinear initial boundary value problems, KAZUAKI TAIRA
224 Computability, enumerability, unsolvability, S.B. COOPER, T.A. SLAMAN & S.S. WAINER (eds)
225 A mathematical introduction to string theory, S. ALBEVERIO, *et al*
226 Novikov conjectures, index theorems and rigidity I, S. FERRY, A. RANICKI & J. ROSENBERG (eds)
227 Novikov conjectures, index theorems and rigidity II, S. FERRY, A. RANICKI & J. ROSENBERG (eds)
228 Ergodic theory of $\mathbf{Z}^d$ actions, M. POLLICOTT & K. SCHMIDT (eds)
229 Ergodicity for infinite dimensional systems, G. DA PRATO & J. ZABCZYK
230 Prolegomena to a middlebrow arithmetic of curves of genus 2, J.W.S. CASSELS & E.V. FLYNN
231 Semigroup theory and its applications, K.H. HOFMANN & M.W. MISLOVE (eds)
232 The descriptive set theory of Polish group actions, H. BECKER & A.S. KECHRIS
233 Finite fields and applications, S.COHEN & H. NIEDERREITER (eds)
234 Introduction to subfactors, V. JONES & V.S. SUNDER
235 Number theory 1993–94, S. DAVID (ed)
236 The James forest, H. FETTER & B. GAMBOA DE BUEN
237 Sieve methods, exponential sums, and their applications in number theory, G.R.H. GREAVES *et al*
238 Representation theory and algebraic geometry, A. MARTSINKOVSKY & G. TODOROV (eds)
240 Stable groups, FRANK O. WAGNER
241 Surveys in combinatorics, 1997, R.A. BAILEY (ed)
242 Geometric Galois actions I, L. SCHNEPS & P. LOCHAK (eds)
243 Geometric Galois actions II, L. SCHNEPS & P. LOCHAK (eds)
244 Model theory of groups and automorphism groups, D. EVANS (ed)
245 Geometry, combinatorial designs and related structures, J.W.P. HIRSCHFELD *et al*
246 p-Automorphisms of finite p-groups, E.I. KHUKHRO
247 Analytic number theory, Y. MOTOHASHI (ed)
248 Tame topology and o-minimal structures, LOU VAN DEN DRIES
249 The atlas of finite groups: ten years on, ROBERT CURTIS & ROBERT WILSON (eds)
250 Characters and blocks of finite groups, G. NAVARRO
251 Gröbner bases and applications, B. BUCHBERGER & F. WINKLER (eds)
252 Geometry and cohomology in group theory, P. KROPHOLLER, G. NIBLO & R. STÖHR (eds)
253 The q-Schur algebra, S. DONKIN
254 Galois representations in arithmetic algebraic geometry, A.J. SCHOLL & R.L. TAYLOR (eds)
255 Symmetries and integrability of difference equations, P.A. CLARKSON & F.W. NIJHOFF (eds)
256 Aspects of Galois theory, HELMUT VÖLKLEIN *et al*
257 An introduction to noncommutative differential geometry and its physical applications 2ed, J. MADORE
258 Sets and proofs, S.B. COOPER & J. TRUSS (eds)
259 Models and computability, S.B. COOPER & J. TRUSS (eds)
260 Groups St Andrews 1997 in Bath, I, C.M. CAMPBELL *et al*
261 Groups St Andrews 1997 in Bath, II, C.M. CAMPBELL *et al*
262 Analysis and logic, C.W. HENSON, J. IOVINO, A.S. KECHRIS & E. ODELL
263 Singularity theory, BILL BRUCE & DAVID MOND (eds)
264 New trends in algebraic geometry, K. HULEK, F. CATANESE, C. PETERS & M. REID (eds)
265 Elliptic curves in cryptography, I. BLAKE, G. SEROUSSI & N. SMART
267 Surveys in combinatorics, 1999, J.D. LAMB & D.A. PREECE (eds)
268 Spectral asymptotics in the semi-classical limit, M. DIMASSI & J. SJÖSTRAND
269 Ergodic theory and topological dynamics, M.B. BEKKA & M. MAYER
270 Analysis on Lie groups, N.T. VAROPOULOS & S. MUSTAPHA
271 Singular perturbations of differential operators, S. ALBEVERIO & P. KURASOV
272 Character theory for the odd order theorem, T. PETERFALVI
273 Spectral theory and geometry, E.B. DAVIES & Y. SAFAROV (eds)
274 The Mandlebrot set, theme and variations, TAN LEI (ed)
275 Descriptive set theory and dynamical systems, M. FOREMAN *et al*
276 Singularities of plane curves, E. CASAS-ALVERO
277 Computational and geometric aspects of modern algebra, M.D. ATKINSON *et al*
278 Global attractors in abstract parabolic problems, J.W. CHOLEWA & T. DLOTKO
279 Topics in symbolic dynamics and applications, F. BLANCHARD, A. MAASS & A. NOGUEIRA (eds)
280 Characters and automorphism groups of compact Riemann surfaces, THOMAS BREUER
281 Explicit birational geometry of 3-folds, ALESSIO CORTI & MILES REID (eds)
282 Auslander-Buchweitz approximations of equivariant modules, M. HASHIMOTO
283 Nonlinear elasticity, Y. FU & R.W. OGDEN (eds)
284 Foundations of computational mathematics, R. DEVORE, A. ISERLES & E. SÜLI (eds)
285 Rational points on curves over finite, fields, H. NIEDERREITER & C. XING
286 Clifford algebras and spinors 2ed, P. LOUNESTO
287 Topics on Riemann surfaces and Fuchsian groups, E. BUJALANCE *et al*
288 Surveys in combinatorics, 2001, J. HIRSCHFELD (ed)
289 Aspects of Sobolev-type inequalities, L. SALOFF-COSTE
290 Quantum groups and Lie theory, A. PRESSLEY (ed)
291 Tits buildings and the model theory of groups, K. TENT (ed)
292 A quantum groups primer, S. MAJID
293 Second order partial differential equations in Hilbert spaces, G. DA PRATO & J. ZABCZYK
294 Introduction to the theory of operator spaces, G. PISIER
295 Geometry and Integrability, LIONEL MASON & YAVUZ NUTKU (eds.)
296 Lectures on invariant theory, IGOR DOLGACHEV
297 The homotopy category of simply connected 4-manifolds, H.-J. BAUES
299 Kleinian Groups and Hyperbolic 3-Manifolds Y. KOMORI, V. MARKOVIC, C. SERIES (eds.)
300 Introduction to Möbius Differential Geometry, UDO HERTRICH-JEROMIN
301 Stable Modules and the D(2)-Problem, F.E.A. JOHNSON
302 Discrete and Continuous Nonlinear Schrödinger Systems, M. J. ABLORWITZ, B. PRINARI, A. D. TRUBATCH
303 Number Theory and Algebraic Geometry, MILES REID ALEXEI SKOROBOGATOV (eds.)
304 Groups St Andrews 2001 in Oxford Vol. 1, COLIN CAMPBELL, EDMUND ROBERTSON, GEOFF SMITH (eds.)
305 Groups St Andrews 2001 in Oxford Vol. 2, C. M. CAMPBELL, E.F. ROBERTSON, G.C. SMITH (eds.)
307 Surveys in Combinatorics 2003, C. D. WENSLEY (ed.)
309 Corings and Comodules, TOMASZ BRZEZINSKI & ROBERT WISBAURER
310 Topics in Dynamics and Ergodic Theory, SERGEY BEZUGLYI & SERGIY KOLYADA (eds.)
312 Foundations of Computational Mathematics, Minneapolis 2002, FELIPE CUCKER *et al* (eds.)

London Mathematical Society Lecture Note Series. 318

Perturbation of the Boundary in Boundary-Value Problems of Partial Differential Equations

Dan Henry

with editorial assistance from

Jack Hale

and

Antônio Luiz Pereira

CAMBRIDGE
UNIVERSITY PRESS

CAMBRIDGE UNIVERSITY PRESS
Cambridge, New York, Melbourne, Madrid, Cape Town,
Singapore, São Paulo, Delhi, Mexico City

Cambridge University Press
The Edinburgh Building, Cambridge CB2 8RU, UK

Published in the United States of America by Cambridge University Press, New York

www.cambridge.org
Information on this title: www.cambridge.org/9780521574914

First published 2005

A catalogue record for this publication is available from the British Library

Library of Congress Cataloguing in Publication Data

Henry, D.
Perturbation of the boundary – value problems of partial differential equations / D. Henry.
p. cm. – (London Mathematical Society lecture note series ; 318)
Includes bibliographical references and index.
ISBN 0-521-57491-9 (pbk. : alk. paper)
1. Boundary value problems. 2. Perturbation (Mathematics) 3. Differential equations, Partial. I. Title. II. Series.
QA379.H46 2005
515′.353 – dc22 2004054767

ISBN 978-0-521-57491-4 Paperback

Contents

Introduction

Perturbation of the boundary (or of the domain of definition of a boundary value problem) is a rather neglected mathematical topic, though it has attracted occasional interest: Rayleigh [33] in 1877 (the first edition), Hadamard [9] in 1908, Courant and Hilbert [5] in the German edition of 1937, Polya and Szëgo [31] in 1951, Garabedian and Schiffer [7] in 1952, and some more recent work [3, 14, 6, 13, 26, 21–23, 42, 35, 32, 27, 10, 11]. The list is far from complete, but is notably sparse.

There seem to be two related reasons for this neglect: (1) the subject is too easy; (2) it is too difficult. If you are interested only in Fundamental Questions, this is certainly a trivial topic. One perturbs the region by applying a diffeomorphism near the identity; but you can change variables via this diffeomorphism to keep the region fixed, and are then only perturbing the coefficients in a fixed region. It is simply the chain rule. However, if you try to carry out this trivial change of variables, you will become mired in such long and difficult calculations that you'll be tempted to quit. If you persist, and are fortunate, and are extremely careful, there may be a miraculous simplification at the end. On experiencing this miracle for the second time I became suspicious – theorems 2.2, 2.4 show how to go directly to the "miracle," bypassing the computational morass. (Peetre [27] also found part of this result, and it is implicit in Courant and Hilbert [5: vol. 1 p. 260] for variational problems.) It is, at the end, merely the chain rule, and we may then apply standard tools (implicit function theorem, Liapunov, Schmidt method, transversality theorem) to problems of perturbation of the boundary. But the standard tools are not enough – new problems arise requiring, for example, a more general form of the transversality theorem for problems with Fredholm index $-\infty$ (ch. 5). Open problems abound. The calculus developed in chapter 2 applies – in principle – to almost any boundary or initial boundary-value problem, but often leads immediately to difficult unsolved problems. To avoid excessive depression of author and reader, we will concentrate on questions we can answer – generally boundary-value problems for scalar second-order elliptic equations. The subject is not, after all, entirely trivial.

I have worked on perturbation of the boundary sporadically since about 1973, after reading Joseph's article [13]. The formulas of theorems 2.2, 2.4 date from 1975 – in a more complicated version – and most of the examples of chapters 3 and 4 (and a few from chapter 6) were developed in 1975–1981 and exposed in seminars at the University of Kentucky, Brown University and the University of São Paulo. In 1982, I had the opportunity to develop this topic at some length [11] at the University of Brasilia. Later that year I generalized the transversality theorem (lectures at São Carlos, September 1982), which solved some problems left open in [11] and raised a host of new open problems. These demanded some tedious calculation – such as those in chapter 7 – which were only completed recently, and then only in special cases: much remains to be done. If the word "I" seems to appear excessively here, it is because very few other people have worked on these problems in the past twelve years with a comparable approach, and my work has been independent of these few (aside from some of the examples cited below). There are, of course, other notions of a "small change in the domain" besides "image under diffeomorphism near the identity" (see, for example, [3, 26, 10]); the advantage of using such regular perturbations will hopefully become more clear as we proceed.

In the past five years, my work has been supported by FAPESP, and of course by IME-USP. It would have been nice to have this ready for the fiftieth anniversary of the University of São Paulo, but as usual I missed the deadline. Still, better late than never: Happy 51st birthday!

Most of the following (Chapters 1 to 7) was written in 1985. But at the end of that year, I found a way to circumvent Horrible Chapter Seven, the method of rapidly oscillating solutions. My course was clear: everything should be rewritten with the new method! Unfortunately, I couldn't seem to find the time and energy needed for the task. Finally, Jack Hale and Antonio Luiz Pereira persuaded me to publish it as it stands, with only minor corrections. (Except that the correction of Theorem 7.6.10 is not so minor, and this required rewriting Example 6.8.) A new chapter was added, on the method of rapidly oscillating solutions, with a new example (8.5: generic simplicity of solutions of a system) to show the power of the method. And, with a few years perspective, Chapter 7 does not seem so horrible.

Thanks Jack and Antonio Luiz for getting it moving!

Chapter 1

Geometrical Preliminaries

In this chapter we introduce some notation, find convenient representations for regions in $\mathbb{R}^n$ with smooth boundary (1.2–1.7) and develop some tools needed later: approximation by smooth regions (1.8), smooth extension of functions (1.9), smooth deformations of regions (1.10), differentiation of integrals (1.11), and differential operators on a hypersurface (1.12). In short, a mixed bag of topics which will be needed later, and it may be wise to omit details until the need becomes apparent.

We treat only smoothly-bounded regions. Initial-boundary-value problems lead naturally to the study of regions with corners; but our examples will be elliptic boundary-value problems, and smooth regions provide sufficient variety.

1.1 Some Notation

For a function f defined near $x \in \mathbb{R}^n$, the mth derivative at x, $D^m f(x)$, may be considered as a homogeneous polynomial of degree m ($h \mapsto D^m f(x)h^m$) on $\mathbb{R}^n$, or as a symmetric m-linear form, or as the collection of partial derivatives

$$D^m f(x) = \left\{ \left(\frac{\partial}{\partial x}\right)^\alpha f(x) : |\alpha| = m \right\},$$

depending on convenience. Then the norm $|D^m f(x)|$ may denote

$$\max_{|\alpha|=m} \left| \left(\frac{\partial}{\partial x}\right)^\alpha f(x) \right| \text{ or } \max_{|h|\leq 1} |D^m f(x)h^m|.$$

The last version has the advantage of being independent of rotation of coordinates in $\mathbb{R}^n$, but the norms are equivalent.

If Ω is an open set in $\mathbb{R}^n$ and $m \geq 0$ an integer, $C^m(\Omega)$ is the space of m-times continuously and bounded differentiable functions on Ω whose derivatives extend continuously to the closure $\overline{\Omega}$, with the usual norm

$$\|\phi\|_{C^m(\Omega)} = \max_{0\le j\le m} \sup_{x\in\Omega} |D^j\phi(x)|.$$

The space of values is some normed linear space E, and is not clear which "E" is meant, we may write $C^m(\Omega, E)$.

- $C^m_{unif}(\Omega)$ is the closed surface of $C^m(\Omega)$ consisting of functions whose m^{th} derivative is uniformly continuous; if Ω is bounded, this is $C^m(\Omega)$.
- $C^{m,\alpha}(\Omega)$ is the subspace of $C^m_{unif}(\Omega)$ consisting of functions whose m^{th} derivative is Hölder continuous with exponent α $(0 < \alpha \le 1)$, provided with the norm

$$\|\phi\|_{C^{m,\alpha}(\Omega)} = \max\left(\|\phi\|_{C^m(\Omega)}, H^{\Omega}_{\alpha}(D^m\phi)\right)$$

where

$$H^{\Omega}_{\alpha}(f) = \sup\{|f(x) - f(y)|/|x-y|^{\alpha} : x \ne y \in \Omega\}$$

(This space is "boundedly closed" in $C^0(\Omega)$; that is, a bounded sequence in $C^{m,\alpha}(\Omega)$ which converges uniformly (in $C^0(\Omega)$) has its limits in $C^{m,\alpha}(\Omega)$.)
- $C^{m,\alpha+}(\Omega)$ is the closed subspace of $C^{m,\alpha}(\Omega)$, $0 < \alpha < 1$, consisting of functions $\phi \in C^{m,\alpha}(\Omega)$ such that

$$|D^m\phi(x) - D^m\phi(y)|/|x-y|^{\alpha} \to 0 \quad \text{as } x - y \to 0 \quad (x, y \in \Omega),$$

provided with the same $C^{m,\alpha}(\Omega)$ norm.

It is sometimes convenient to write

$$C^{m,0} \text{ for } C^m, \qquad C^{m,0+} \text{ for } C^m_{unif}$$

so we allow $0 \le \alpha \le 1$ in $C^{m,\alpha}$ and $0 \le \alpha < 1$ in $C^{m,\alpha+}$. $C^{m,\alpha}_{loc}(\Omega)[C^{m,\alpha+}_{loc}(\Omega)]$ is the space of functions whose restrictions to any Ω_0 (with $\overline{\Omega}_0$ compact in Ω) are in $C^{m,\alpha}(\Omega_0)[C^{m,\alpha+}(\Omega_0)$, respectively].

$$C^{\infty}(\Omega) = \bigcap_m C^m(\Omega), \quad C^{\infty}_{loc}(\Omega) = \bigcap_m C^{\infty}_{loc}(\Omega)$$

$C^{\omega}(\Omega)$ = analytic functions defined on an open set of $\mathbb{R}^n$ containing $\overline{\Omega}$ (hence, extending to analytic functions on an open set in $\mathbb{C}^n$, when we consider $\mathbb{R}^n$ as $\{z \in \mathbb{C}^n : \operatorname{Im} z = 0\}$.)

Definition 1.2. An open set $\Omega \subset \mathbb{R}^n$ has *C^m-regular boundary* [or $C^{m,\alpha}$ or $C^{m,\alpha+}$ or $C^{m,\alpha}_{loc}$ or $C^{m,\alpha+}_{loc}$ or C^{∞} or C^{∞}_{loc} or C^{ω}; regular boundary], if there exists $\phi \in C^m(\mathbb{R}^n, \mathbb{R})$ [or $C^{m,\alpha}$ or $C^{m,\alpha+}$ or ...] which is at least in $C^1_{unif}(\mathbb{R}^n, \mathbb{R})$ such that

$$\Omega = \{x \in \mathbb{R}^n : \phi(x) > 0\}$$

and $\phi(x) = 0$ implies $|\text{grad}\phi(x)| \geq 1$. We may also say Ω is a C^m region or Ω is C^m *regular* [or $C^{m,\alpha}$ or ...].

Theorem 1.3. *$\Omega \subset \mathbb{R}^n$ has C^m-regular boundary (or $C^{m,\alpha}$ or $C^{m,\alpha+}$) – at least C^1_{unif} – if and only if there are positive constants r, M such that, given any open ball $B \subset \mathbb{R}^n$ of radius r, after appropriate rotation and translation of coordinates, we have*

$$\Omega \cap B = \{x \in \mathbb{R}^n | x_n > \psi(\hat{x})\} \cap B, \quad \hat{x} = (x_1, \ldots, x_{n-1})$$
$$\partial\Omega \cap B = \{x \in \mathbb{R}^n | x_n = \psi(\hat{x})\} \cap B$$

for some $\psi \in C^m(\mathbb{R}^{n-1}, \mathbb{R})$ (or $C^{m,\alpha}$ or $C^{m,\alpha+}$) with norm $\leq M$. Note the conditions are trivial if $B \cap \partial\Omega = \emptyset$.

Remark. This implies easily that our definition of C^m-regular boundary is equivalent to that used by F. Browder and Agmon-Douglis-Nirenberg in their studies of elliptic boundary value problems. We will see that Def. 1.2 is very convenient for discussing perturbations of the boundary (as in 1.8 below). Our Definition 1.2 applied only when $\partial\Omega$ is at least uniformly C^1. The condition of Thm. 1.3 is more general, in that we may permit (for example) ψ to be merely Lipschitz continuous (in $C^{0,1}(\mathbb{R}^{n-1})$) which gives the "minimally smooth" domains of Stein's extension theorem [38, Sec. 6.3]. Some of our results apply to regions with convex corners, transversal intersections of smooth regions, as described in the remark following Thm. 1.9.

Proof. Suppose $\Omega = \{x \mid \phi(x) > 0\}$ is C^m (or $C^{m,\alpha}$ or $C^{m,\alpha+}$) regular, $\phi(x) = 0 \Rightarrow |\text{grad}\,\phi(x)| \geq 1$, $L = \sup|\text{grad}\,\phi|$, and choose $r > 0$ so $|x - y| \leq 6Lr \Rightarrow |D\phi(x) - D\phi(y)| < 1/2$.

Let B be a ball of radius r in $\mathbb{R}^n$ which meets $\partial\Omega$; we may assume the center of the ball is 0. If $0 \in \partial\Omega$, choose the positive x_n-axis along the inward normal (i.e., grad ϕ (0)). Otherwise let p be a point of $\partial\Omega \cap B$ closest to 0 and choose the x_n- axis to contain p and be directed into Ω. Then $p = (0, p_n)$, $|p_n| < r$, and $\frac{\partial\phi}{\partial x_n}(p) = |D\phi(p)| \geq 1$ (possibly $p = 0$). Also $\phi\,(0, s)$ has the same sign as $s - p_n$ in $-r < s < r$, and for $|x - p| \leq 6Lr$ we have $\frac{\partial\phi}{\partial x_n}(x) > 1/2$.

Let $\hat{x} \in \mathbb{R}^{n-1}$, $|\hat{x}| \leq 2r$; then $|\phi(\hat{x}, x_n) - \phi(0, x_n)| \leq 2Lr$ and $\pm\phi(\hat{x}, p_n \pm 4Lr) \geq \pm\phi(0, p_n \pm 4Lr) - 2Lr > 0$ so there exists unique $\psi(\hat{x}) \in (p_n - 4Lr, p_n + 4Lr)$ with $\phi(\hat{x}, \psi(\hat{x})) = 0$. By the implicit function theorem, ψ is C^m (or $C^{m,a}$ or $C^{m,\alpha+}$, respectively) and $|D\psi(\hat{x})| = |-(\partial\phi/\partial\hat{x})/(\partial\phi/\partial x_n)| \leq 2M$ for $|\hat{x}| \leq 2r$. Choose some (fixed) $C^\infty\theta : \mathbb{R} \to [0, 1]$ with $\theta(t) = 1$ for $t \leq 1$, $\theta(t) = 0$ for $t \geq 3/2$, and let $\psi_0(\hat{x}) = \theta(|\hat{x}|/r)\psi(\hat{x})$, or zero for $|\hat{x}| \geq 3r/2$. Then $\psi_0 \in C^m(\mathbb{R}^{n-1})$ (or $C^{m,\alpha}$ or $C^{m,\alpha+}$) with norm bounded by a multiple (depending only on n, m, α, r) of the norm of ϕ, and

$$\Omega \cap B = \{x \in B \mid x_n > \psi_0(\hat{x})\}, \quad \partial\Omega \cap B = \{x \in B \mid x_n = \psi(\hat{x})\}.$$

For the converse, we need the following

Lemma 1.4. *Let $\sigma > \sqrt{n}, r > 0$, and for each $k = (k_1, \ldots, k_n) \in \mathbb{Z}^n$, let B_k be the open ball in $\mathbb{R}^n$ with radius r and center $(r/\sigma)k$, while $B_k^{1/2}$ is the concentric ball with radius $r/2$. Then every point of $\mathbb{R}^n$ is contained in some $B_k^{1/2}$, and no point is contained in more than $(2\sigma + 1)^n$ of the balls B_k.*

Proof of the Lemma. The result is invariant under a homothety ($x \mapsto cx, c =$ constant > 0) so it suffices to treat the case $r = \sigma$. If $x \in \mathbb{R}^n$ there exists $k \in \mathbb{Z}^n$ so $x - k \in [-½, ½)^n$, hence $|x - k| \leq \sqrt{n}/2 < \sigma/2$ so $x \in B_k^{1/2}$.

Suppose $x \in B_k$; then for each $j = 1, \ldots n, |x_j - k_j| < \sigma$ so k_j is an integer in $(x_j - \sigma, x_j + \sigma)$. But $(x_j - \sigma, x_j + \sigma)$ contains no more than $2\sigma + 1$ integers, so there are at most $(2\sigma + 1)^n$ choices of $k \in \mathbb{Z}^n$ such that $x \in B^k$.

Completion of Proof of (1.3). Assume r, M given satisfying the requirements of the theorem; we must find $\phi : \mathbb{R}^n \to \mathbb{R}$ which satisfies the conditions (1.2). Choose $\sigma = \sqrt{2n+1}$ in the lemma. There is a C^∞ partition of unity $\{\phi_k\}_{k \in \mathbb{Z}^n}$

$$\operatorname{supp} \phi_k \subset B_k^{1/2}, \quad \phi_k \geq 0, \quad \sum_k \phi_k(x) \equiv 1$$

with $\|\phi_k\|C^m$ (or $\|\phi_k\|C^{m,\alpha}$) uniformly $\leq K = K(n, r, m, \alpha)$.

If $B_k \cap \partial\Omega \neq \phi$, there is a function $S_k(x)$ of class C^m(or $C^{m,\alpha}$or $C^{m,\alpha+}$) – $S_k(x) = x_n - \psi(\hat{x})$ after rotation of coordinates – such that

$$\Omega \cap B_k = \{x \in B_k : S_k(x) > 0\},$$
$$\partial\Omega \cap B_k = \{x \in B_k : S_k(x) = 0\},$$

with $\|S_k\|_{C^m \text{ (or } C^{m,\alpha})} \leq M$ and

$$S_k(x) = 0 \Rightarrow |DS_k(x)| \geq 1.$$

If $B_k \subset \Omega$, let $S_k = 1$; if $B_k \cap \overline{\Omega} = \emptyset$, let $S_k = -1$. Define $\psi(x) = \sum_k \phi_k(x) S_k(x)$. Then ψ is in $C^m(\mathbb{R}^n, \mathbb{R})$ [or $C^{m,\alpha}$or $C^{m,\alpha+}$], $\psi > 0$ in Ω, $\psi = 0$ on $\partial\Omega$, $\psi < 0$ outside $\overline{\Omega}$. If $x \in \partial\Omega$ then $\phi_k(x)S_k(x) = 0$ for each k and at x, grad $(\phi_k S_k) = \phi_k$ grad S_k which either vanishes or has the direction of the inward normal so $|\text{grad } \psi(x)| = \sum_k \phi_k(x) |\text{ grad } S_k(x)| \geq 1$.

The following "normal coordinates" are sometimes useful.

Theorem 1.5. *Let $\Omega \subset \mathbb{R}^n$ have C^m-regular boundary (or $C^{m,\alpha}$ or $C^{m,\alpha+}$, $2 \leq m \leq \infty$). There exists $r > 0$ so that if*

$$B_r(\partial\Omega) = \{x : \operatorname{dist}(x, \partial\Omega) < r\}$$
$$\pi(x) = \text{the point of } \partial\Omega \text{ nearest to } x$$
$$t(x) = \pm\operatorname{dist}(x, \partial\Omega) \quad (\text{“}+\text{” outside, “}-\text{” inside})$$

then $t(\cdot) : B_r(\partial\Omega) \to (-r, r)$, $\pi(\cdot) : B_r(\partial\Omega) \to \partial\Omega$ are well-defined, π is a C^{m-1} (or $C^{m-1,\alpha}$ or $C^{m-1,\alpha+}$) retraction onto $\partial\Omega$ ($\pi(x) = x$ when $x \in \partial\Omega$) and t has the same smoothness as $\partial\Omega$ (C^m or $C^{m,\alpha}$ or $C^{m,\alpha+}$). Further

$$x \mapsto (t(x), \pi(x)) : B_r(\partial\Omega) \to (-r, r) \times \partial\Omega$$

is a C^{m-1} (or $C^{m-1,\alpha}$ or $C^{m-1,\alpha+}$) diffeomorphism with inverse

$$(t, \xi) \mapsto \xi + tN(\xi) : (-r, r) \times \partial\Omega \to B_r(\partial\Omega)$$

where $N(\xi)$ is the unit outward normal to $\partial\Omega$ at ξ.

$$t(\cdot) \text{ is the unique solution of } |\nabla t(x)| = 1 \text{ in } B_r(\partial\Omega),$$

with $t = 0$ on $\partial\Omega$, $\partial t/\partial N > 0$ on $\partial\Omega$.

Extending the normal field N to a neighborhood of $\partial\Omega$ by

$$N(\xi + tN(\xi)) = N(\xi) \quad -r < t < r,$$

we have $N(x) = \operatorname{grad} t(x)$ on $B_r(\partial\Omega)$. Also $K(x) = DN(x) = D^2 t(x)$, restricted to the tangent space at $x \in \partial\Omega$, is the curvature of $\partial\Omega$. It is sometimes convenient to call $K(x)$ the curvature, though it is degenerate ($K(x)N(x) = 0$) in the normal direction.

Remark. The fact that $t(\cdot)$ has the same smoothness as $\partial\Omega$ – does not lose a derivative, as happens with $\pi(\cdot)$ – seems to have been noted first by Gilbarg and Trudinger [8].

The best (largest) choice of r is $r = 1/\max|k|$, where k is the sectional curvature of the boundary in any (tangent) direction at any point of $\partial\Omega$.

Corollary 1.6. *A C^m-regular region $\Omega \subset \mathbb{R}^n$, $m \geq 2$, may be represented by $\{x | \phi(x) > 0\}$ where ϕ is C^m and $|\nabla\phi(x)| \equiv 1$ on a neighborhood of $\partial\Omega$. In this case, ϕ is unique on a neighborhood of $\partial\Omega$.*

Proof of (1.5). By the inverse function theorem, the C^{m-1} map $(t, \xi) \mapsto \xi + tN(\xi) = x$ has a C^{m-1} inverse $t = t(x)$, $\xi = \pi(x)$, on some neighborhood of $\partial\Omega$. On the other hand, for each $x \in \mathbb{R}^n$, $\xi \mapsto \frac{1}{2}|x - \xi|^2$ ($\xi \in \partial\Omega$) has a minimum and if ξ is a minimizing point then $x - \xi \perp T_\xi(\partial\Omega)$ so $x = \xi + tN(\xi)$

for some real t with $t = \pm\text{dist}(x, \partial\Omega)$. We show, for some $r > 0$, that $x = \xi + tN(\xi)$, $t = \pm\text{dist}(x, \partial\Omega)$ has a unique solution ξ whenever $x \in B_r(\partial\Omega)$ so $\xi = \pi(x)$ is the (unique) nearest points.

Suppose $\Omega = \{x : \phi(x) > 0\}$, $\phi(x) = 0 \Rightarrow |\nabla\phi(x)| \geq 1$ and $r = 1/\sup|D^2\phi(x)|$. If $\text{dist}(x, \partial\Omega) < r$ but there are two "nearest points" $\xi_1 \neq \xi_2$, $x = \xi_j + tN(\xi_j)$, then $t^2 = |\xi_1 - \xi_2 + tN(\xi_1)|^2 = |\xi_1 - \xi_2|^2 + 2t(\xi_1 - \xi_2).N(\xi_1) + t^2$. Now $\phi(\xi_2) = \phi(\xi_1) = 0 = \nabla\phi(\xi_1)\cdot(\xi_2 - \xi_1) + \frac{1}{2}\int_0^1 D^2\phi(\theta\xi_2 + (1-\theta\xi_1))\cdot(\xi_2 - \xi_1)^2 d\theta$ and $N(\xi_1) = -\nabla\phi(\xi_1)/|\nabla\phi(\xi_1)|$, so for $|t| < r$

$$0 < |\xi_1 - \xi_2|^2 = 2tN(\xi_1)\cdot(\xi_2 - \xi_1) \leq |t| \sup|D^2\phi|\,|\xi_1 - \xi_2|^2 < |\xi_1 - \xi_2|^2,$$

a contradiction. Thus $t(\cdot)$, $\pi(\cdot)$ are well-defined and C^{m-1} on $B_r(\partial\Omega)$.

Extend N to be constant on normal lines, so $N(x) = N(\pi(x))$ is C^{m-1} on $B_r(\partial\Omega)$. It is clear that $t(x + sN(x)) = t(x) + s$ when $t(x)$ and $t(x) + s$ are in $(-r, r)$, so

$$\partial t(x)/\partial x_j = \partial_j t(x + sN(x)) + s\sum_{k=1}^{n} \partial_k t(x + sN(x))\partial_j N_k(x).$$

Let $s = -t(x)$, so $x + sN(x) = \pi(x) \in \partial\Omega$, and note $\nabla t(\pi(x)) = N(\pi(x)) = N(x)$. Then on $B_r(\partial\Omega)$

$$\partial_j t(x) = N_j(x) - t(x)\sum_{k=1}^{m} N_k(x)\partial_j N_k(x) = N_j(x)$$

or $N(x) = \text{grad}\, t(x)$. (Observe that $K(x)N(x) = \partial N/\partial N(x) = 0$.) Clearly N is C^{m-1} so $t(\cdot)$ is C^m.

To identify $DN(\xi) = D^2 t(\xi)$ as the curvature of $\partial\Omega$ at ξ, we may choose coordinates so Ω is locally $\{x_n > \psi(\hat{x})\}$ where $\psi(0) = 0$, $D\psi(0) = 0$ and $K_{ij} = \partial^2\psi/\partial x_i \partial x_j\,(0)$ $(1 \leq i, j \leq n-1)$ is the curvature matrix (on the tangent plane $\mathbb{R}^{n-1} \times 0$). It is easy to show $N(x) = (0, -1) + (K\hat{x}, 0) + o(|x|)$ as $x \to 0$ so $DN(0) = \left(\begin{smallmatrix} K & 0 \\ 0 & 0 \end{smallmatrix}\right)$, and the restriction in the tangent plane is the curvature K of $\partial\Omega$ at 0.

Remark. If S is a C^2 hypersurface, $x_0 \in S$, the usual existence theory [5], [6] for first order scalar PDEs says there is a unique C^1 solution of $|\nabla t| = 1$ near x_0 with $t = 0$, $\partial t/\partial N > 0$ on S near x_0. If S is not C^2 (or $C^{1,1}$) there may be no C^1 solution of this problem. For example if $1 < p < 2$, $S = \{x : x_2 = |x_1|p\}$, there are two "nearest points" to x when $x_1 = 0$, $x_2 > 0$ (near 0) and the gradient of $\text{dist}(x, S)$ has a discontinuous jump as x_1 crosses 0 with $x_2 > 0$.

Theorem 1.7. *Let Γ be a compact subgroup of the orthogonal group $O(n)$ and let $\Omega \subset \mathbb{R}^n$ be a C^m-regular region such that $\gamma(\Omega) = \Omega$ for all $\gamma \in \Gamma$. Then there exists C^m $\phi : \mathbb{R}^n \to \mathbb{R}$ such that for $\Omega = \{x : \phi(x) > 0\}$,*

$\phi(x) = 0 \Rightarrow |\nabla\phi(x)| \geq 1$, and $\phi(\gamma \cdot x) = \phi(x)$ for all $\gamma \in \Gamma, x \in \mathbb{R}^n$. If $\partial\Omega$ is at least C^2, we may choose such ϕ with $|\nabla\phi| = 1$ on a neighborhood of $\partial\Omega$.

Proof. There exists a function ϕ_0 satisfying all requirements except perhaps Γ-invariance. Let $\phi(x)$ be the average of $\gamma \mapsto \phi_0(\gamma x)$ with respect to Haar measure in Γ; then ϕ is certainly C^m and Γ-invariant. Further $x \in \Omega \Rightarrow \gamma x \in \Omega \Rightarrow \phi_0(\gamma x) > 0$ for all $\gamma \in \Gamma$ so $\phi(x) > 0$ in Ω; and similarly $\phi(x) = 0$ on $\partial\Omega$, $\phi(x) < 0$ outside $\overline{\Omega}$. On $\partial\Omega$, $N(x) = -\nabla\phi_0(x)/|\nabla\phi_0(x)|$ and it follows easily that $N(\gamma x) = \gamma N(x)$ for $x \in \partial\Omega, \gamma \in \Gamma$, so

$$\text{grad}(\phi_0(\gamma x)) = \gamma(\text{grad}\ \phi_0(\gamma x) = -N(x)|\text{grad}\ \phi_0(\gamma x)|$$

and (averaging with respect to γ) $|\text{grad}\ \phi(x)| = \text{average}_\gamma |\text{grad}\ \phi_0(\gamma x)| \geq 1$ for $x \in \partial\Omega$. If in fact $|\nabla\phi_0(x)| = 1$ near $\partial\Omega$, then $\phi(x) = \phi_0(x) = \pm\text{dist}(x, \partial\Omega)$ near $\partial\Omega$ so $|\nabla\phi(x)| = 1$ near $\partial\Omega$.

A common technique in analysis is to approximate a given function by a smooth function, to facilitate calculations, and take limits only at the end. Similarly we may approximate a given region by smooth regions. If $\Omega = \{x : \phi(x) > 0\}$ is a C^m_{unif} region, and ψ is C^m-close to ϕ, we show $\{x : \psi(x) > 0\}$ is C^m-close to Ω, that is, it is diffeomorphic to Ω by a diffeomorphism C^m-close to the identity.

Theorem 1.8. *Let $\phi : \mathbb{R}^n \to \mathbb{R}$ be (at least) uniformly $C^1, \phi(x) = 0 \Rightarrow |\text{grad}\ \phi(x)| \geq 1, \Omega = \{x : \phi(x) > 0\}$, and for some $a > 0$,*

$$|\phi(x)| \geq \min\left\{\frac{1}{2}\ \text{dist}(x, \partial\Omega), a\right\} \quad \textit{for all } x.$$

The last condition may always be achieved by modifying ϕ away from $\partial\Omega$. Also, let $r_0 > 0$.

Then there exists $\epsilon_0 > 0$ such that, if $\|\psi - \phi\|_{C^1(\mathbb{R}^n)} < \epsilon_0$, there is a diffeomorphism $h(\cdot, \psi) : \mathbb{R}^n \to \mathbb{R}^n$ supported in $B_{r_0}(\partial\Omega)$ – i.e., $h(x, \psi) = x$ when dist $(x, \partial\Omega) \geq r_0$ – such that $h(\Omega; \psi) = \{x : \psi(x) > 0\}$ and $\|h(\cdot, \psi) - id_{\mathbb{R}^n}\|_{c^1(\mathbb{R}^n)} \to 0$ as $\|\psi - \phi\|_{c^1(\mathbb{R}^n)} \to 0$. If ψ is of class $C^{m,\alpha}$ [or $C^{m,\alpha+}$ or C^∞ or C^ω] then $\{x : \psi(x) > 0\}$ is a $C^{m,\alpha}$ [or $C^{m,\alpha+}$ or C^∞ or C^ω] regular region, assuming $\|\psi - \phi\|_{C^1} < \epsilon_0$.

If ϕ, ψ are $C^{m,\alpha+}$ then $h(\cdot, \psi)$ is a $C^{m,\alpha+}$ diffeomorphism, $\|h(\cdot, \psi) - id_{\mathbb{R}^n}\|_{C^{m,\alpha}} \to 0$ as $\|\psi - \phi\|_{C^{m,\alpha}} \to 0$ and $(x, \psi) \mapsto h(x, \psi) : \mathbb{R}^n \times C^{m,\alpha+}(\mathbb{R}^n) \to \mathbb{R}^n$ is $C^{m,\alpha+}$.

If ϕ, ψ are $C^{m,\alpha}$ then $h(\cdot, \psi)$ is a $C^{m,\alpha}$ diffeomorphism and $(x, \psi) \mapsto h(x, \psi) : \mathbb{R}^n \times C^{m,\alpha} \to \mathbb{R}^n$ is $C^{m,\alpha}$. As $\|\psi - \phi\|_{C^1} \to 0$ with $\|\psi\|_{C^{m,\alpha}}$ bounded, $h(\cdot, \psi)$ remains bounded in $C^{m,\alpha}$ and converges to $id_{\mathbb{R}^n}$ in $C^{k,\beta}$ when $k + \beta < m + \alpha$.

Remark. The hypothesis $\|\psi - \phi\|_{C^{m,\alpha}} \to 0$ does not yield $C^{m,\alpha}$ convergence of $h(\cdot, \psi) \to id$, if ϕ is not $C^{m,\alpha+}$.

Proof. Choose a C^∞ vector field $M(x)$ on a neighborhood of $\partial\Omega$, uniformly close to grad ϕ: for some $r_1 > 0$ and $C \geq 2$

$$\frac{1}{2} \leq |M(x)| \leq C$$

and

$$M(x) \cdot \operatorname{grad} \phi(x) \geq \frac{1}{2}|M(x)|^2 \quad \text{if } \operatorname{dist}(x, \partial\Omega) \leq r_1.$$

Let $r_2 = \min\{a, r_0, \frac{1}{2}r_1\}$ and choose C^∞ $\theta : \mathbb{R}^n \to [0,1]$ so that $\theta = 1$ on $B_{r_2/2}(\partial\Omega)$, $\theta = 0$ outside $B_{r_2}(\partial\Omega)$. We will show that, if $\|\psi - \phi\|_{C^1}$ is small and $x \in B_{r_2}(\partial\Omega)$, there is a unique $t = t(x;\psi)$ near 0 such that

$$\psi(x + tM(x)) = \phi(x).$$

Then define h by

$$h(x;\psi) = \begin{cases} x + \theta(x)t(x;\psi) & M(x) \text{ in } B_{r_2}(\partial\Omega) \\ x & \text{outside } B_{r_2}(\partial\Omega) \end{cases}$$

and the conditions of the theorem are satisfied.

First choose $s_0 > 0$ so small that $s_0 \leq \frac{1}{2}r_2$ and $|D\phi(x) - D\phi(y)| \leq 1/8C$ when $|x - y| \leq C s_0$. Suppose $\|\psi - \phi\|_{C^0} = \sup|\psi - \phi| \leq s_0/32$ and $\sup|D\psi - D\phi| \leq 1/8C$. Then for $-s_0 \leq t \leq s_0$, $\operatorname{dist}(x, \partial\Omega) \leq r_2$,

$$\begin{aligned} \frac{\partial}{\partial t}(\psi(x + tM(x)) - \phi(x)) &= M(x) \cdot D\psi(x + tM(x)) \\ &= M(x) \cdot D\phi(x) + M(x) \cdot (D\psi(x + tM(x)) - D\phi(x)) \\ &\geq \frac{1}{2}|M|^2 - |M|\left(\frac{1}{8C} + \frac{1}{8C}\right) \geq \frac{1}{16} \end{aligned}$$

while for $t = \pm s_0$

$$\begin{aligned} t(\psi(x + tM(x)) - \phi(x)) &\geq t(\phi(x + tM(x)) - \phi(x)) - s_0\|\psi - \phi\|_{C^0} \\ &\geq s_0^2\left(\frac{1}{2}|M|^2 - |M|/8c\right) - \frac{s_0^2}{32} \geq \frac{s_0^2}{32} > 0. \end{aligned}$$

Thus there is a unique $t = t(x,\psi)$ in $(-s_0, s_0)$ with $\psi(x + tM(x)) = \phi(x)$, and it is easy to see

$$|t(x,\psi)| \leq 16\|\psi - \phi\|_{C^0} \quad \text{and} \quad \frac{\partial}{\partial x}t(x;\psi) \to 0$$

uniformly on $B_{r_2}(\partial\Omega)$ as $\|\psi - \phi\|_{C^1} \to 0$. (Here we again use the uniform continuity of $D\psi$.)

The evaluation map $(\psi, z) \mapsto \psi(z) : C^{m,\alpha} \times \mathbb{R}^n \to \mathbb{R}$ [or $C^{m,\alpha+} \times \mathbb{R}^n \to \mathbb{R}$] is of class $C^{m,\alpha}$ [or $C^{m,\alpha+}$]. The proof is similar to that of the case C^m in [1]. Thus, by the implicit function theorem, if ϕ is $C^{m,\alpha}$ [or $C^{m,\alpha+}$], $(x, \psi) \mapsto t(x; \psi) : B_{r_2}(\partial\Omega) \times C^{m,\alpha}$ [or $C^{m,\alpha+}$] $\to \mathbb{R}$ is of class $C^{m,\alpha}$ [or $C^{m,\alpha+}$].

It is useful to define an extension operator E_Ω for open $\Omega \subset \mathbb{R}^n$ such that given $u : \Omega \to \mathbb{R}$ (or $\mathbb{R}^N$ or $\mathbb{C}^N$), $E_\Omega(u)$ is defined on all $\mathbb{R}^n$ with the same smoothness as u and $E_\Omega(u) = u$ on Ω. Stein [38] defines such an operator for regions whose boundary is merely Lipschitzian. When the boundary is smooth, there is a simpler construction which also applies (with some modifications, noted below) when $\overline{\Omega}$ has convex corners.

Theorem 1.9. *Let $\Omega \subset \mathbb{R}^n$ have $C^{m,\alpha}$ (or $C^{m,\alpha+}$) regular boundary, $1 \leq m < \infty$. There is a linear operator E_Ω which carries functions $u : \Omega \to \mathbb{R}$ to functions $E_\Omega(u) : \mathbb{R}^n \to \mathbb{R}$ such that $E_\Omega(u) = u$ on Ω and for may integer k in $0 \leq k \leq m$ and any p, β in $1 \leq p \leq \infty, 0 \leq \beta \leq 1$, with $k + \beta \leq m + \alpha$, we have*

$$\|u\|_{W^{k,p}(\Omega)} \leq \|E_\Omega(u)\|_{W^{k,p}(\mathbb{R}^n)} \leq C_{m,\Omega}\|u\|_{W^{k,p}(\Omega)}$$
$$\|u\|_{C^{k,\beta}(\Omega)} \leq \|E_\Omega(u)\|_{C^{k,\beta}(\mathbb{R}^n)} \leq C_{m,\Omega}\|u\|_{C^{k,\beta}(\Omega)}$$

where $C_{m,\Omega}$ is a constant depending only on m and Ω. If $\partial\Omega$ is $C^{k,\beta+}$ and u is $C^{k,\beta+}$ in Ω, then $E_\Omega(u)$ is $C^{k,\beta+}$ in $\mathbb{R}^n$. For any $r_0 > 0$ we may choose E_Ω such that the support of $E_\Omega(u)$ is in an r_0-neighborhood of the support of u ($\subset \overline{\Omega}$), and such that $E_\Omega(u)$ outside Ω depends only on the values of u in an r_0-neighborhood of $\partial\Omega$. In this case, the constant $C_{m,\Omega}$ depends also on r_0. The constants $C_{m,h(\Omega)}$ are bounded when h varies in a $C^{m,\alpha}$-small neighborhood of the inclusion $i_\Omega : \Omega \to \mathbb{R}^n$. If $\partial\Omega$ is C^∞_{loc} and $u = U|_\Omega$ for some $U \in C^\infty_c(\mathbb{R}^n)$ then $E_\Omega(u) \in C^\infty_c(\mathbb{R}^n)$.

Proof. First we consider a half-space, $\mathbb{R} \times \mathbb{R}^{n-1}$. Let $a_j, b_j \leq -1$ $(j = 1, 2, \ldots)$ be real such that $a_j \to 0$ rapidly, $\sum |a_j||b_j|^k < \infty$ and $\sum_1^\infty a_j b_j^k = 1$ for every $k = 0, 1, 2, \ldots$. Then if $B_j(t) = t$ for $t \geq 0$, $B_j(t) = b_j t$ for $t \leq 0$, $v \mapsto V$ defined by

$$V(t, \xi) = \sum_{j=1}^{\infty} a_j v(B_j(t), \xi) \quad (t \in \mathbb{R}, \xi \in \mathbb{R}^{n-1})$$

is the desired extension operator for $\Omega = \mathbb{R}_+ \times \mathbb{R}^{n-1}$. Specifically let $b_j = 1 - 2^j$ (as in [35]) and define the a_j by $\sum_1^\infty a_j z^j = A(z)$

$$A(z) = z \prod_{k=1}^{\infty}(1 - z/2^k) \Big/ \prod_{1}^{\infty}(1 - 1/2^k).$$

Then A is an entire analytic function of order 0, $A(1) = 1$, $A(2^j) = 0$ for $j = 1, 2, 3, \ldots$ and so

$$\sum_{j=1}^{\infty} a_j b_j^k = \sum_{i=0}^{k} \binom{k}{i} (-1)^i A(2^i) = 1 \quad \text{for each } k = 0, 1, 2, \ldots$$

In the general case, let $M : \mathbb{R}^n \to \mathbb{R}^n$ be C^∞ (each derivative bounded on $\mathbb{R}^n$) which is uniformly close to the outward unit normal on $\partial\Omega$, recall $\partial\Omega$ is at least uniformly C^1 (Def. 1.1). Then there exists $r_1 > 0$ so that $(t, \xi) \mapsto \xi + tM(\xi) : (-r_1, r_1) \times \partial\Omega \to \mathbb{R}^n$ is a $C^{m,\alpha}$ (or $C^{m,\alpha+}$) diffeomorphism onto a neighborhood V_1 of $\partial\Omega$, and $V_1 \supset B_{r_2}(\partial\Omega)$ for some $r_2 > 0$. Chose $0 < r_3 < r_2$ and some C^∞ $\theta : \mathbb{R}^n \to [0, 1]$ such that $\theta = 1$ in $B_{r_3}(\partial\Omega)$, $\theta = 0$ outside $B_{r_2}(\partial\Omega)$. Define

$$E_\Omega(u)(x) = \begin{cases} u(x) & \text{in } \Omega \\ \theta(x) \sum_{k=1}^{\infty} a_k(\theta u)(\xi + b_k t M(\xi)) & \text{for } x = \xi + tM(\xi), \\ 0 & \text{outside } \Omega \cup V_1. \end{cases}$$

Writing $\partial/\partial M$ for $(\partial/\partial t)_{\xi=\text{const}}$ in the coordinates $x = \xi + tM(\xi)$ near $\partial\Omega$, it follows $(\partial/\partial M)^j E_\Omega u(\xi \pm 0 \cdot M(\xi)) = (\partial/\partial M)^j u(\xi - 0 \cdot M(\xi))$, $\xi \in \partial\Omega$, for each j and E_Ω is the desire extension operator.

Remarks. We may treat similarly regions with convex corners, of the form $\Omega = \{x : \phi_j(x) > 0 \text{ for } j = 1, \ldots, \nu\}$ where each ϕ_j is $C^{m,\alpha}$ (or $C^{m,\alpha+}$) and for each subset $\{j_1, \ldots, j_k\} \subset \{1, \ldots, \nu\}$, $(0, \ldots, 0)$ is a regular value of $(\phi_{j_1}, \ldots, \phi_{j_k})$. There are also some uniformity conditions, which we avoid by assuming $\partial\Omega$ is compact. Then for each point x of $\overline{\Omega}$, there is a neighborhood of x in $\overline{\Omega}$, $C^{m,\alpha}$ (or $C^{m,\alpha+}$) diffeomorphic to an open set in $\overline{\mathbb{R}}_+^k \times \mathbb{R}^{n-k}$ (and taking x to 0), for some integer k, $0 \le k \le \min(n, \nu)$. We say x is a kth order corner ($k = 0$ for an interior point) and exactly k of the ϕ_j vanish at x. We may define an extension operator for $\mathbb{R}_+^k \times \mathbb{R}^{n-k}$ by extending one variable at a time as above. The result is

$$V(t, \xi) = \sum_{i_1, \ldots, i_k \ge 1} a_{i_1} \cdots a_{i_k} v(B_{i_1}(t_1), \ldots, B_{i_k}(t_k), \xi)$$

for $t \in \mathbb{R}^k$, $\xi \in \mathbb{R}^{n-k}$. A partition of unity and the usual tricks give the extension operator for Ω.

Another type of extension problem occurs when we have a function defined initially only on $\partial\Omega$. As above this reduces to extension from $\mathbb{R}^{n-1} \times 0$ to $\mathbb{R}^n$, supposing $\partial\Omega$ is smooth. (When $\partial\Omega$ has corners, there will also be some compatibility conditions to satisfy.) Such extensions relative to the Sobolev spaces are fairly complicated, but fortunately in our applications, this extension problem occurs only in the trivial case C^m.

Corollary 1.10. *Let $\Omega \subset \mathbb{R}^n$ have $C^{m,\alpha}$ [or $C^{m,\alpha+}$] boundary ($m \geq 1$) at least $C^1_{unif} = C^1$. Let $h \in C^{m,\alpha}((-r,r) \times \Omega, \mathbb{R}^n)$ [or $C^{m,\alpha+}$] for some $r > 0$ and $h(0,x) = x$ for $x \in \Omega$, i.e., $h(0,\cdot)$ is the inclusion i_Ω of Ω in $\mathbb{R}^n$ and $h(t,\cdot) : \Omega \to \mathbb{R}^n$ is an imbedding for small t. Then there exists $H \in C^{m,\alpha}((-r,r) \times \mathbb{R}^n, \mathbb{R}^n)$ [or $C^{m,\alpha+}$] such that $H(t,x) = h(t,x)$ for $x \in \Omega$, $H(0,\cdot) =$ identity on $\mathbb{R}^n$, and for some $r_0 > 0$, $H(t,\cdot)$ is a diffeomorphism on $\mathbb{R}^n$ with inverse $H^{-1}(t,\cdot)$ for $-r_0 < t < r_0$ and $H^{-1} \in C^{m,\alpha}((-r_0,r_0) \times \mathbb{R}^n)$ [or $C^{m,\alpha+}$, respectively]. Similar results hold when $t \mapsto h(t,\cdot) \in C^{m,\alpha}$ is $C^{k,\beta}$ with $k + \beta \leq m + \alpha$. If both $(t,x) \mapsto h(t,x)$, $\frac{\partial h}{\partial t}(t,x)$ are in $C^{m,\alpha}$ [or $C^{m,\alpha+}$] on $(-r,r) \times \Omega$, $h(0,\cdot) = i_\Omega$, there exists $V \in C^{m,\alpha}((-r_0,r_0) \times \mathbb{R}^n, \mathbb{R}^n)$ [or $C^{m,\alpha+}$] such that*

$$\frac{\partial}{\partial t}h(t,x) = V(t,h(t,x)) \text{ for } x \in \Omega,\ -r_0 < t < r_0$$
$$h(0,x) = x.$$

Conversely, given such V, the function h defined by this differential equation has $(t,x) \mapsto h(t,x)$, $\frac{\partial h}{\partial t}(t,x)$ of class $C^{m,\alpha}$ [or $C^{m,\alpha+}$, respectively].

Remark. This corollary provides alternative representations of a "smoth curve of regions $t \mapsto \Omega(t)$," $\Omega(t) = h(t,\Omega)$. We will ordinarily use the first form (a curve of imbeddings of Ω in $\mathbb{R}^n$), but the other representations are also convenient in some cases. In light of Th. 1.8, we may equally use $\Omega(t) = \{x : \phi(x,t) > 0\}$ or (as in the proof of that theorem) $\partial\Omega(t) = \{\xi + S(\xi,t)M(\xi) \mid \xi \in \partial\Omega\}$, a graph over $\partial\Omega$, given a smooth vector field M transverse to $\partial\Omega$, where $S(\xi,t)$ is a smooth real-valued function on $\partial\Omega \times (-r,r)$ which vanishes when $t = 0$. The last representation has the advantage of uniqueness given M; the others are degenerate, with many imbeddings (or diffeomorphisms or ...) yielding the same geometric result. However, this degeneracy does not seem to cause problems. Some analogous results occur as lemmas in Examples 3.1, 3.2 below.

Ana Maria Micheletti [22] shows the set of regions C^m-diffeomorphic to a given bounded C^m region $\Omega \subset \mathbb{R}^n$ can be considered a complete metric space relative to the "Courant distance:"

$$d(\Omega_1,\Omega_2) = \inf \sum_{j=1}^{N} \left(\|f_j - id\|_{C^m} + \left\|f_j^{-1} - id\right\|_{C^m}\right)$$

considered over all $N \geq 1$ and $f_j \in C^m(\mathbb{R}^n,\mathbb{R}^n)$ $(1 \leq j \leq N)$ which are diffeomorphisms with compact support ($f_j(x) = x$ outside a compact set) and such that $f_1 \circ f_2 \circ \ldots \circ f_N$ carries Ω_1 onto Ω_2. We will not use this metric space explicitly, but our representations above (for $\alpha = 0$: recall Ω is bounded here) all yield continuous curves $t \mapsto \Omega(t)$ in this metric.

Proof of the Corollary. Let E_Ω be the extension operator provided by Thm. 1.9, and define $H(t,\cdot) = id_{\mathbb{R}^n} + E_\Omega(h(t,\cdot) - i_\Omega)$. Since E_Ω is linear, we also have $\partial_t^i H(t,\cdot) = E_\Omega \partial_t^j h(t,\cdot)$ for $i \le j \le m$ and it follows that $H \in C^{m,\alpha}((-r,r) \times \mathbb{R}^n, \mathbb{R}^n)$ [or $C^{m,\alpha+}$] with $H(0,x) = x$ on $\mathbb{R}^n$. Since $\sup_x |\frac{\partial}{\partial x} H(x,t) - I| < 1$ for t near 0, the inverse $H^{-1}(t,\cdot)$ exists and is mostly easily studied as $(t,x) \to (t, H^{-1}(t,\cdot)(x))$, which is the inverse of $(t,x) \mapsto (t, H(t,x))$. The inverse-function theorem in the class $C^{m,\alpha}$ [or $C^{m,\alpha+}$] shows $(t,x) \mapsto H^{-1}(t,\cdot)(x)$ is $C^{m,\alpha}$ [or $C^{m,\alpha+}$] for t near 0. If also $(t,x) \mapsto \frac{\partial h}{\partial t}(t,x)$ is $C^{m,\alpha}$ [or $C^{m,\alpha+}$] then $(t,x) \mapsto H^{-1}(t,\cdot)(x)$, $\frac{\partial}{\partial t} H(t,x)$ and $V(t,x) = (\partial_t H)(t, H^{-1}(t,\cdot)(x))$ are $C^{m,\alpha}$ [or $C^{m,\alpha+}$], and $\frac{\partial}{\partial t} H(t,x) = V(t, H(t,x))$ for t near 0, $x \in \mathbb{R}^n$.

Theorem 1.11. *Let $\Omega \subset \mathbb{R}^n$ be a C^1_{unif} region and $h \in C^1_{unif}((-r,r) \times \Omega, \mathbb{R}^n)$ with $h(0,\cdot) = i_\Omega$ [$h(0,x) = x$ for $x \in \Omega$]. Also let $f : \mathbb{R} \times \mathbb{R}^n \to \mathbb{R}$ and $\partial f/\partial t$ be continuous (at least near $t = 0$), with compact support. Then if $\Omega(t) = h(t,\Omega)$, $t \mapsto \int_{\Omega(t)} f(t,x)dx$ is C^1 near $t = 0$ and*

$$\frac{d}{dt}\int_{\Omega(t)} f(t,x)dx = \int_{\Omega(t)} \frac{\partial f}{\partial t}(t,x)dx + \int_{\partial\Omega(t)} f(t,x) V \cdot N dA_x$$

where V is the velocity field (so $\partial h/\partial t(t,x) = V(t, h(t,x))$) and N the unit outward normal.

Let Ω be a C^2 region and assume in addition to the previous hypotheses, that $\partial^2 h/\partial x^2$ and $\partial^2 h/\partial t \partial x$ are continuous and $\partial f/\partial x$ is continuous (f is C^1). Then for t near 0, $t \mapsto \int_{\partial\Omega(t)} f(t,x) dA_x$ is C^1 and

$$\frac{d}{dt}\int_{\partial\Omega(t)} f(t,x)dA_x = \int_{\partial\Omega(t)} \left(\frac{\partial f}{\partial t} + V \cdot N N \cdot \frac{\partial f}{\partial x} + HV \cdot N f\right) dA$$

where, as before, V is the velocity field, N the outward unit normal, and $H = \operatorname{div} N$ is the mean curvature of $\partial\Omega(t)$ (the sum of the principal curvatures).

Proof. It suffices to prove these results when f and h are smooth (say, C^2) and to compute the derivative at $t = 0$. Then

$$h(t,x) = x + tV(x) + O(t^2),$$
$$\frac{\partial h}{\partial x}(t,x) = I + tV(x) + O(t^2)$$

and

$$\int_{\Omega(t)} f(t,y)dy = \int_{\Omega} f(t,h(t,x))\det\left(\frac{\partial h}{\partial x}(t,x)\right)dx$$
$$= \int_{\Omega} \{f(0,x) + t(f_t + f_x \cdot V + f \operatorname{div} V) + O(t^2)\}\, dx$$

so the derivative at $t = 0$ is $\int_{\Omega}(f_t + \operatorname{div}(fV)) = \int_{\Omega} f_t + \int_{\partial\Omega} fV \cdot N$.

For the second case, we may suppose the normal field $N(t,x)$ for $\partial\Omega(t)$ is extended as a C^1 unit-vector field near $\partial\Omega(t)$, and then extended in any C^1 manner throughout $\mathbb{R}^n$. Then

$$\int_{\partial\Omega(t)} f(t,x)dA_x = \int_{\Omega(t)} \operatorname{div}(f(t,x)N(t,x))dx$$

and we may apply the first case, merely recalling that $\partial N/\partial t \cdot N = 0$ on $\partial\Omega(t)$.

We will sometimes need to use differential operators (gradient, divergence and Laplacian) in a hypersurface $S \subset \mathbb{R}^n$. The following definitions are all equivalent to the corresponding formulas of Riemannian geometry or tensor analysis, in the metric induced in S from (Euclidean) $\mathbb{R}^n$. These formulas are (of course) intrisic to S and say nothing about the surrounding $\mathbb{R}^n$; but our interest is precisely in this relation to a neighborhood of S (see Th. 1.13).

Definition 1.12. Let S be a C^1 hypersurface in $\mathbb{R}^n$ and let $\varphi : S \to \mathbb{R}$ be C^1 (so it may be extended to be C^1 on a neighborhood of S); then $\nabla_S\varphi$ is the tangent vector field on S such that, for each C^1 curve $t \mapsto x(t) \in S$, we have

$$\frac{d}{dt}\varphi(x(t)) = \nabla_S\varphi(x(t)) \cdot \dot{x}(t).$$

Let S be a C^2 hypersurface in $\mathbb{R}^n$ and let $\mathbf{a} : S \to \mathbb{R}^n$ be a C^1 vector field – it is a tangent-vector field if $\mathbf{a} \cdot N \equiv 0$ on S, where N is a unit normal field on S. Then $\operatorname{div}_S \mathbf{a} : S \to \mathbb{R}$ is the continuous function such that, for every $C^1\varphi : S \to \mathbb{R}$ with support compact in S,

$$\int_S (\operatorname{div}_S \mathbf{a})\varphi = -\int_S \mathbf{a} \cdot \nabla_S\varphi.$$

Note $\operatorname{div}_S \mathbf{a}$ depends only on the tangential component of $\mathbf{a}|S$. Finally, if $u : S \to \mathbb{R}$ is C^2, then $\Delta_S u = \operatorname{div}_S(\nabla_s u)$ or equivalently, for all C^1 $\varphi : S \to \mathbb{R}$ with compact support,

$$\int_S \varphi\Delta_S u = -\int_S \nabla_S\varphi \cdot \nabla_S u$$

Theorem 1.13. *(i) If S is a C^1 hypersurface and $\varphi : \mathbb{R}^n \to \mathbb{R}$ is C^1 on a neighborhood of S, then on S*

$$\begin{aligned}\nabla_S\varphi(x) &= \text{the component of grad } \varphi(x) \text{ tangent to } S \text{ at } x \\ &= \nabla\varphi(x) - N(x)\partial\varphi/\partial N(x)\end{aligned}$$

where N is an unit-normal field on S. $\nabla_S\varphi$ depends only on the restriction $\varphi|S$.
(ii) If S is a C^2 hypersurface, $\mathbf{a} : \mathbb{R}^n \to \mathbb{R}^n$ is C^1 on a neighborhood of S, $N : \mathbb{R}^n \to \mathbb{R}^n$ is a C^1 unit-vector field on a neighborhood of S which is a normal field at points of S near $x_0 \in S$ and $H =$ div N is the mean curvature of S (near x_0), then

$$\mathrm{div}_S\mathbf{a} = \mathrm{div}\,\mathbf{a} - H\mathbf{a}\cdot N - \frac{\partial}{\partial N}(a\cdot N)$$

on S (near x_0), where div $\mathbf{a} = \sum_{i=1}^n \frac{\partial a_i}{\partial x_i}$. div$_S\mathbf{a}$ *depends only on the tangential component of* $\mathbf{a}$ *at point of S.*
(iii) If S is a C^2 hypersurface, $u : \mathbb{R}^n \to \mathbb{R}$ is C^2 on a neighborhood of S, and N is a normal-vector field for S (near x_0) in the sense of (ii) above, then

$$\Delta_S u = \Delta u - H\frac{\partial u}{\partial N} - \frac{\partial^2 u}{\partial N^2} + \nabla_S u \cdot \frac{\partial N}{\partial N}$$

on S near x_0. We may choose N so that $\frac{\partial N}{\partial N} = 0$ on S and then the final term is omitted. $\Delta_S u$ depends only on the restriction $u|S$.

Proof. (i) is trivial; we prove (ii), and (iii) may be proved in the same way (or see [11]).

Use the "normal coordinates" of Th. 1.5, and let S_t be the normal translation of S, $S_t = \{\xi + tN(\xi) \mid \xi \in S\}$; we may suppose φ has small support, so we need only work with a small piece of S. Let $(S, S_t) = \{\xi + \theta N(\xi) \mid 0 < \theta < t, \xi \in S\}$ and by the divergence theorem

$$\int_{(S,S_t)} (\varphi\,\mathrm{div}\,\mathbf{a} + \nabla\varphi\cdot\mathbf{a}) = \int_{S_t} \varphi\mathbf{a}\cdot N - \int_S \varphi\mathbf{a}\cdot N.$$

The derivative at $t = 0$ is then

$$\int_S (\varphi\,\mathrm{div}\,\mathbf{a} + \nabla\varphi\cdot\mathbf{a}) = \int_S \frac{\partial}{\partial N}(\varphi\mathbf{a}\cdot N) + H\varphi\mathbf{a}\cdot N$$

so

$$\int_S \left\{\mathbf{a}\cdot\nabla_S\varphi + \varphi(\mathrm{div}\,\mathbf{a} - H\mathbf{a}\cdot N - \frac{\partial}{\partial N}(\mathbf{a}\cdot N))\right\} = 0$$

for all such φ (with small support). This is the formula claimed for a particular choice of the normal field (with $\partial N/\partial N = 0$), and the general case follows easily.

Remark. $\Delta_S u$ may also be determined from

$$\int_{S\cap B_\epsilon(x)} udA/|S\cap B_\epsilon(x)| = u(x) + \frac{\epsilon^2}{2(n+1)}\Delta_S u(x) + o(\epsilon^2) \text{ as } \epsilon \to 0^+.$$

Alternatively if S is written as a graph over the tangent plane $T_x(S)$, S (near x) is $\{x + \xi - \nu_x(\xi)N(x) \mid \xi \in T_x(S) \text{ near } 0\}$ where $\nabla_x(\cdot)$ is C^2, $\nu_x(\xi) = O(|\xi|^2)$ as $\xi \to 0$, and if Δ_ξ denotes the Laplacian relative to (induced) Cartesian coordinates in $T_x(\partial\Omega)$, then

$$\Delta_\xi u(x + \xi - \nu_x(\xi)N(x))|_{\xi=0} = \Delta_S u(x).$$

Chapter 2

Differential Calculus of Boundary Perturbations

We will develop a kind of differential calculus in which the independent variable is the domain of definition of a boundary-value problem. There are conceptual difficulties in this task which have already been encountered in continuum mechanics. There are two customary ways to describe the motion (or deformation of the region):

(i) the Lagrangian description, in which we label each "particle," for example by giving its position at some time t_0;

(ii) the Eulerian description, in which we write the velocity and other variables as functions of time and position in a fixed coordinate system.

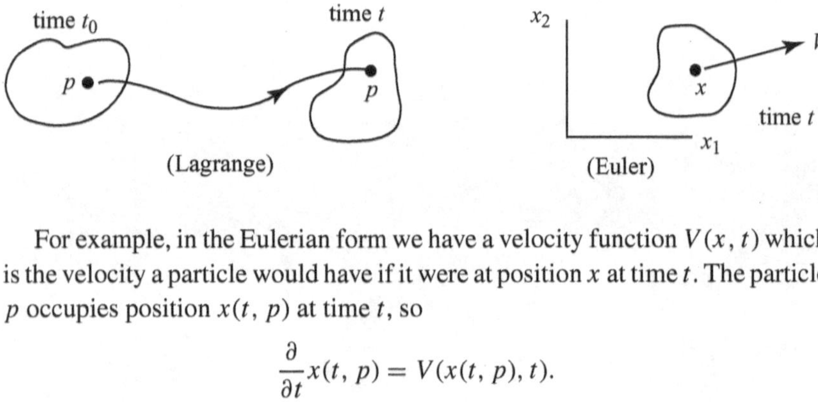

For example, in the Eulerian form we have a velocity function $V(x, t)$ which is the velocity a particle would have if it were at position x at time t. The particle p occupies position $x(t, p)$ at time t, so

$$\frac{\partial}{\partial t}x(t, p) = V(x(t, p), t).$$

The Eulerian form is frequently (not always) more natural and simpler for computations, but we use the Lagrangian form to prove theorems. In fact, we use both methods – they are essentially equivalent – and the results of this chapter show how to pass from one to the other.

We will use an artificial "time" t to parameterize the regions; when there is a natural time-dependence in the problem this could cause confusion and another letter ($\neq t$) should be used.

Consider a formal non-linear differential operator $u \mapsto v$,

$$v(y) = f\left(y, u(y), \frac{\partial u}{\partial y_1}, \dots, \frac{\partial u}{\partial y_n}, \frac{\partial^2 u}{\partial y_1^2}, \frac{\partial^2 u}{\partial y_1 \partial y_2}, \dots\right), \quad y \in \mathbb{R}^n$$

where f is a given function and u, f might have several components. For the sake of less cumbersome notation, we define a constant matrix coefficient differential operator L

$$Lu(y) = \left(u(y), \frac{\partial u}{\partial y_1}(y), \dots, \frac{\partial u}{\partial y_n}(y), \frac{\partial^2 u}{\partial y_1^2}(y), \dots\right)$$

with as many terms as are needed, so our nonlinear operator becomes

$$u \mapsto f(\cdot, Lu(\cdot)).$$

More precisely, suppose $Lu(\cdot)$ has values in $\mathbb{R}^p$ and $f(y, \lambda)$ is defined for (y, λ) in some open set $G \subset \mathbb{R}^n \times \mathbb{R}^p$. For subsets $\Omega \subset \mathbb{R}^n$ define F_Ω by

$$F_\Omega(u)(y) = f(y, Lu(y)), \quad y \in \Omega$$

for sufficiently smooth functions u on Ω such that $(y, Lu(y)) \in G$ for all $y \in \overline{\Omega}$. For example, if f is continuous, Ω is bounded and L involves derivatives of order $\leq m$, the domain of F_Ω is an open set – perhaps empty – in $C^m(\Omega)$, and the values of F_Ω are in $C^0(\Omega)$. [Other function spaces could be used, with the obvious modifications.] If Ω is bounded and the domain of F_Ω is nonempty then also the domain of $F_{h(\Omega)}$ is nonempty in $C^m(h(\Omega))$ for all $h : \Omega \to \mathbb{R}^n$ in some neighborhood of the inclusion i_Ω. We will usually work with bounded domains, but this is not essential; in Example 4 (Capacity) below, we consider deformations of $\mathbb{R}^n \backslash \Omega$ where Ω is bounded, applying diffeomorphisms with compact support (equal to the identity at large distances). There may be some technical problems when $\partial\Omega$ is unbounded, but for many purposes it is sufficient to restrict attention to diffeomorphisms with compact support.

Let $h : \Omega \to \mathbb{R}^n$ be a C^m imbedding, i.e., a C^m diffeomorphism to its image $h(\Omega)$. We define the composition map (or pull-back) h^* by

$$h^*u(x) = u(h(x)), \quad x \in \Omega$$

when u is a given function on $h(\Omega)$, Then h^* is an isomorphism of $C^m(h(\Omega))$ onto $C^m(\Omega)$, with inverse $h^{*-1} = (h^{-1})^*$, and we use the same notation for the pull-back in other function spaces, Hölder spaces $h^* : C^{k,\beta}(h(\Omega)) \to C^{k,\beta}$ $(0 \leq k + \beta \leq m)$, Sobolev spaces $h^* : W^{k,p}(h(\Omega)) \to W^{k,p}(\Omega)$ $(0 \leq k \leq m, 1 \leq p \leq \infty)$, etc. If $u = col(u_1, \dots, u_q)$ then $h^*u = col(h^*u_1, \dots, h^*u_1)$. An appropriate pull-back for the normal-vector field N (actually, covector) will be discussed below (2.3); our problems do not have sufficient geometric structure to make it worthwhile burdening the pull-back h^* with more responsibilities.

For such an imbedding h of a bounded region Ω, $F_{h(\Omega)}$ has an open domain $\mathcal{D}(F_{h(\Omega)} \subset C^m(h(\Omega))$ and

$$F_{h(\Omega)} : \mathcal{D}(F_{h(\Omega)}) \subset C^m(h(\Omega) \to C^0(h(\Omega))$$

is what we term the *Eulerian form* of the formal nonlinear differential operator $v \mapsto f(\cdot, Lv(\cdot))$ on $h(\Omega)$, while

$$h^* F_{h(\Omega)} h^{*-1} : h^*\mathcal{D}(F_{h(\Omega)}) \subset C^m(\Omega) \to C^0(\Omega)$$

is the *Lagrangian form;* the same terms are used relative to other function spaces.

The advantage of the Lagrangian form is that it acts in spaces which don't depend on h, facilitating use of the implicit function theorem (for example). But then we need to know the *smoothness* of

$$(h, u) \mapsto h^* F_{h(\Omega)} h^{*-1}(u)$$

and we must be able to *calculate derivatives* with respect to h. (Since h^* is linear, the derivative with respect to u presents no problems.)

Smoothness. First note (with $v = h^{*-1}u$, $y = h(x)$, $x \in \Omega$)

$$\begin{aligned} h^* F_{h(\Omega)} h^{*-1} u(x) &= F_{h(\Omega)} v(h(x)) = f(y, Lv(y)) \\ &= f(h(x), h^* L h^{*-1} u(x)). \end{aligned}$$

A typical constant coefficient differential operator is

$$\begin{aligned} (\partial/\partial y)^\alpha &= \prod_{i=1}^{n} (\partial/\partial y_j)^{\alpha j}, \\ \alpha &= (\alpha_1, \ldots, \alpha_n), \alpha_j = \text{integer} \geq 0. \end{aligned}$$

From the chain rule

$$h^* \left(\frac{\partial}{\partial y}\right)^\alpha h^{*-1} u(x) = \left(\frac{\partial}{\partial y}\right)^\alpha v(y) = \prod_{j=1}^{n} \left(\sum_{i=1}^{n} (h_x^{-1})_{ij} \frac{\partial}{\partial x_i} \right)^{\alpha_j} u(x).$$

Thus, for example

$$(h, u) \mapsto (h^* L h^{*-1}) u : \text{Diff}^m(\Omega) \times C^m(\Omega) \to C^0(\Omega)$$

is analytic (in fact, rational), where $\text{Diff}^m(\Omega)$ is the open subset of maps in $C^m(\Omega, \mathbb{R}^n)$ which are diffeomorphisms to their images (= imbeddings), supposing Ω is a bounded C^m region. If in addition f is C^k or analytic on G, then

$$(h, u) \mapsto h^* F_{h(\Omega)} h^{*-1}(u) = f(h(\cdot), h^* L h^{*-1} u)$$
$$: \mathrm{Diff}^m(\Omega) \times C^m(\Omega) \to C^0(\Omega)$$

is C^k or analytic, respectively, on its (open) domain of definition. Other function spaces may be (and will be) employed, and smoothness results are equally simple.

Calculation of Derivative. We will compute the Gâteaux derivative of $h \mapsto h^* F_{h(\Omega)} h^{*-1}(u)$, i,e., the t-derivative along a C^1 curve $t \mapsto h(t, \cdot)$ of imbeddings. In the natural Eulerian form, we would compute

$$\frac{\partial}{\partial t} F_{\Omega(t)}(v)(y) = \frac{\partial}{\partial t} f(y, Lv(y))$$

with y fixed in $\Omega(t) = h(t, \Omega)$; but $y = h(t, x)$ for some $x \in \Omega = \Omega(0)$, and to keep y fixed, x must move. If $U(x, t) = h_x^{-1} h_t$ and $x(t)$ solves the differential equation $dx/dt = -U(x, t)$, then $\frac{d}{dt} h(t, x(t)) = 0$. Thus the partial derivative in t in the Eulerian form with $y = h(x, t)$ fixed, corresponds to the *anti-convective derivative* D_t in the reference region Ω (Lagrangian form)

$$D_t = \frac{\partial}{\partial t} - U(x, t) \cdot \frac{\partial}{\partial x}, \qquad U = h_x^{-1} h_t.$$

More precisely, we have the following.

Lemma 2.1. *Suppose* $\psi : \mathbb{R}^n \times \mathbb{R} \to \mathbb{R}$ *is* C^1, $h : \mathbb{R}^n \times \mathbb{R} \to \mathbb{R}^n$ *is* C^1 *and* $\det h_x(x, t) \neq 0$ *in the region of interest; then*

$$D_t(h^*(t)\psi(\cdot, t)) = h^*(t) \frac{\partial \psi}{\partial t}(\cdot, t)$$

where $h^*(t)$ *is the pull-back by* $h(\cdot, t)$.

Proof. If $y = h(x, t)$, by the chain rule,

$$\begin{aligned} D_t(h^*(t)\psi(\cdot, t))(x) &= \left(\partial_t - \sum_1^n U_j \frac{\partial}{\partial x_j}\right) \psi(h(x, t), t) \\ &= \frac{\partial \psi}{\partial t}(y, t) + \sum_1^n \frac{\partial \psi}{\partial y_k} \left(\frac{\partial h_k}{\partial t} - \sum_{j=1}^n U_j \frac{\partial h_k}{\partial x_j}\right) \\ &= \frac{\partial \psi}{\partial t}(y, t) = h(t)^* \frac{\partial \psi}{\partial t}(\cdot, t). \end{aligned}$$

Remark. The above lemma (and the results below) are stated "pointwise," but will later be interpreted as equalities of curves (parametrized by t) with values in certain function spaces. The interpretation is immediate for the spaces C^m, but in general we argue by continuity, starting from C^m approximations.

Theorem 2.2. *Suppose $f(t, y, \lambda)$ is C^1 on an open set in $\mathbb{R} \times \mathbb{R}^n \times \mathbb{R}^p$, L is a constant-coefficient differential operator of order $\leq m$ with $Lv(y) \in \mathbb{R}^p$ (where defined) and for open sets $Q \subset \mathbb{R}^n$ and C^m functions v on Q, let $F_Q(t)v$ be the function*

$$y \mapsto f(t, y, Lv(y)), \quad y \in Q,$$

where defined.

Suppose $t \mapsto h(t, \cdot)$ is a curve of imbeddings of an open set Ω in $\mathbb{R}^n$, $\Omega(t) = h(t, \Omega)$ and for $|j| \leq m$, $|k| \leq m+1$, $(t, x) \mapsto \partial_t \partial_x^j h(t, x)$, $\partial_x^k h(t, x)$, $\partial_t \partial_x^j u(t, x)$, $\partial_x^k u(t, x)$ are continuous on $\mathbb{R} \times \Omega$ near $t = 0$, and

$$h(t, \cdot)^{*-1} u(t, \cdot)$$

is in the domain of $F_{\Omega(t)}(t)$. Then at points of Ω,

$$D_t\big(h^* F_{\Omega(t)}(t) h^{*-1}\big)(u) = \big(h^* \dot{F}_{\Omega(t)}(t) h^{*-1}\big)(u) + \big(h^* F'_{\Omega(t)}(t) h^{*-1}\big)(u) D_t u$$

where D_t is the anti-convective derivative defined above,

$$\dot{F}_Q(t) v(y) = \frac{\partial f}{\partial t}(t, y, Lv(y))$$

and

$$F'_Q(t) v \cdot w(y) = \frac{\partial f}{\partial \lambda}(t, y, Lv(y)) Lw(y), \quad y \in Q;$$

is the linearization of $v \mapsto F_Q(t)v$.

Proof. Define $v = h^{*-1} u$, i.e., $u(t, x) = v(t, h(t, x))$; by the lemma,

$$D_t(h^* L h^{*-1})(u) = D_t(h^* L v) = h^* L \frac{\partial v}{\partial t} = (h^* L h^{*-1}) D_t u$$

so

$$\begin{aligned} D_t\big(h^* F_{\Omega(t)}(t) h^{*-1} u\big) &= D_t h^* f(t, \cdot, Lv) \\ &= h^* \left(\frac{\partial f}{\partial t}(t, \cdot, Lv) + \frac{\partial f}{\partial \lambda}(t, \cdot, Lv) L \frac{\partial v}{\partial t} \right) \end{aligned}$$

which is the result claimed.

Remark. Suppose we deal with a linear operator

$$A = \sum_{|\alpha| \leq m} a_\alpha(y) \left(\frac{\partial}{\partial y} \right)^\alpha,$$

not explicitly dependent on t, and $h(x,t) = x + tV(x) + o(t)$ as $t \to 0$. Then at $t = 0$

$$\frac{\partial}{\partial t}(h^* A_{h(\Omega)} h^{*-1} u)\big|_{t=0} = V \cdot \nabla(Au) + A(\partial u/\partial t - V \cdot \nabla u)$$
$$= A\frac{\partial u}{\partial t} + [V \cdot \nabla, A]u,$$

The commutator of $V \cdot \nabla$ and A is still of order m – assuming smooth coefficients – and this is the result we would obtain by first changing variables (x in place of $h(t,x) = y$) and then taking the derivative at $t = 0$. If we applied the chain rule directly, we would never see a derivative of order $(m+1)$ and probably not notice the simple commutator structure, especially if (as usual) we compute the derivative at a solution u of $Au = 0$. This is the reason for the superiority of our general approach compared to a "bare-hands" change of variable (and the source of the "miracle" mentioned in the Introduction). This point was also noted by Peetre [27], and related to a Lie derivative. For operators in variational form, Courant [5, Vol. 1, p. 260] gives an equivalent formula.

We must also treat boundary conditions, and a quite general form of boundary condition is

$$b(t, y, Lv(y), MN_{\Omega(t)}(y)) = 0 \qquad \text{for } y \in \partial\Omega(t),$$

where L,M are constant-coefficient differential operators and $N_{\Omega(t)}(y)$ is the outward unit normal for $y \in \partial\Omega(t)$, extended smoothly as a unit vector field on a neighborhood of $\partial\Omega(t)$. For example, the Neumann problem requires $N_{\Omega(t)} \cdot$ grad $v = 0$ on $\partial\Omega(t)$, while some boundary conditions of the theory of elasticity involve the curvature of the boundary which may be expressed using the derivative of the normal. In these cases – and any "natural" boundary conditions – the particular extension of $N_{\Omega(t)}$ away from the boundary is irrelevant. Rather than hypothesizing this, we choose some extension of N_Ω in the reference region and then define $N_{\Omega(t)} = N_{h(t,\Omega)}$ by

$$h^* N_{h(\Omega)}(x) = N_{h(\Omega)}(h(x)) = {}^T h_x^{-1} N_\Omega(x) / \| {}^T h_x^{-1} N_\Omega(x) \| \tag{2.1}$$

for x near $\partial\Omega$, where ${}^T h_x^{-1}$ is the inverse-transpose ("contra-gradient") of the Jacobian matrix $h_x = [\partial h_i / \partial x_j]_{i,j=1}^n$ and $\| \,.\, \|$ is the Euclidean norm. This is the extension understood in the above boundary condition: $b(t, y, Lv(y), MN_{\Omega(t)}(y))$ is defined for $y \in \Omega$ near $\partial\Omega$ and has limit zero (in some sense, depending on the function spaces employed) as $y \to \partial\Omega(t)$.

Note that $h \mapsto h^* N_{h(\Omega)}(x)$ is analytic for each x, but the smoothness in x depends on the smoothness of $\partial\Omega$ (and N_Ω and h). If $\Omega = \{x : \varphi(x) > 0\}$ and

$N_\Omega = -\operatorname{grad}\varphi(x)/\|\operatorname{grad}\varphi(x)\|$ near $\partial\Omega$, and if $\psi_h(h(x)) = \varphi(x)(\psi_h = h^{*-1}\varphi)$ then according to the extension (2.1) above

$$N_h(\Omega)(y) = -\operatorname{grad}\psi_h(y)/\|\operatorname{grad}\psi_h(y)\| \qquad \text{for } y \text{ near } \partial h(\Omega).$$

At points of $\partial h(\Omega)$, the unit normal vector is determined geometrically, compatibly with (2.1).

Lemma 2.3. *Let Ω be a C^2-regular region, $N_\Omega(\cdot)$ a C^1 unit-vector field defined on a neighborhood of $\partial\Omega$ which is the outward normal on $\partial\Omega$, and for C^2 imbeddings $h : \Omega \to \mathbb{R}^n$ define $N_{h(\Omega)}$ on a neighborhood of $h(\partial\Omega) = \partial h(\Omega)$ by (2.1) above. Suppose $h(t,\cdot)$ is an imbedding for each t, defined by*

$$\frac{\partial}{\partial t}h(t,x) = V(t,h(t,x)) \quad \text{for } x \in \Omega,\ h(0,x) = x,$$

$(t,y) \mapsto V(t,y)$ is C^2 and $\Omega(t) = h(t,\Omega)$, $N_{\Omega(t)} = N_{h(t,\Omega)}$. Then for x near $\partial\Omega$, $y = h(t,x)$ near $\partial\Omega(t)$, we may compute the derivative $(\frac{\partial}{\partial t})_{y=\text{const.}}$ and, if $y \in \partial\Omega(t)$,

$$\begin{aligned}\left(\frac{\partial}{\partial t}\right) N_{\Omega(t)}(y) &= D_t(H^* N_{h(\Omega)})(x)\\ &= -(\nabla_{\partial\Omega(t)}\sigma + \sigma\,\partial N_{\Omega(t)}/\partial N_{\Omega(t)})(y)\end{aligned}$$

$\sigma = V \cdot N_{\Omega(t)}$ is the normal velocity and $\nabla_{\partial\Omega(t)}\sigma = \nabla\sigma - N_{\Omega(t)}\frac{\partial\sigma}{\partial N_{\Omega(t)}}$ is the component of the gradient tangent to $\partial\Omega(t)$.

Remark. We may, if desired, choose N_Ω so $\partial N_\Omega/\partial N_\Omega \equiv 0$ near $\partial\Omega$ (as in Th. 1.5); but in general $\partial N_{\Omega(t)}/\partial N_{\Omega(t)}$ will be a nontrivial vector field orthogonal to $N_{\Omega(t)}$.

The formula is claimed only for $y \in \partial\Omega(t)$, but to make sense of the derivative we compute initially for u near $\partial\Omega(t)$.

Proof. It suffices to find the derivative at $t = 0$ (since we can transfer the "origin" Ω of (2.1) to any $\Omega(t)$; see below). Thus let $h(t,x) = x + tV(x) + o(t)$, $V(x) = V(0,x)$ so $N_{\Omega(t)}(h(t,x))$ has i^{th} component

$$N_i - t\sum_j \frac{\partial V_j}{\partial x_i}N_j + tN_i\sum_{j,k} N_j N_k \frac{\partial V_j}{\partial x_k} + o(t)$$

and $D_t(h^* N_{\Omega(t)})|_{t=0} = -V_j\partial N_i/\partial x_j - N_j\partial V_j/\partial x_i + N_i N_j N_k \partial V_j/\partial x_k$, summing over repeated indices. If $V = \sigma N + \tau$, $\tau \cdot N \equiv 0$, we find

$$D_t(h^* N_{\Omega(t)})_i|_{t=0} = -\partial\sigma/\partial x_i + N_i N_k \partial\sigma/\partial x_k - \sigma N_j \partial N_i/\partial x_j + q_i$$

with $q_i = -\tau_j \partial N_i/\partial x_j + \tau_j \partial N_j/\partial x_i - \tau_j N_i N_k \partial N_j/\partial x_k$ and we have used $\tau \cdot N \equiv 0$, $N \cdot N \equiv 1$. Now $q_i N_i \equiv 0$ and for any $\rho \perp N$ we have $\rho_i q_i = \rho_i \tau_j (\partial N_j/\partial x_i - \partial N_i/\partial x_j)$ which vanishes on $\partial\Omega$: this depends only on $N|\partial\Omega$, not on the extension, and for the extension of Th. 1.5, $[\partial N_i/\partial x_j]$ is a symmetric matrix. Thus $q = 0$ on $\partial\Omega$, which proves the result.

Theorem 2.4. *Let $b(t, y, \lambda, \mu)$ be a C^1 function on an open set of $\mathbb{R} \times \mathbb{R}^n \times \mathbb{R}^p \times \mathbb{R}^q$ and let L, M be constant-coefficient differential operators (and order $\leq m$) of appropriate dimensions so $b(t, y, Lv(y), MN(y))$ makes sense. Assume Ω is a C^{m+1} region, $N_\Omega(x)$ is a C^m unit-vector field near $\partial\Omega$ which is the outward unit normal on $\partial\Omega$, and define $N_{h(\Omega)}$ by*

$$N_{h(\Omega)}(h(x)) = {}^T h_x^{-1} N_\Omega(x) / \| {}^T h_x^{-1} N_\Omega(x) \|$$

when $h : \Omega \mapsto \mathbb{R}^n$ is a C^{m+1} smooth imbedding. Also define $B_{h(\Omega)}(t)$ by

$$B_{h(\Omega)}(t) V(y) = b\big(t, y, Lv(y), MN_{h(\Omega)}(y)\big)$$

for $y \in h(\Omega)$ near $\partial h(\Omega)$.

If $t \mapsto h(t, \cdot)$ is a curve of C^{m+1} imbeddings of Ω and for $|j| \leq m$, $|k| \leq m+1$, $(t, x) \mapsto (\partial_t \partial_x^j h, \partial_x^k h, \partial_t \partial_x^j u, \partial_x^k u)(t, x)$ are continuous on $\mathbb{R} \times \Omega$ near $t = 0$, then at points of Ω near $\partial\Omega$

$$\begin{aligned} D_t\big(h^* B_{h(\Omega)}(t) h^{*-1}\big)(u) = {} & \big(h^* \dot{B}_{h(\Omega)} h^{*-1}\big)(u) + \big(h^* B'_{h(\Omega)} h^{*-1}\big)(u) \cdot D_t u \\ & + \big(\big(h^* \partial B_{h(\Omega)}/\partial N\big) h^{*-1}\big)(u) \cdot D_t\big(h^* N_{h(\Omega)}\big) \end{aligned}$$

where $h = h(t, \cdot)$; $\dot{B}_{h(\Omega)}$, $B'_{h(\Omega)}$ are defined as in Th. 2.2,

$$\big(\partial B_{h(\Omega)}/\partial N\big)(v) \cdot n(y) = \frac{\partial h}{\partial \mu}\big(t, y, Lv(y), MN_{h(\Omega)}(y)\big) \cdot Mn(y)$$

and $D_t(h^* N_{h(\Omega)})|_{\partial\Omega}$ is computed in the previous lemma.

Proof. The same calculation as in Th. 2.2

Change of Origin. In the above, the "origin" or reference region is Ω, but we may easily transfer the origin to any Ω_1 diffeomorphic to Ω. Let $H_1 : \Omega \to \Omega_1$ be the diffeomorphism and for every imbedding $h : \Omega \to \mathbb{R}^n$ define the imbedding $h_1 = h \circ H_1^{-1} : \Omega_1 \to \mathbb{R}^n$. Similarly define $x_1 = H_1(x)$, $u_1 = H^{*-1} u$, $N_{\Omega_1}(x_1) = N_{H_1(\Omega)}(H_1 x) = {}^T H_{1,x}^{-1} N_\Omega(x) / \| \cdots \|$ and then $h(\Omega) = h_1(\Omega_1)$,

$$h^* F_{h(\Omega)} h^{*-1} u(x) = h_1^* F_{h_1(\Omega_1)} h_1^{*-1} u_1(x_1)$$

$$h^* B_{h(\Omega)} h^{*-1} u(x) = h_1^* B_{h_1(\Omega_1)} h_1^{*-1} u_1(x_1)$$

using the normal

$$N_{h_1(\Omega_1)}(h_1(x_1)) = {}^T h_{1,x_1}^{-1} N_{\Omega_I}(x_1)/\|\cdots\| = {}^T h_x^{-1} N_\Omega(x)/\|\cdots\|$$
$$= N_{h(\Omega)}(h(x)).$$

This "change of origin" is used frequently in the sequel, as it permits us to compute derivatives in h at $h = i_\Omega$, where the formulas are simpler.

Chapter 3

Examples Using the Implicit Function Theorem

We prove differentiability with respect to the domain for various quantities (e.g., eigenvalues) associated with a boundary-value problem, and compute the derivative, and sometimes a second derivative, using the formulas of Chapter 2. Examples 3.1, 3.2 and 3.5 are treated in some detail, to show the calculations are not merely formal.

Example 3.1. Torsional Rigidity. The resistance to torsion of a cylindrical rod depends not only on the elastic constants of the material, but also on the geometry of the cross-section $\Omega \subset \mathbb{R}^2$, through the torsional rigidity $R(\Omega)$.

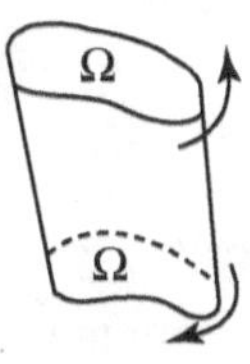

$$R(\Omega) = \int_\Omega |\nabla u|^2 \text{ where } u : \Omega \to \mathbb{R} \text{ is the solution of}$$

$$\Delta u = -1 \text{ in } \Omega, \quad u = 0 \text{ on } \partial\Omega.$$

It is convenient to use $R(\Omega) = \int_\Omega u$, noting that $\int_\Omega u = \int_\Omega -u\Delta u = \int_\Omega |\nabla u|^2$.

Let $0 < \alpha < 1$ and let $\Omega \subset \mathbb{R}^2$ be a bounded $C^{2,\alpha}$-regular region. $\mathrm{Diff}^{2,\alpha}(\Omega)$ is the open set of $C^{2,\alpha}(\Omega, \mathbb{R}^2)$ consisting of imbeddings; if $h \in \mathrm{Diff}^{2,\alpha}(\Omega)$, then $h(\Omega)$ is also $C^{2,\alpha}$-regular.

Suppose $v : h(\Omega) \to \mathbb{R}$ solves $\Delta v = -1$ in $h(\Omega)$, $v = 0$ on $\partial h(\Omega)$; then $R(h(\Omega)) = \int_{h(\Omega)} v$. Also $u = h^* v$ solves

$$(h^* \Delta h^{*-1}) u = -1 \text{ in } \Omega, \quad u = 0 \text{ on } \partial\Omega,$$

and

$$R(h(\Omega)) = \int_\Omega u(x) |\det h'(x)| dx.$$

Consider the map $\Phi : (u, h) \mapsto (h^* \Delta h^{*-1})u$ from $C^{2,\alpha}(\Omega)_0 \times \mathrm{Diff}^{2,\alpha}(\Omega)$ to $C^{0,\alpha}(\Omega)$, where $C^{2,\alpha}(\Omega)_0 = \{u \in C^{2,\alpha}(\Omega) | u = 0 \text{ on } \partial\Omega\}$. Φ is an analytic map, and we apply the implicit function theorem for h near i_Ω (= the inclusion of Ω in $\mathbb{R}^2$).

Let u_0 be the solution of $\Delta u_0 = -1$ in Ω, $u_0 = 0$ on $\partial\Omega$, so $\Phi(u_0, i_\Omega) = -1$. Now $u \mapsto \Phi(u, i_\Omega) = \Delta u : C^{2,\alpha}(\Omega)_0 \to C^{0,\alpha}(\Omega)$ is an isomorphism, by familiar results of elliptic theory. So for every h in a $C^{2,\alpha}(\Omega, \mathbb{R}^2)$-neighborhood of i_Ω, there is a unique solution $u_h \in C^{2,\alpha}(\Omega)_0$ of $\Phi(u_h, h) = -1$ and $h \mapsto u_h$ is analytic, with $u_h = u_0$ when $h = i_\Omega$. It follows that $h \mapsto R(h(\Omega)) = \int_\Omega u_h(x) \det h'(x) dx$ is also analytic and we will compute its derivative at $h = i_\Omega$.

It is convenient to have more smoothness for the initial calculations. If Ω is $C^{m,\alpha}$ regular ($m \geq 2$), the same argument shows $h \mapsto (u_h, R(h(\Omega))) : C^{m,\alpha}(\Omega, \mathbb{R}^2) \to C^{m,\alpha}(\Omega)_0 \times \mathbb{R}$ is analytic on a neighborhood of i_Ω. Suppose $t \mapsto h(t, \cdot) \in C^{m,\alpha}(\Omega, \mathbb{R}^2)$ is a C^1 curve near $t = 0$ with $h(t, x) = x + tV(x) + c(t)$ as $t \to 0$. Then $t \mapsto u_{h(t,\cdot)} \in C^{m,\alpha}(\Omega)_0$ is also C^1 near $t = 0$ and, say,

$$u_{h(t,\cdot)} = u_0 + tu_1 + o(t) \quad \text{as } t \to 0,$$

so

$$\frac{d}{dt} R(h(t, \Omega))|_{t=0} = \int_\Omega (u_1 + u_0 \operatorname{div} V).$$

Now $t \mapsto D_t u_{h(t,\cdot)} \in C^{m-1,\alpha}(\Omega)$ is continuous and (if $m \geq 3$)

$$D_t \big(h^* \Delta h^{*-1} u_{h(t,\cdot)}\big) = (h^* \Delta h^{*-1}) D_t u_{h(t,\cdot)}$$

is an equation of continuous functions of t with values in $C^{m-3,\alpha}(\Omega)$. The equation is valid if everything is smooth (Theorem 2.2) and by continuity, is valid along our solution curve $t \mapsto u_{h(t,\cdot)}$ (if $m \geq 3$). At $t = 0$ we have $\Delta v_1 = 0$, where $v_1 = D_t u_{h(t,\cdot)}|_{t=0} = u_1 - V \cdot \nabla u_0$. When $x \in \partial\Omega$, $u_{h(t,\cdot)}(x) = 0$ so $u_1 = u_0 = 0$ on $\partial\Omega$. Also

$$\begin{aligned}
\frac{d}{dt} R(h(t, \Omega))|_{t=0} &= \int_{\Omega_0} (u_1 - V \cdot \nabla u_0 + \operatorname{div} (V u_0)) = \int_\Omega v_1 \\
&= \int_\Omega -v_1 \Delta u_0 + u_0 \Delta v_1 = \int_{\partial\Omega} -v_1 \frac{\partial u_0}{\partial N} + u_0 \frac{\partial v_1}{\partial N} \\
&= \int_{\partial\Omega} V \cdot N \left(\frac{\partial u_0}{\partial N}\right)^2.
\end{aligned}$$

This was proved for the case $m \geq 3$, but we know the derivative exists in the setting of $C^{2,\alpha}$ data; we prove by continuity that the final equation

$$\frac{d}{dt}R(h(t,\Omega))|_{t=0} = \int_{\partial\Omega} V \cdot N\left(\frac{\partial u_0}{\partial N}\right)^2$$

also holds for $C^{2,\alpha}$ data.

We may assume $h(t,\Omega) = \{x \mid \varphi(t,x) > 0\}$ for some φ such that $t \mapsto \varphi(t,\cdot) \in C^{2,\alpha}(\mathbb{R}^2)$ is C^1, $\varphi(t,x) = 0 \Rightarrow |\varphi_x(t,x)| \geq 1$. There exists $\varphi(s,t,x)$ which is C^∞ for $s \neq 0$, $\varphi(0,t,x) = \varphi(t,x)$, and $(s,t) \mapsto \varphi(s,t,\cdot)$, $\partial_t\varphi(s,t,\cdot) \in C^{2,\alpha}(\mathbb{R}^2)$ are continuous near (0,0). Let $\Omega(s,t) = \{x | \varphi(s,t;x) > 0\}$, and let $u(s,t,\cdot)$ be the solution of $\Delta_x u = -1$ in $\Omega(s,t)$, $u = 0$ on $\partial\Omega(s,t)$; then for $s \neq 0$

$$\frac{\partial}{\partial t}R(\Omega(s,t)) = \int_{\partial\Omega(s,t)} V \cdot N_{\Omega(s,t)}\left(\frac{\partial u}{\partial N}\right)^2$$

and we want the limit as $s \to 0$. By Theorem 1.8, $\Omega(s,t)$ converges to $h(t,\Omega)$ in $C^{2,\alpha}$ as $s \to 0$, so the implicit function argument above proves $R(\Omega(s,t)) \to R(h(t,\Omega))$ as $s \to 0$. To see the limit of the integral has the expected form, we write it as an integral over $\partial\Omega$. By Theorem 1.8, $\Omega(s,t) = h(s,t,\Omega)$ for an appropriate $C^{2,\alpha}$ diffeomorphism $h(s,t,\cdot) : \mathbb{R}^n \to \mathbb{R}^n$, and $u(s,t,h(s,t,x)) \equiv u_{h(s,t,\cdot)}(x)$ for $x \in \Omega$ [u_h is the function provided by the previous argument]. On $\partial\Omega(s,t)$, $|\frac{\partial u(s,t,\cdot)}{\partial N}|^2 = |\nabla u(s,t,\cdot)|^2$; and at $h(s,t,x)$, $x \in \partial\Omega$, this is $|h(s,t,\cdot)^*\nabla h^*(s,t,\cdot)^{-1}u_{h(s,t,\cdot)}(x)|^2$. Further $V \cdot N_{\Omega(s,t)} = \frac{\varphi_t}{|\varphi_x|}(s,t,h(s,t,x))$ and the corresponding elements of area for $\partial\Omega(s,t)$, $\partial\Omega$ are in the proportion

$$\frac{dA_{(s,t)}}{dA_{(0,0)}} = \left(\frac{\det({}^T E\,{}^T h_x h_x E)}{\det({}^T E E)}\right)^{1/2}$$

where $E = [e_1, \ldots, e_{n-1}]$ is an $n \times (n-1)$ matrix whose columns are a basis for the tangent space to $\partial\Omega$ (for example $e_j = \partial x(\xi)/\partial\xi_j$, when $\partial\Omega$ is locally the image of $\xi \mapsto x(\xi) : \mathbb{R}^{n-1} \to \mathbb{R}^n$). Substitution in the integral shows we may allow $s \to 0$ and then

$$\frac{\partial}{\partial t}R(\Omega(t)) = \int_{\partial\Omega(t)} V \cdot N_{\Omega(t)}\left(\frac{\partial u}{\partial N}\right)^2 dA$$

where $\Delta u = -1$ in $\Omega(t)$, $u = 0$ on $\partial\Omega(t)$, and $\Omega(t) = h(t,\Omega)$, $t \mapsto h(t,\cdot)$ is a C^1-curve of $C^{2,\alpha}$-imbeddings of the $C^{2,\alpha}$ region Ω.

The Fréchet derivative (at $h = i_\Omega$) is

$$\dot{h} \mapsto \int_{\partial\Omega} \dot{h} \cdot N_\Omega\left(\frac{\partial u_0}{\partial N_\Omega}\right)^2 dA : C^{2,\alpha}(\Omega,\mathbb{R}^2) \to \mathbb{R}.$$

Observe that $\frac{d}{dt}$ (area $\Omega(t)) = \int_{\partial\Omega(t)} V \cdot N_{\Omega(t)} dA$ (by Theorem 1.11 with $f \equiv 1$) so Ω is a critical point of $R(\Omega)$, with fixed area, if and only if $(\partial u_0/\partial N_\Omega)^2 =$ constant on $\partial\Omega$. Since $u_0 > 0$ in Ω, by the maximum principle, this says

$$\Delta u_0 = -1 \text{ in } \Omega, \qquad u_0 = 0 \text{ on } \partial\Omega, \qquad \frac{\partial u_0}{\partial N} = \text{constant} < 0 \text{ on } \partial\Omega$$

if and only if Ω is a critical point of the torsional rigidity with fixed area. This is evidently the case when Ω is a disc in $\mathbb{R}^2$, since u_0 is radially symmetric. Conversely, by a result of Serrin [36], the only (connected, bounded, C^2) regions for which such a solution u_0 exists are discs, so the disc is the only critical point. In fact, the disc maximizes the torsional rigidity for given area. This may be shown by an argument like that of Faber and Krahn (as given in Garabedian [6, p. 413–416]), or see the article of Pólya [30].

The argument above is in Lagrangian form – as it must be to use the implicit function theorem. But once we know a (sufficiently smooth) solution exists, subsequence calculations are simpler in the Eulerian form. Suppose $h(t,x)$, $\frac{\partial h}{\partial t}(t,x)$ are $C^{m,\alpha}$ on $\mathbb{R} \times \Omega$ (near $t = 0$) with $h(0,x) = x$, $m \geq 2$; there is (by Corollary 1.10) an extension H to $\mathbb{R} \times \mathbb{R}^n$ with $(t,x) \mapsto H(t,x)$, $\frac{\partial H}{\partial t}(t,x)$ of class $C^{m,\alpha}$, $H(0,x) = x$ for all $x \in \mathbb{R}^n$, $H(t,x) = h(t,x)$ for $x \in \Omega$, t near 0 and $\frac{\partial}{\partial t} H(t,x) = V(t, H(t,x))$ where $(t,y) \mapsto V(t,y)$ is $C^{m,\alpha}$. Let $u_{h(t,\cdot)}$ be the function provided by the implicit function argument above, and define $U(t,x) = (E_\Omega u_{h(t,\cdot)})(x)$ where E_Ω is the extension operator (Theorem 1.9); then $(t,x) \mapsto U(t,x)$ is $C^{m,\alpha}$ on $\mathbb{R} \times \mathbb{R}^n$ (near $t = 0$). Finally, define $v : \mathbb{R} \times \mathbb{R}^n \to \mathbb{R}$ (near $0 \times \mathbb{R}^n$) by $v(t, H(t,x)) = U(t,x)$ or $v(t,y) = U(t, H(t,\cdot)^{-1}(y))$, so v is also $C^{m,\alpha}$. It follows that $(t,y) \mapsto \Delta_y v(t,y)$ is $C^{m-2,\alpha}$, and

$$\begin{cases} \Delta_y v(t,y) = -1 & \text{in } \Omega(t) = h(t,\Omega) \\ v(t,y) = 0 & \text{on } \partial\Omega(t) = h(t,\partial\Omega). \end{cases}$$

(To prove *directly* the smoothness of v seems a very difficult problem.) In this form, the calculations are quite straightforward. Thus $R(\Omega(t)) = \int_{\Omega(t)} v(t,y)dy$ so (recalling $v = 0$ on the boundary)

$$\frac{d}{dt} R(\Omega(t)) = \int_{\Omega(t)} \frac{\partial v}{\partial t}.$$

But $\Delta_y(\frac{\partial v}{\partial t}) = 0$ in $\Omega(t)$ and so (assuming $m \geq 3$, so $\frac{\partial v}{\partial t}(t,\cdot)$ is at least C^2)

$$\begin{aligned} \frac{d}{dt} R(\Omega(t)) &= \int_{\Omega(t)} \left\{ -\frac{\partial v}{\partial t} \Delta_y v + \Delta_y \left(\frac{\partial v}{\partial t} \right) v \right\} \\ &= \int_{\partial\Omega(t)} -\frac{\partial v}{\partial t} \frac{\partial v}{\partial N_y} \end{aligned}$$

Now if $y = y(t)$ solves $\dot{y}(t) = V(t, y(t))$ and $y(t_0) \in \partial\Omega(t_0)$, for some t_0, it follows $y(t) \in \partial\Omega(t)$ for all t so $v(t, y(t)) \equiv 0$; therefore $\frac{\partial v}{\partial t} + V\frac{\partial v}{\partial y} = 0$ on $\partial\Omega(t)$. Also $v(t, \cdot) = 0$ on $\partial\Omega(t)$ implies $\frac{\partial v}{\partial y} = \frac{\partial v}{\partial N} N_{\Omega(t)}$ hence

$$\frac{d}{dt}R(\Omega(t)) = \int_{\partial\Omega(t)} V \cdot N_{\Omega(t)}\left(\frac{\partial v}{\partial N_{\Omega t}}\right)^2.$$

The calculations in this form are so simple that we will go on to compute the second derivative. Writing $\sigma = V \cdot N_{\Omega(t)}$ (defined on a neighborhood of $\partial\Omega(t)$, for each t) and recalling $|\nabla v|^2 = (\partial v/\partial N_{\Omega(t)})^2$ on $\partial\Omega(t)$, we have (for $m \geq 4$)

$$\frac{d^2}{dt^2}R(\Omega(t)) = \int_{\partial\Omega(t)} \left\{\frac{\partial}{\partial t}(\sigma|\nabla v|^2) + \sigma\frac{\partial}{\partial N}(\sigma|\nabla v|^2) + H\sigma^2|\nabla v|^2\right\}$$

where $H = \operatorname{div} N_{\Omega(t)}$ is the mean curvature (Theorem 1.11). Writing $\dot{v}$ for $\frac{\partial v}{\partial t}$, this becomes

$$\begin{aligned}\frac{d^2}{dt^2}R(\Omega(t)) &= \int_{\partial\Omega(t)} \left(\frac{\partial v}{\partial N}\right)^2\left(\frac{\partial\sigma}{\partial t} + \sigma\frac{\partial\sigma}{\partial N} + H\sigma^2\right)\\ &\quad + \int_{\partial\Omega(t)} 2\sigma\frac{\partial v}{\partial N}\left(\sigma\frac{\partial^2 v}{\partial N^2} + \frac{\partial\dot{v}}{\partial N}\right)\end{aligned}$$

($\Delta\dot{v} = 0$ in $\Omega(t)$, $\dot{v} = \sigma\frac{\partial v}{\partial N}$ on $\partial\Omega(t)$; $\sigma = V \cdot N_{\Omega(t)}$) and we observe

$$\frac{d^2}{dt^2}\text{ area }\Omega(t) = \int_{\partial\Omega(t)}\left(\frac{\partial\sigma}{\partial t} + \sigma\frac{\partial\sigma}{\partial N} + H\sigma^2\right).$$

Suppose $\Omega(0)$ is the unit disc in $\mathbb{R}^2$ and keep area $\Omega(t) =$ constant. At $t = 0$, $v = \frac{1}{4}(1 - r^2)$ in polar coordinates and $V \cdot N = \sigma = \sum_{-\infty}^{\infty}\sigma_k e^{ik\theta}$ on $\partial\Omega(0)$, with $\sigma_0 = 0$ since $\sigma_0 = \frac{1}{2\pi}\int_{\partial\Omega(0)} V \cdot N = \frac{1}{2\pi}\frac{d}{dt}\text{ area }\Omega(t)|_{t=0} = 0$. Also at $t = 0$, $\dot{v}$ is the solution of $\Delta\dot{v} = 0$ in $r < 1$, $\dot{v} = \frac{1}{2}\sigma$ on $r = 1$, or $\dot{v}(r, \theta) = \frac{1}{2}\sum_{-\infty}^{\infty}\sigma_k r^{|k|}e^{ik\theta}$, while $\partial v/\partial N = \partial v/\partial r = -\frac{1}{2}$ on $r = 1$. Substitution in the formula above yields

$$\begin{aligned}\frac{d^2}{dt^2}R(\Omega(t))|_{t=0} &= \int_0^{2\pi} d\theta\left(\frac{1}{2}\sigma^2 - \sigma\frac{\partial\dot{v}}{\partial r}\bigg|_{r=1}\right)\\ &= -2\pi\sum_2^{\infty}|\sigma_k|^2(k-1).\end{aligned}$$

This is independent of σ_1 (or $\sigma_{-1} = \overline{\sigma}_1$) which is reasonable as

$$2\pi\sigma_{\pm 1} = \int_{\partial\Omega(0)} e^{\mp i\theta}\sigma = \frac{d}{dt}\iint_{\Omega(t)}(x \mp iy)\,dx\,dy\bigg|_{t=0}$$

gives the velocity of the center of mass at $t = 0$ and does not affect the shape of $\Omega(t)$. The other coefficients are all negative – hardly surprising, as the disc maximizes $R(\Omega)$ with fixed area.

Example 3.2. Simple Eigenvalues of the Dirichlet Problem for the Laplacian. Suppose $\Omega \subset \mathbb{R}^n$ is a C^2-regular bounded region and λ_0 is a simple eigenvalue of the Dirichlet problem in Ω,

$$\Delta u + \lambda u = 0 \text{ in } \Omega, \qquad u = 0 \text{ on } \partial\Omega, \qquad u \not\equiv 0$$

and let u_0 be the corresponding eigenfunction with $\int_\Omega u_0^2 = 1$. For all $h \in C^2(\Omega, \mathbb{R}^n)$ in some C^2-neighborhood of i_Ω, $h(\Omega)$ is also a C^2 region and there is a unique simple eigenvalue $\lambda(h(\Omega))$ near λ_0, $h \mapsto \lambda(h(\Omega))$ is analytic, and its derivative at i_Ω is

$$\dot{h} \mapsto -\int_{\partial\Omega} \dot{h} \cdot N_\Omega \left(\frac{\partial u_0}{\partial N_\Omega}\right)^2.$$

The derivative was computed formally, in some special cases, by Rayleigh [33, p. 338, eq. 11] and for general two-dimensional regions by Hadamard [9], but the first adequate proof is probably that of Garabedian and Schiffer [7].

We work in the Sobolev spaces $H^m(\Omega)$ and $H_0^1(\Omega) = \{u \in H^1(\Omega) \mid u = 0 \text{ on } \partial\Omega\}$. Define $F : H^2 \cap H_0^1(\Omega) \times \mathbb{R} \times \text{Diff}^2(\Omega) \to L_2(\Omega) \times \mathbb{R}$ by

$$F(u, \lambda, h) = \left(h^*(\Delta + \lambda)h^{*-1}u, \int_\Omega u^2 \det h'\right).$$

Then $F(u_0, \lambda_0, i_\Omega) = (0, 1)$ and whenever $F(u, \lambda, h) = (0, 1)$, $v = h^{*-1}u : h(\Omega) \to \mathbb{R}$ satisfies

$$\Delta v + \lambda v = 0 \text{ in } h(\Omega), \qquad v = 0 \text{ on } \partial h(\Omega), \qquad \int_{h(\Omega)} v^2 = 1.$$

Also F is analytic and the derivative

$$\frac{\partial F}{\partial(u, \lambda)}(u_0; \lambda_0, i_\Omega) : (v, \mu) \mapsto \left((\Delta + \lambda_0)v + \mu u_0, 2\int_\Omega u_0 v\right)$$

is an isomorphism. Indeed, since λ_0 is a simple eigenvalue, the image $(\Delta + \lambda_0)(H^2 \cap H_0^1(\Omega)) = \{u_0\}^\perp$, so for any $(f, g) \in L_2(\Omega) \times \mathbb{R}$

$$(\Delta + \lambda_0)v + \mu u_0 = f, \qquad 2\int_\Omega u_0 v = g$$

has a unique solution $(v, \mu) \in H^2 \cap H_0^1(\Omega) \times \mathbb{R}$:

$$\mu = \int_\Omega u_0 f, \qquad v = v_0 + \frac{1}{2} g u_0$$

where

$$(\Delta + \lambda_0)v_0 = f - \mu u_0, \qquad v_0 \perp u_0, \qquad v_0 \in H^2 \cap H_0^1(\Omega).$$

Thus the implicit function theorem applies and there are analytic functions $h \mapsto u_h, \lambda_h$, for $\|h - i_\Omega\|_{C^2}$ small, such that $F(u_h, \lambda_h, h) = (0, 1)$.

If $\partial\Omega$ is C^m regular ($m \geq 2$), the same argument applies to $F : H^m \cap H_0^1(\Omega) \times \mathbb{R} \times \mathrm{Diff}^m(\Omega) \to H^{m-2}(\Omega) \times \mathbb{R}$ to give analytic $h \mapsto (u_h, \lambda_h) \in H^m \cap H_0^1(\Omega) \times \mathbb{R}$ for $\|h - i_\Omega\|_{C^m(\Omega)}$ small. If E_Ω is the extension map of Theorem 1.9, $h \mapsto E_\Omega(u_h) \in H^m(\mathbb{R}^n)$ is analytic, as is $h \mapsto H = id_{\mathbb{R}^n} + E_\Omega(h - i_\Omega) \in C^m(\mathbb{R}^n, \mathbb{R}^n)$, and each H^{-1} is C^m, for $\|h - i_\Omega\|_{C^m}$ small. Let $v_h = (H^{-1})^* E_\Omega(u_h) \in H^m(\mathbb{R}^n)$; then $\Delta v_h + \lambda_h v_h \in H^{m-2}(\mathbb{R}^n)$ and vanishes in $h(\Omega)$, $v_h = 0$ on $\partial h(\Omega)$, and $\int_{h(\Omega)} v_h^2 = 1$. We show $h \mapsto v_h \in H^i(\mathbb{R}^n)$ is C^{m-i} near $h = i_\Omega$ for $0 \leq j \leq m$.

Lemma. *Suppose $m \geq 1$ and let X be a convex set in $C^m(\mathbb{R}^n, \mathbb{R}^n)$ consisting of C^m maps H with $H(x) = x$ for $|x| \geq R_0$ (some fixed R_0) which are diffeomorphisms with $D^j H(x), D^j H^{-1}(x)$ $(0 \leq j \leq m)$ uniformly bounded. For each $j, 0 \leq j \leq m, H \mapsto H^{-1} : X \subset C^m \to C^{m-j}(\mathbb{R}^n, \mathbb{R}^n)$ is a C^j map and for $1 \leq p < \infty$,*

$$(f, H) \mapsto f \circ H^{-1} : W^{m,p}(\mathbb{R}^n) \times X \to W^{m-j,p}(\mathbb{R}^n)$$

is also C^j.

Proof. The evaluation map $(H, x) \mapsto H(x) : X \times \mathbb{R}^n \to \mathbb{R}^n$ is C^m [1] so $(H, x) \mapsto (H, H(x))$ is C^m and has C^m inverse $(H, y) \mapsto (H, H^{-1}(y))$. We are concerned only with a compact set $\{|y| \leq R_0\}$ so $(H, y) \mapsto H^{-1}(y)$ of class C^m implies $H \mapsto H^{-1}|\overline{B}_{R_0} : X \to C^{m-j}(\overline{B}_{R_0}, \mathbb{R}^n)$ is C^j, by the converse Taylor theorem [1]. Again by the converse Taylor theorem, it is enough to prove:

> If $f_\nu \to f$ in $L_p(\mathbb{R}^n)$, $g_\nu \mapsto g$ uniformly on $\mathbb{R}^n$ and g_ν, g and C^1 diffeomorphisms with $|Dg_\nu|, |D(g_\nu^{-1})|$ uniformly bounded and g_ν, g equal to the identity outside a fixed compact set, then $f_\nu \circ g_\nu \to f \circ g$ in $L_p(\mathbb{R}^n)$.

But $f_\nu \to f$ in measure, and our hypotheses ensure $f_\nu \circ g_\nu - f \circ g_\nu \to 0$ in measure and also $f \circ g_\nu \to f \circ g$ in measure so $f_\nu \circ g_\nu \to f \circ g$ in measure. Further $\int_E |f_\nu|^p \to 0$ as measure $E \to 0$, uniformly in ν, which implies $\int_E |f_\nu \circ g_\nu|^p \leq \text{const.} \int_{g_\nu(E)} |f_\nu|^p \to 0$ as meas $E \to 0$, and the proof is complete.

Thus if Ω is C^3 regular, $t \mapsto h(t, \cdot) \in C^3(\Omega, \mathbb{R}^n)$ is C^1 with $h(0, \cdot) = i_\Omega$ then $t \mapsto v_{h(t,\cdot)} = v(t, \cdot) \in H^2(\mathbb{R}^n)$ is C^1, $t \mapsto \lambda_{h(t,\cdot)} = \lambda(t)$ is C^1

$$\Delta v(t, \cdot) + \lambda(t) v(t, \cdot) = 0 \text{ in } \Omega(t) = h(t, \Omega)$$

$$v(t, \cdot) = 0 \text{ on } \partial\Omega(t), \qquad \int_{\Omega(t)} v(t, \cdot)^2 = 1$$

with $\Omega(0) = \Omega$, $v(0,\cdot) = u_0$, $\lambda(0) = \lambda_0$. Let $\dot{v} = \frac{\partial}{\partial t} v|_{t=0}$ and then $(\Delta + \lambda_0)\dot{v} + \dot{\lambda}(0) u_0 = 0$ in Ω, $\dot{v} + \dot{h}\cdot\nabla u_0 = 0$ on $\partial\Omega$ so

$$\begin{aligned}\dot{\lambda}(0) &= \int_\Omega \{-u_0(\Delta+\lambda_0)\dot{v} + (\Delta+\lambda_0)u_0\dot{v}\}\\ &= \int_{\partial\Omega}\left\{-u_0\frac{\partial\dot{v}}{\partial N} + \frac{\partial u_0}{\partial N}\dot{v}\right\}\\ &= \int_{\partial\Omega} -\dot{h}\cdot\nabla u_0\frac{\partial u_0}{\partial N} = -\int_{\partial\Omega}\dot{h}\cdot N\left(\frac{\partial u_0}{\partial N}\right)^2\end{aligned}$$

where $\dot{h} = \frac{\partial h}{\partial t}\big|_{t=0}$. More generally

$$\frac{d}{dt}\lambda(\Omega(t)) = \int_{\partial\Omega(t)} V\cdot N_{\Omega(t)}\left(\frac{\partial v(t,\cdot)}{\partial N_{\Omega(t)}}\right)^2$$

where $\frac{\partial h}{\partial t}(t,x) = V(t,h(t,x))$, $\Omega(t) = h(t,\Omega)$ and we deal with sufficiently smooth (C^3) deformation of a C^3 domain Ω. But, as in the previous example, we may approximate C^2 regions by C^3 (or C^∞ or analytic) regions and then take limits to conclude this is the expression for the derivative also in the C^2 setting. Note that, when $t \mapsto h(t,\cdot) \in C^2(\Omega,\mathbb{R}^n)$ is C^1 (or C^∞), our previous argument says $t \mapsto v(t,\cdot) \in H^2(\mathbb{R}^n)$ is continuous, though it is differentiable into $H^1(\mathbb{R}^n)$, and the calculation above would then involve distributions in $H^{-1}(\mathbb{R}^n)$. We avoid working with distributions by only taking limits in the equation for $\frac{d}{dt}\lambda(\Omega(t))$.

Now observe that $\Omega \mapsto \lambda(\Omega)$ has a critical point (with vol (Ω) fixed) if and only if the corresponding eigenfunction v_Ω satisfies $(\partial v_\Omega/\partial N_\Omega)^2 = \text{constant} \neq 0$ on $\partial\Omega$. If we are considering the first (principal) eigenvalue λ_0 and Ω is connected, then v_Ω does not change sign in Ω, so the condition is: $\partial v_\Omega/\partial N_\Omega = \text{constant} \neq 0$ on $\partial\Omega$. By arguments of Serrin and Gidas-Ni-Nirenberg, it can be shown [11] that this holds if and only if Ω is a ball. The ball is the only critical point of $\Omega \mapsto \lambda_0(\Omega)$, the principal eigenvalue of bounded connected C^2 regions Ω with given volume. If fact, the ball minimizes the principal eigenvalue for given volume. This was conjectured by Rayleigh [33] on the basis of various examples and a formal calculation of the second derivative (see below) and proved by Faber and Krahn, whose argument is in [6, p. 413–416].

We now calculate the second derivative of $\Omega \mapsto \lambda(\Omega)$ at the point $\Omega =$ unit ball of $\mathbb{R}^n$, for volume-preserving deformations. Specifically let $h : \mathbb{R}\times\Omega_0 \to \mathbb{R}^n$ be smooth near $t = 0$ ($\Omega_0 =$ unit ball) and such that $h(0,\cdot) = i_{\Omega_0}$ and vol $h(t,\Omega_0) =$ vol Ω_0. Writing $\frac{\partial h}{\partial t}(t,x) = V(t,h(t,x))$, $\Omega(t) = h(t,\Omega)$, etc., we differentiate

$$\frac{d}{dt}\lambda(\Omega(t)) = \int_{\partial\Omega(t)} V \cdot N_{\Omega(t)} \left(\frac{\partial v}{\partial N_{\Omega(t)}}\right)^2$$

as in Example 3.1 to conclude

$$\frac{d^2}{dt^2}\lambda(\Omega(t))|_{t=0} = -\int_{\partial\Omega_0} 2\sigma \frac{\partial u_0}{\partial N}\left(\frac{\partial \dot{v}}{\partial N} - \sigma H \frac{\partial u_0}{\partial N}\right)$$

where $\sigma = V \cdot N_\Omega$ and $\dot{v} = \frac{\partial}{\partial t} v|_{t=0}$ satisfies

$$(\Delta + \lambda_0)\dot{v} = 0 \text{ in } \Omega_0, \qquad \dot{v} + \dot{h} \cdot \nabla u_0 = 0 \text{ on } \partial\Omega_0$$

and $\int_{\Omega_0} \dot{v} u_0 = 0$ (recall $\frac{d}{dt}\lambda(\Omega(t))|_{t=0} = 0$). Now $u_0(x) = k_n(x)^{-\nu} J_\nu(\sqrt{\lambda_0}|x|)$ where $\nu = (n-2)/2$, J_ν is the ν-order Bessel function, $\lambda_0 > 0$ with $J_\nu(\sqrt{\lambda_0}) = 0$ and k_n is chosen so $\int_{\Omega_0} u_0^2 = 1$ and $\frac{\partial u_0}{\partial N} > 0$ (say) on $\partial\Omega_0$, so $\partial u_0/\partial N = \sqrt{2\lambda_0/|S^{n-1}|}$. [This λ_0 is a simple eigenvalue for the unit ball, which depends on a nontrivial fact ("Bourget's hypothesis") about Bessel functions: see Watson, *Theory of Bessel Functions*, Cambridge U. Press.] Expand σ in spherical harmonics: $\sigma(x) = \sum_0^\infty P_j(x)$ when $|x| = 1$, where $P_j(x)$ is a homogeneous polynomial of degree j with $\Delta P_j(x) = 0$. (See Stein [38] for basic properties of harmonic polynomials.) Since the volume is fixed

$$0 = \frac{d}{dt}\text{vol } \Omega(t)|_{t=0} = \int_{\partial\Omega_0} \sigma = \int_{\partial\Omega_0} P_0,$$

so $P_0 = 0$. Then $\dot{v}(x) = -\sqrt{2\lambda_0/|S^{n-1}|}\sum_1^\infty f_j(\sqrt{\lambda_0}|x|)/f_j(\sqrt{\lambda_0})P_j(\frac{x}{|x|})$ where $f_j(z) = z^{-\nu}J_{\nu+j}(z)$, and substitution (with $H = n-1$) yields

$$\left.\frac{d^2}{dt^2}\lambda(\Omega(t))\right|_{t=0} = \frac{4\lambda_0}{|S^{n-1}|}\sum_{j=1}^{\infty}\left(\int_{\partial\Omega_0} P_j^2\right)Q_j$$

$$Q_j = \sqrt{\lambda_0}J'_{\nu+j}(\sqrt{\lambda_0})/J_{\nu+j}(\sqrt{\lambda_0}) + \nu + 1.$$

Since $J_\nu(\lambda_0) = 0$, it follows (by the recursion relations for J) that $Q_1 = 0$ so the second derivative does not depend on P_1. This is understandable as P_1 gives the initial velocity of the center of mass, and says nothing about change of shape.

Rayleigh [33] computed this derivative formally when $n = 2$, and showed (if λ_0 is the principal eigenvalue) that $Q_j > 0$ for all $j \geq 2$. The same argument works for $2 \leq n \leq 5$, but in general we refer to the argument of Polya and

Szëgo [31, p. 135], who also treated $n = 2$, but whose argument shows, for every $n \geq 2$, that $Q_j > 0$ for all $j \geq 2$.

As a more specific example, suppose $\Omega(t)$ is the ellipsoid with semi-axes $e^{K_1 t}, \ldots, e^{K_n t}$, i.e.,

$$\Omega(t) = \left\{ x \in \mathbb{R}^n | \sum_{j=1}^{n} (x_j e^{-K_j t})^2 < 1 \right\},$$

and $K_1 + \cdots + K_n = 0$ so vol $\Omega(t) =$ constant. Then $\Omega(t) = h(t, \Omega_0)$, $\Omega_0 =$ unit ball, $h(t, x) = (e^{K_j t} x_j)_{j=1}^n V(t, y) = (K_j, y_j)_{j=1}^n$ and

$$N_{\Omega(t)}(y) = \left(\frac{e^{-2K_j t} y_j}{\sqrt{\sum_i e^{-4K_i t} y_i^2}} \right)_{j=1}^n$$

so, at $t = 0$, $\sigma(y) = V \cdot N_{\Omega(t)}|_{t=0} = \sum_1^n K_j y_j^2$. Since $\sum_1^n K_j = 0$, σ is a second order spherical harmonic and the above formula gives

$$\frac{d^2}{dt^2} \lambda(\Omega(t))|_{t=0} = \frac{8\lambda_0(\lambda_0 - n) \sum_1^n K_j^2}{n^2(n+2)}.$$

If $n = 2$, $K_1 = -K_2 = 1$ then $\Omega(t)$ is the ellipse with semi-axes e^t, e^{-t} and $t \mapsto \lambda(\Omega(t))$ is even – since $\Omega(-t)$ is obtained from $\Omega(t)$ by rotation – so

$$\lambda(\Omega(t)) = \lambda_0 + \frac{1}{2}\lambda_0(\lambda_0 - 2)t^2 + O(t^4).$$

In terms of the eccentricity $\epsilon(\sqrt{1 - \epsilon^2}) = e^{-2(t)})$ we have

$$\lambda = \lambda_0 + \frac{1}{32}\lambda_0(\lambda_0 - 2)(\epsilon^4 + \epsilon^6) + O(\epsilon^8)$$

as the eigenvalue near λ_0 in the ellipse of area π and eccentricity ϵ. D. Joseph [13] computed the series for this eigenvalue but, after reduction to area π, his series has $\frac{3}{2}\lambda_0 - 5$ in place of $\lambda_0 - 2$ in the coefficient of ϵ^6 (our series agrees to order ϵ^4). When λ_0 is the principal eigenvalue, the coefficent of ϵ^6 should be increased about 3%, which has only slight effect on the values in the table [13, p. 337].

Example 3.3. Capacity. Physically, the capacity $K(\Omega)$ of a bounded region (conductor) $\Omega \subset \mathbb{R}^3$ is the amount of electrical charge in Ω required to raise the equilibrium electrical potential of the conductor to 1, relative to potential zero at infinity. At large distances, the resulting potential ϕ looks like a point charge at the origin, $\phi(x) \approx K(\Omega)/4\pi|x|$ as $|x| \to \infty$.

For a bounded C^2 region $\Omega \subset \mathbb{R}^n$ $(n \geq 3)$ let ϕ be the solution of

$$\Delta\phi = 0 \text{ in } \mathbb{R}^n \backslash \Omega, \qquad \phi = 1 \text{ on } \partial\Omega, \qquad \phi(x) \to 0 \text{ as } |x| \to \infty.$$

Then there is a constant $K(\Omega) \geq 0$ so

$$-\phi(x)/E_n(x) = (n-2)|S^{n-1}||x|^{n-2}\phi(x) \to K(\Omega)$$

as $|x| \to \infty$ (E_n is the fundamental solution of the Laplacian) and we call $K(\Omega)$ the *capacity* of Ω. In this case we also have

$$|S^{n-1}||x|^{n-1}\frac{x}{|x|} \cdot \nabla\phi(x) \to -K(\Omega) \qquad \text{as } |x| \to \infty.$$

[The case $n = 2$ is somewhat different, requiring logarithmic potential.] Other representations of the capacity follow from the divergence theorem:

$$\begin{aligned}\int_{\mathbb{R}^n\backslash\Omega} |\nabla\phi|^2 &= \int_{\mathbb{R}^n\backslash\Omega} \operatorname{div}(\phi\nabla\phi) = -\int_{\delta\Omega} \phi\frac{\partial\phi}{\partial N} \\ &= -\int_{\partial\Omega} \frac{\partial\phi}{\partial N} = -\int_{|x|=R} \frac{\partial\phi}{\partial N} \ (R \text{ large}) = K(\Omega).\end{aligned}$$

The charge-density on $\partial\Omega$ is $\sigma = -\partial\phi/\partial N$ ($\sigma > 0$, by the maximum principle) and then $K(\Omega) = \int_{\partial\Omega}\sigma$ is the total charge on $\partial\Omega$ (or in $\overline{\Omega}$). It is also easy to show

$$K(\Omega) = \min\left\{\int_{\mathbb{R}^n} |\nabla\psi|^2 : \psi \in H^1(\mathbb{R}^n),\ \psi = 1 \text{ on } \Omega\right\}$$

and this is often taken as the definition of capacity when Ω is not smooth. However we assume Ω is smooth (C^2) and show $K(\Omega)$ is an analytic function of Ω, for C^2 deformations, and compute the derivative:

$$\frac{\partial}{\partial h}K(h(\Omega))|_{h=i}\dot{h} = \int_{\partial\Omega} \dot{h} \cdot N_\Omega\left(\frac{\partial\phi}{\partial N_\Omega}\right)^2, \qquad \dot{h} \in C^2(\mathbb{R}^n, \mathbb{R}^n).$$

(As usual, N_Ω is the unit outward normal from Ω, although the capacitary potential ϕ is defined in $\mathbb{R}^n\backslash\Omega$.)

Query: Is the ball the only critical point of the capacity, with given volume?

Working in an unbounded region raises some technical problems. For this example, we could avoid these problems by using inversion in a sphere to reduce to a bounded domain. But we will use a method which generalizes.

Choose p in $3 < p < \infty$ (or more generally, $p > n/(n-2)$) so $|x|^{2-n}$ will be p-power integrable near ∞. Also choose some large R so that $\overline{\Omega}$ (and any relevant perturbations) are in $B_R = \{x : |x| < R\}$, and choose $C^\infty \psi_0$ so that $\psi_0 = 1$ on $\partial\Omega$ and $\psi_0 = 0$ outside B_R. Define

$$\begin{aligned}\mathrm{Diff}_R^2(\mathbb{R}^n) = \{&\text{all } C^2 \text{ diffeomorphisms } h : \mathbb{R}^n \to \mathbb{R}^n \text{ such that} \\ &h(x) = x \text{ for } |x| \geq R\}\end{aligned}$$

$$\begin{aligned}X_R &= \{\psi \in W^{2,p} \cap W_0^{1,p}(\mathbb{R}^n \backslash \Omega) | \Delta\psi = 0 \text{ outside } B_R\} \\ Y_R &= \{f \in L_p(\mathbb{R}^n \backslash \Omega) | f = 0 \text{ outside } B_R\},\end{aligned}$$

with the induced norms. X_R, Y_R are Banach spaces while Diff_R^2 is the translation of of an open set in a Banach space. Note for $h \in \mathrm{Diff}_R^2(\mathbb{R}^n)$ that $(h^*\Delta h^{*-1})\psi(x) = \Delta\psi(x)$ when $|x| > R$, and also note $\psi \mapsto \Delta\psi : X_R \to Y_R$ is an isomorphism.

Apply the implicit function theorem to the analytic map $(h, \psi) \mapsto (h^*\Delta h^{*-1})(\psi + \psi_0) : \mathrm{Diff}_R^2 \times X_R \to Y_R$ to prove, for each $h \in \mathrm{Diff}_R^2$ in a C^2-neighborhood of $id_{\mathbb{R}^n}$, there is a unique $\psi_h \in X_R$ with $(h^*\Delta h^{*-1})(\psi_h + \psi_0) = 0$ in $\mathbb{R}^n \backslash \Omega$, and $h \mapsto \psi_h$ is analytic. Then $\phi_h = h^{*-1}(\psi_h + \psi_0) : \mathbb{R}^n \backslash h(\Omega) \to \mathbb{R}$ is a harmonic function which vanishes at ∞ and $\phi_h = 1$ on $\partial h(\Omega)$, so ϕ_h is the capacitary potential for $h(\Omega)$ and

$$K(h(\Omega)) = \int_{\mathbb{R}^n \backslash h(\Omega)} |\nabla\phi_h|^2 = \int_{\mathbb{R}^n \backslash \Omega} |h^*\nabla h^{*-1}(\psi_h + \psi_0)|^2 \det h'$$

is also analytic in h. (We could write this as an integral over $B_R \backslash \Omega$, plus another over ∂B_R; to avoid convergence problems at ∞.)

Assuming a bit more smoothness, say C^3, we compute the derivative. Omitting details, say $h(t, x) = x + tV(x) + o(t)$, $\phi_{h(t,\cdot)} = \phi_0 + t\phi_1 + o(t)$ is the capacitary potential for $\omega(t) = h(t, \Omega)$, so $\Delta\phi_0 = 0$, $\Delta\phi_1 = 0$ in $\mathbb{R}^n \backslash \Omega$; ϕ_0, ϕ_1 vanish at ∞, $O(|x|^{2-n})$; $\phi_0 = 1$ and $\phi_1 + V \cdot \nabla_{\phi_0} = 0$ on $\partial\Omega$. Then

$$\begin{aligned}\left.\frac{d}{dt}\right|_{t=0} K(\Omega(t)) &= \left.\frac{d}{dt}\right|_{t=0} \int_{\mathbb{R}^n \backslash \Omega(t)} |\nabla\phi_{h(t,\cdot)}|^2 \\ &= \int_{\mathbb{R}^n \backslash \Omega} 2\nabla\phi_0 \cdot \nabla\phi_1 + \int_{\partial\Omega} -V \cdot N(\nabla\phi_0)^2 \\ &= -\int_{\partial\Omega} \left(2\phi_1 \frac{\partial\phi_0}{\partial N} + V \cdot N\left(\frac{\partial\phi_0}{\partial N}\right)^2\right) \\ &= \int_{\partial\Omega} V \cdot N\left(\frac{\partial\phi_0}{\partial N}\right)^2.\end{aligned}$$

Alternatively,

$$\begin{aligned}\left.\frac{d}{dt}\right|_{t=0} K(\Omega(t)) &= -\left.\frac{d}{dt}\right|_{t=0}\int_{\partial\Omega(t)} N_{\Omega(t)}\cdot\nabla\phi_{h(t,\cdot)}\\ &= -\int_{\partial\Omega}\left(\dot{N}\cdot\nabla\phi_0 + N\cdot\nabla\phi_1 + V\cdot N\frac{\partial^2\phi_0}{\partial N^2} + HV\cdot N\frac{\partial\phi_0}{\partial N}\right)\\ &= -\int_{\partial\Omega}\frac{\partial\phi_1}{\partial N}\\ &= -\int_{\partial\Omega}\phi_0\frac{\partial\phi_1}{\partial N} = -\int_{\partial\Omega}\phi_1\frac{\partial\phi_0}{\partial N} = \int_{\partial\Omega} V\cdot N\left(\frac{\partial\phi_0}{\partial N}\right)^2,\end{aligned}$$

using Theorems 1.11, 1.12 and Lemma 2.3.

Example 3.4. Green's Function. The first general treatment of perturbation of the boundary is due to Hadamard [9] who took as the fundamental problem calculation of the change in Green's function. This is also the basis of the work of Schiffer and Garabedian [7], among others. Our method allows a more direct approach to Green's function (and Neumann's and other corresponding kernels). We treat only the case of the Dirichlet problem for a second-order uniformly elliptic operator L

$$L = \sum_{i,j=1}^{n} a_{ij}(x)\frac{\partial^2}{\partial x_i\partial x_j} + \sum_{j=1}^{n} b_j(x)\frac{\partial}{\partial x_j} + C(x),$$

but other cases may be treated similarly [11]. See Peetre [27] for another approach.

We assume everything is smooth, and uniqueness holds for the Dirichlet problem for L in $\Omega\subset\mathbb{R}^n$, Ω smooth and bounded, $n\geq 3$. Then the Green function $G_\Omega(x,y)$ exists and may be defined by

$$L_y^* G_\Omega(x,y) = 0 \text{ for } x\neq y \text{ in } \Omega,\qquad G_\Omega(x,y)=0 \text{ for } y\in\partial\Omega,$$

$$G_\Omega(x,y) = C_n(x)\left(\sum_{j,k} a^{jk}(x)(y_j-x_j)(y_k-x_k)\right)^{(2-n)/2} + O(|x-y|^{3-n})$$

as $y\to x (x\in\Omega)$, where $[a^{jk}]$ is the inverse of the symmetric positive-definite matrix $[a_{jk}]$ and

$$\frac{1}{C_n(x)} = (2-n)|S^{n-1}|\sqrt{\det[a_{jk}(x)]}.$$

Then for any C^2 $\phi:\overline{\Omega}\to\mathbb{R}$ and $x\in\Omega$,

$$\phi(x) = \int_\Omega G_\Omega(x,y)L\phi(y)dy + \int_{\partial\Omega} dA_y\,\phi(y)\left(\sum a_{jk}(x)N_jN_k\right)\frac{\partial G_\Omega}{\partial N(y)}(x,y).$$

Say $f \in C^\infty(\mathbb{R}^n, \mathbb{R})$ and $h(t,x) = x + tV(x) + o(t)$ is a smooth curve of smooth imbeddings of Ω. For each t near 0 there is a solution $u(t,\cdot)$ of $Lu(t,\cdot) = f$ in $\Omega(t) = h(t,\Omega)$, $u(t,\cdot) = 0$ on $\partial\Omega(t)$, and we may consider $u(t,x)$ extended smoothly to a neighborhood of $\{0\} \times \overline{\Omega}$. Calculating in the Eulerian mode, with $\dot{u} = \partial u/\partial t$ and $\partial h/\partial t = V(t,h)$, we find $L\dot{u} = 0$ in $\Omega(t)$, $\dot{u} + V \cdot \nabla u = 0$ on $\partial\Omega(t)$ so (if $G_{\Omega(t)}$ is the appropriate Green function and $x \in \Omega(t)$),

$$\dot{u}(t,x) = -\int_{\partial\Omega(t)} dA_y \left(\sum_{j,k} a_{jk} N_j N_k\right) \frac{\partial G_{\Omega(t)}(x,y)}{\partial N_{\Omega(t)}(y)} V \cdot N_{\Omega(t)}(t,y) \frac{\partial u(t,y)}{\partial N_{\Omega(t)}}.$$

Also $u(t,y) = \int_{\Omega(t)} G_{\Omega(t)}(y,z) f(z) dz$, so

$$\dot{u}(t,x) = \frac{d}{dt} \int_{\Omega(t)} G_{\Omega(t)}(x,z) f(z) dz = \int_{\Omega(t)} \dot{G}_{\Omega(t)}(x,z) f(z) dz$$

where we define (for x, z in Ω)

$$\dot{G}_{\Omega(t)}(x,z) = -\int_{\partial\Omega(t)} dA_y \left(\sum a_{jk}(y) N_j N_k\right) V \cdot N_{\Omega(t)} \frac{\partial G_{\Omega(t)}(x,y)}{\partial N(y)} \times \frac{\partial G_{\Omega(t)}(y,z)}{\partial N(y)}.$$

By the implicit function theorem, we may easily prove differentiability of $t \mapsto G_{\Omega(t)}(x,z)$ for $x \neq z$ in $\Omega(t)$, so in fact

$$\dot{G}_{\Omega(t)}(x,z) = \frac{\partial}{\partial t} G_{\Omega(t)}(x,z), \quad \text{even for } x = z.$$

Example 3.5. Simple Eigenvalues of $\Delta u + (\lambda + C(x))u = 0$ in Ω, $\partial u/\partial N + \beta(x)u = 0$ on $\partial\Omega$. The previous examples used the Dirichlet boundary condition, but we may also treat other boundary conditions using Theorem 2.4.

Suppose $\beta, C : \mathbb{R}^n \to \mathbb{R}$ are C^2 functions and $\Omega \subset \mathbb{R}^n$ is a bounded C^2 region. Assume λ_0 is a simple eigenvalue of the above (Robin) problem in Ω, with eigenfunction u_0, $\int_\Omega u_0^2 = 1$. We show for every $h : \Omega \to \mathbb{R}^n C^2$ close to i_Ω, the corresponding problem in $h(\Omega)$ has a unique eigenvalue λ_h near λ_0, λ_h is a simple eigenvalue, $h \mapsto \lambda_h$ is differentiable and the derivative at $h = i_\Omega$ is

$$\dot{h} \mapsto \int_{\partial\Omega} \dot{h} \cdot N_\Omega \left\{ |\nabla_{\partial\Omega} u_0|^2 - \left(\lambda_0 + c + \beta^2 - H\beta - 2\frac{\partial\beta}{\partial N_\Omega}\right) u_0^2 \right\}$$

where $H = \operatorname{div} N_\Omega$ is the mean curvature of $\partial\Omega$ and $\nabla_{\partial\Omega} u_0$ is the tangential component of the gradient of u_0.

Let $H^{1/2}(\partial\Omega)$ be the space of restrictions to $\partial\Omega$ of functions in $H^1(\partial\Omega)$, i.e., the quotient $H^1(\partial\Omega)/H_0^1(\partial\Omega)$. Consider the map

$$F : (u, \lambda, h) \mapsto \left(h^*(\Delta + c + \lambda)h^{*-1}u,\, h^*(N_{h(\Omega)} \cdot \nabla + \beta)h^{*-1}u|\partial\Omega,\, \int_\Omega u^2 \right)$$

$$: H^2(\Omega) \times \mathbb{R} \times \mathrm{Diff}^2(\Omega) \to L_2(\Omega) \times H^{1/2}(\partial\Omega) \times \mathbb{R}.$$

Then $F(u_0, \lambda_0, i_\Omega) = (0, 0, 1)$, and $F(u, \lambda, h) = (0, 0, 1)$ implies $v \equiv h^{*-1}u \in H^2(h(\Omega))$ satisfies

$$(\Delta + c + \lambda)v = 0 \text{ in } h(\Omega), \qquad \frac{\partial v}{\partial N} + \beta v = 0 \text{ on } \partial h(\Omega),$$

$$\int_{h(\Omega)} \frac{v^2}{|\det h'|} = 1,$$

so λ is an eigenvalue of the Robin problem in $h(\Omega)$.

Now $(u, \lambda, h) \mapsto h^*\Delta h^{*-1}u + \lambda u + (c \circ h) \cdot u$ is C^2 in the relevant spaces, while

$$(h, u) \mapsto h^*(N_{h(\Omega)} \cdot \nabla + \beta)h^{*-1}u|_{\partial\Omega} = h^*N_{h(\Omega)} \cdot h^*\nabla h^{*-1}u|_{\partial\Omega} + (\beta \circ h)u|_{\partial\Omega}$$

is C^1, for $(h, u) \mapsto (\beta \circ h)u : \mathrm{Diff}^2 \times H^2 \to H^1(\Omega)$ is C^1. Thus F is a C^1 map and we could have assumed c is C^1. The smoothness of F is limited only by that of c and β: if c is C^r, β is C^{r+1}, then F is C^r. Also, if $\partial\Omega$ is $C^m (m \geq 2)$ we could work with the corresponding F on $H^m \times \mathbb{R} \times \mathrm{Diff}^m \to H^{m-2} \times H^{m-3/2} \times \mathbb{R}$, which is a C^r map provided c is C^{r+m-2}, β is C^{r+m-1}.

Since λ_0 is assumed to be a simple eigenvalue,

$$\frac{\partial F}{\partial(u, \lambda)}(u_0, \lambda_0, i_\Omega) : (\dot{u}, \dot{\lambda}) \mapsto ((\Delta + c + \lambda_0)\dot{u} + \dot{\lambda}u_0,$$
$$\left(\frac{\partial}{\partial N} + \beta\right)\dot{u},\, 2\int_\Omega (u_0\dot{u}))$$
$$: H^2(\Omega) \times \mathbb{R} \to L_2(\Omega) \times H^{1/2}(\partial\Omega) \times \mathbb{R}$$

is an isomorphism, so the implicit function theorem says $F(u, \lambda, h) = (0, 0, 1)$ has a unique solution $u = u_h$, $\lambda = \lambda_h$ near u_0, λ_0 for every h C^2-near i_Ω, and $h \mapsto (u_h, \lambda_h)$ is C^1.

To compute the derivative, we assume more smoothness, say Ω is C^3, β is C^3, c is C^2, and we work in $H^3(\Omega)$. The same argument gives $h \mapsto (u_h, \lambda_h) : \mathrm{Diff}^3(\Omega) \to H^3(\Omega) \times \mathbb{R}$ of class C^1. We will calculate in the Lagrangian mode, for a change, which (in this problem) is about the same level of difficulty as the Eulerian.

Let $u(t,\cdot) = u_{h(t,\cdot)}$, $\lambda(t) = \lambda_{h(t,\cdot)}$, where $h(t,x)$ is smooth with $h(t,x) = x + tV(x) + o(t)$ as $t \to \infty$. Now for t near 0, $h^*(\Delta + c + \lambda(t))h^{*-1}u(t,\cdot) = 0$ in Ω and

$$D_t h^*(\Delta + c + \lambda(t))h^{*-1}u(t,\cdot) = \dot\lambda(t)u(t,\cdot) + h^*(\Delta + c + \lambda)h^{*-1}D_t u$$

holds by continuity (a continuous function of t with values in $L_2(\Omega)$), so at $t = 0$, with $\dot u = \frac{\partial}{\partial t}u(t,\cdot)|_{t=0}$,

$$0 = \dot\lambda(0)u_0 + (\Delta + c + \lambda_0)(\dot u \cdot V \cdot \nabla u_0) \text{ in } \Omega.$$

Also by Theorem 2.4

$$\begin{aligned} D_t h^*\big(N_{h(\Omega)} \cdot \nabla + \beta\big)h^{*-1}u(t,\cdot) = h^*&\big(N_{h(\Omega)} \cdot \nabla + \beta\big)h^{*-1}D_t u(t,\cdot) \\ &+ D_t\big(h^* N_{h(\Omega)}\big) \cdot h^*\nabla h^{*-1}u(t,\cdot) \end{aligned}$$

holds near $\partial\Omega$, when everything is smooth, and by continuity also holds for our solution curve as an equality of continuous functions of t with values in H^1 on a neighborhood of $\partial\Omega$. Taking $t \to 0$ and restricting to $\partial\Omega$, we find

$$\begin{aligned} -V \cdot \nabla\left(\frac{\partial u_0}{\partial N} + \beta u_0\right) &= \left(\frac{\partial}{\partial N} + \beta\right)(\dot u - V \cdot \nabla u_0) \\ &\quad - \left(\nabla_{\partial\Omega}\sigma + \sigma\frac{\partial N}{\partial N}\right) \cdot \nabla u_0, \qquad \sigma = V \cdot N_\Omega, \end{aligned}$$

by Lemma 2.3. Since $\partial u_0/\partial N + \beta u_0 = 0$ on $\partial\Omega$, the left-side of this equation involves only the normal derivative, and we have

$$\begin{aligned} &\left(\frac{\partial}{\partial N} + \beta\right)(\dot u - V \cdot \nabla u_0) \\ &\quad = -\sigma\frac{\partial}{\partial N}\left(\frac{\partial u_0}{\partial N} + \beta u_0\right) + \left(\nabla_{\partial\Omega} + \sigma\frac{\partial N}{\partial N}\right) \cdot \nabla u_0 \\ &\quad = \operatorname{div}_{\partial\Omega}(\sigma\nabla_{\partial\Omega}u_0) + \sigma u_0\left(\lambda_0 + c + \beta^2 - \beta H - \frac{\partial\beta}{\partial N}\right) \end{aligned}$$

on $\partial\Omega$, were we have used Theorem 1.12 and the equations satisfied by u_0.

Now multiplying the earlier equation by u_0 and integrating over Ω,

$$\begin{aligned} 0 &= \dot\lambda(0)\int_\Omega u_0^2 + \int_\Omega(\Delta + c + \lambda_0)(\dot u - V \cdot \nabla u_0) \\ &= \dot\lambda(0) + \int_{\partial\Omega} u_0\left(\frac{\partial}{\partial N} + \beta\right)(\dot u - V \cdot \nabla u_0) \\ &= \dot\lambda(0) + \int_{\partial\Omega}\sigma u_0^2\left(\lambda_0 + c + \beta^2 - \beta H - \frac{\partial\beta}{\partial N}\right) + \int_{\partial\Omega} u_0 \operatorname{div}_{\partial\Omega}(\sigma\nabla_{\partial\Omega}u_0). \end{aligned}$$

The last integral is $-\int_{\partial\Omega}\sigma|\nabla_{\partial\Omega}u_0|^2$, which gives the result claimed. We may take limits in the final expression – as in Example 3.1 – to see the result also holds if $\partial\Omega$ is C^2, c is C^1 and β is C^2.

Now suppose $n=2$ and β, c are constant – we may suppose $c=0$. We will compute the second derivative at $t=0$ of a simple eigenvalue $\lambda(t)$ of the Robin problem in the ellipse $\Omega(t)$ with semi-axes e^t, e^{-t}. If $\beta=0$ we always have the eigenvalue zero; but we assume $\lambda(0)\neq 0$ in case $\beta=0$. It is clear that $\lambda(-t)=\lambda(t)$, (as in Example 3.2) so $\dot\lambda(0)=0$, and $t\mapsto\lambda(t)$ is analytic. In polar coordinates, at $t=0$, the eigenvalue $\lambda(0)=\lambda_0$ is a root of $\sqrt{\lambda_0}J_0'(\sqrt{\lambda_0})+\beta J_0(\sqrt{\lambda_0})=0$ and the eigenfunction is $u_0(r)=cJ_0(\sqrt{\lambda_0}r)$ with c chosen so $\int_{\Omega(0)}u_0^2=1$, hence $\pi u_0(1)^2=\lambda_0/(\lambda_0+\beta^2)$. If $u(t,\cdot)$ is the eigenfunction in $\Omega(t)$, working in the Eulerian mode, and $\dot u=\frac{\partial}{\partial t}u(t,\cdot)|_{t=0}$ we have

$$(\Delta+\lambda_0)\dot u=0 \text{ in the unit disc } \Omega(0)=\{r<1\}$$

and

$$\left(\frac{\partial}{\partial N}+\beta\right)\dot u+\dot N\cdot\nabla u_0+V\cdot\nabla\left(\frac{\partial u_0}{\partial N}+\beta u_0\right)=0 \qquad \text{on } r=1.$$

Now $V(t,y)=\mathrm{col}(y_1,-y_2)$ so $V\cdot N_{\Omega(0)}=\cos 2\theta$ on $r=1$. Since u_0 is radial and $\dot N\cdot N=0$, the boundary condition is

$$\left(\frac{\partial}{\partial r}+\beta\right)\dot u|_{r=1}=-\cos 2\theta(u_0''(1)+\beta u_0'(1))=\mu u_0(1)\cos 2\theta$$

where $\mu=\lambda_0+\beta^2-\beta$. Further, assuming $\int_{\Omega(t)}u(t,\cdot)^2\equiv 1$, we have $\int_{\Omega(0)}u_0\dot u=0\,(\int_{\partial\Omega}V\cdot N=0)$ so $\dot u=BJ_2(\sqrt{\lambda_0}r)\cos 2\theta$ where the constant B is given by

$$BJ_2(\sqrt{\lambda_0})=\frac{u_0(1)\mu(2\beta-\lambda_0)}{2(\mu-\beta)}.$$

Let $\sigma=V\cdot N_{\Omega(t)}$, $Q(t,\cdot)=|\nabla u(t,\cdot)|^2+u(t,\cdot)^2(\beta\,\mathrm{div}\,N_{\Omega(t,\cdot)}-2\beta^2-\lambda(t))$ so $\frac{d}{dt}\lambda(t)=\int_{\partial\Omega(t)}Q\sigma$ and

$$\frac{d^2}{dt^2}\lambda(t)=\int_{\partial\Omega(t)}\left(\frac{\partial\sigma}{\partial t}+\sigma\frac{\partial\sigma}{\partial N}+H\sigma^2\right)Q+\int_{\partial\Omega(t)}\sigma^2\frac{\partial Q}{\partial N}+\sigma\frac{\partial Q}{\partial t}.$$

At $t=0$, $Q=u_0'(r)^2+u_0(r)^2(\beta/r-2\beta^2-\lambda_0)$ is constant on $\{r=1\}=\partial\Omega(0)$

so the first integral vanishes (area $\Omega(t) = \text{constant}$). Also $\partial Q/\partial N = \partial Q/\partial r|_{r=1} = \beta(4\mu - 1)u_0(1)^2$ at $t = 0$,

$$\begin{aligned} \operatorname{div} \dot{N}|_{t=0} &= 4\cos 2\theta, \\ \left.\frac{\partial Q}{\partial t}\right|_{r=1,t=0} &= 2\nabla u_0 \cdot \nabla \dot{u} + 2u_0\dot{u}(\beta - 2\beta^2 - \lambda_0) + \beta u_0^2 \operatorname{div} N \\ &= u(1)^2 \cos 2\theta\{4\beta + \mu(\lambda_0\mu - 4\beta\mu + 2\beta^2)/(\mu - \beta)\}. \end{aligned}$$

Substitution yields

$$\ddot{\lambda}(0) = \frac{\lambda_0}{\mu^2 - \beta^2}[\lambda_0\mu^2 + \mu(3\beta - 2\beta^2) - 3\beta^2], \qquad \mu = \lambda_0 + \beta^2 - \beta,$$

which (a minor miracle) is the same expression obtained in [11] by a different method. Note that, as $\beta \to \pm\infty$ with λ_0 bounded, this approaches $\lambda_0(\lambda_0 - 2)$ and $J_0(\sqrt{\lambda_0}) \to 0$, so we recover the result for the Dirichlet problem (Example 3.2).

Example 3.6. Simple Eigenvalue of a General Dirichlet Problem. Suppose $Q \subset \mathbb{R}^n$ is open and for multi-indices $\alpha = (\alpha_1, \dots, \alpha_n)$ with $|\alpha| = \sum_1^n \alpha_j \leq 2m$ we have $a_\alpha : Q \to \mathbb{C}$ of class $C^{|\alpha|}$, while a_0 is C^1, $|\sum_{|\alpha|=2m} a_\alpha(x)\xi^\alpha| \geq c_0|\xi|^{2m}$ for $x \in Q$, $\xi \in \mathbb{R}^n$ and some constant $c_0 > 0$, and $\sum_{|\alpha|=2m} a_\alpha(\partial/\partial x)^\alpha$ is properly elliptic (true, for example, if $n \geq 3$ or if the a_α are real when $|\alpha| = 2m$).

Let Ω be a bounded C^{2m}-regular region with $\overline{\Omega} \subset Q$, and suppose $\lambda = \lambda_0$ is a simple eigenvalue of

$$(*)_\Omega \qquad \begin{aligned} &\sum_{|\alpha|\leq 2m} a_\alpha(x)\left(\frac{\partial}{\partial x}\right)^\alpha u = \lambda u \text{ in } \Omega, \\ &u = \frac{\partial u}{\partial N} = \cdots = \frac{\partial^{m-1}u}{\partial N^{m-1}} = 0 \text{ on } \partial\Omega \end{aligned}$$

with eigenfunction u_0. Then u_0 is unique, up to a complex-constant multiplier, and there is an eigenfunction v_0 (equally unique) of the adjoint problem

$$\begin{aligned} &\sum_{|\alpha|\leq m} \left(-\frac{\partial}{\partial x}\right)^\alpha (a_\alpha(x)v) = \lambda v \text{ in } \Omega, \\ &v = \frac{\partial v}{\partial N} = \cdots = \frac{\partial^{m-1}v}{\partial N^{m-1}} = 0 \text{ on } \partial\Omega \end{aligned}$$

with $\int_\Omega u_0 v_0 \neq 0$. We may and shall assume $\int_\Omega u_0 v_0 = 1$.

Then if $t \mapsto h(t, \cdot) \in C^{2m}(\Omega, \mathbb{R}^n)$ is a C^1 function of the form $h(t, x) = x + tV(x) + o(t)$ as $t \to 0$, $\Omega(t) = h(t, \Omega)$, there is a unique eigenvalue $\lambda(t)$ near λ_0 of the corresponding problem $(*)_{\Omega(t)}$ for t near 0, $\lambda(0) = \lambda_0$, $t \mapsto \lambda(t)$

is C^1, and

$$\frac{d}{dt}\lambda(t)|_{t=0} = (-1)^{m+1}\int_{\partial\Omega}\left(\sum_{|\alpha|=2m} a_\alpha N_\Omega^\alpha\right) V\cdot N_\Omega \frac{\partial^m u_0}{\partial N^m}\frac{\partial^m v_0}{\partial N^m}.$$

The existence and differentiability of the eigenvalue and eigenfunction are proved by applying the implicit function theorem to

$$(u,\lambda,h)\mapsto \left(h^*\left(\sum_\alpha a_\alpha\partial^\alpha\right)h^{*-1}u-\lambda u, \int_\Omega uv_0\right)$$
$$: H^{2m}\cap H_0^m(\Omega,\mathbb{C})\times\mathbb{C}\times \mathrm{Diff}^{2m}(\Omega)\to L_2(\Omega,\mathbb{C})\times\mathbb{C}.$$

We will compute the derivative, assuming all convenient smoothness.

From $h^*(\sum_\alpha a_\alpha\partial^\alpha-\lambda)h^*u=0$ and Theorem 2.2, we find

$$\sum_\alpha a_\alpha\partial^\alpha(D_tu)_{t=0}=\dot\lambda(0)u_0 \qquad \text{in } \Omega.$$

Also $w\equiv D_tu|_{t=0}$ satisfies $\partial^j w/\partial N^j=0$ on $\partial\Omega$ for $0\le j\le m-2$ while $\partial^{m-1}w/\partial N^{m-1}=-V\cdot N_\Omega\partial^m u_0/\partial N^m$ on $\partial\Omega$. Therefore

$$\dot\lambda(0)=\dot\lambda(0)\int_\Omega u_0v_0=\int_\Omega\left\{v_0\sum_\alpha a_\alpha\partial^\alpha w-\sum_\alpha(-\partial)^\alpha(a_\alpha v_0)w\right\}$$
$$=(-1)^m\int_{\partial\Omega}\sum_{|\alpha|=2m}a_\alpha N^\alpha\frac{\partial^m v_0}{\partial N^m}\frac{\partial^{m-1}w}{\partial N^{m-1}}$$

which is the result claimed.

We show that a simple eigenvalue may be moved by perturbing the boundary, assuming uniqueness in the Cauchy problem for $\sum_\alpha a_\alpha\partial^\alpha$ and its adjoint – which certainly holds if the a_α are analytic (Holmgren) or if $m=1$ (see, for example, [2] or [12, Theorem 8.9.1]).

Suppose to the contrary that λ_0 is a simple eigenvalue of $(^*)_\Omega$ which does not move or, more generally, that Ω is a critical point of the eigenvalue so $\dot\lambda(0)=0$ for every choice of V in the expression above. Then $\frac{\partial^m u_0}{\partial N^m}\frac{\partial^m v_0}{\partial N^m}=0$ on $\partial\Omega$, so on some non-empty open set in $\partial\Omega$, at least one of these (say u_0) vanishes. By uniqueness in the Cauchy problem, $u_0\equiv 0$ throughout the adjacent component of Ω; in fact $u_0v_0=0$ throughout Ω. But $\int_\Omega u_0v_0=1$, a contradiction. If Ω is connected, we can move a simple eigenvalue by small perturbations of $\partial\Omega$ restricted to a small neighborhood of a given point. Of course if uniqueness fails, one might possibly have an eigenfunction (u_0 or v_0 or both) which vanishes throughout a neighborhood of $\partial\Omega$, and then the eigenvalue would not move. (Pliš gave an example of a fourth order elliptic equation with C^∞ coefficients on $\mathbb{R}^3$ which has nontrivial kernel containing C^∞ functions with compact support.)

Chapter 4

Bifurcation Problems

The following examples are again based on the implicit function theorem, but joined with the method of Liapunov-Schmidt, which gives them a slightly different flavor. We use the domain of definition as a bifurcation parameter. An introduction – and much more – to bifurcation theory can be found in *Methods of Bifurcation Theory*, S. N. Chow and J. K. Hale (Springer-Verlag).

Example 4.1. Multiple Eigenvalues of the Dirichlet Problem for the Laplacian. Let Ω be a bounded C^2 region in $\mathbb{R}^n$ and let $\lambda = \lambda_0$ be an m-fold eigenvalue of

$$\Delta u + \lambda u = 0 \text{ in } \Omega, \quad u = 0 \text{ on } \partial\Omega, \quad u \not\equiv 0,$$

with $m > 1$. For regions C^2-close to Ω, there will be m eigenvalues (counting multiplicity) close to λ_0 and we study their dependence on the region.

Suppose $\{\phi_j\}_{j=1}^m$ is an orthonormal basis of the eigenfunctions in Ω with eigenvalue λ_0. if h is near i_Ω in $C^2(\Omega, \mathbb{R}^n)$ and λ is an eigenvalue of the Dirichlet problem in $h(\Omega)$, there is a solution $v \not\equiv 0$ of $\Delta v + \lambda v = 0$ in $h(\Omega)$, $v = 0$ on $\partial h(\Omega)$, so $h^* v = u \in H^2 \cap H_0^1(\Omega)$ satisfies $h^*(\Delta + \lambda)h^{*-1}u = 0$ in Ω. Define the (real) analytic map

$$\begin{aligned} F &: (u, \lambda, h) \mapsto h^*(\Delta + \lambda)h^{*-1}u \\ &: H^2 \cap H_0^1(\Omega) \times \mathbb{R} \times \mathrm{Diff}^2(\Omega) \to L_2(\Omega). \end{aligned}$$

(Recall the eigenvalues are necessarily real, so we may restrict attention to real eigenfunctions.) Let $P = \sum_1^m \phi_j \otimes \phi_j$ be the orthogonal projection onto span $\{\phi_1, \ldots, \phi_m\}$

$$P\psi = \sum_1^m \phi_j \int_\Omega \phi_j \psi.$$

We seek λ (near λ_0) and $u \neq 0$ such that $F(u, \lambda, h) = 0$ or equivalently

$$u = v + w \neq 0, \quad v = Pu, \quad w = (I - P)u \in \mathcal{N}(P),$$

such that $PF(v + w, \lambda, h) = 0$ and $(I - P)F(v + w, \lambda, h) = 0$. The second equation may be written

$$0 = (\Delta + \lambda)w + (I - P)(h^* \Delta h^{*-1} - \Delta)(v + w) \in \mathcal{N}(P).$$

Since $(\Delta + \lambda)|\mathcal{N}(P) : \mathcal{N}(P) \cap H^2 \cap H_0^1(\Omega) \to \mathcal{N}(P) \subset L_2(\Omega)$ is an isomorphism for $\lambda = \lambda_0$, it is also an isomorphism for λ near λ_0 and has inverse Q_λ which is an analytic function of λ. Then the above equation becomes

$$w = -Q_\lambda(I - P)(h^* \Delta h^{*-1} - \Delta)(v + w), \quad \lambda \text{ near } \lambda_0,$$

which is solvable for w if also $\|h - i_\Omega\|_{C^2}$ is small. The solution $w = S(\lambda, h)v$ is linear in $v \in \mathcal{R}(P)$ and analytic in (λ, h) near (λ_0, i_Ω) with $\|S(\lambda, h)\|_{\mathcal{L}(\mathcal{R}(P), H^2 \cap H_0^1)} = O(\|h - i_\Omega\|_{C^2})$ as $h \to i_\Omega$. There remains another equation, in $\mathcal{R}(P)$. Introducting coordinates in this m-dimensional space, $v = \sum_1^m c_j \phi_j$ for some $c_j \in \mathbb{R}$, and we conclude: $u = \sum_1^m c_j(\phi_j + S(\lambda, h)\phi_j) \neq 0$ and $h^{*-1}u$ is an eigenfunction of the problem in $h(\Omega)$ with eigenvalue λ (λ near λ_0, h near i_Ω) if and only if the c_j are not all zero and $\sum_{k=1}^m M_{jk}(\lambda, h)c_k = 0$ for $1 \leq j \leq m$, where $M_{jk}(\lambda, h) = \int_\Omega \phi_j(h^* \Delta h^{*-1} + \lambda)(\phi_k + S(\lambda, h)\phi_k)$. Thus λ is an eigenvalue if and only if det $[M_{jk}(\lambda, h))]_{j,k=1}^m = 0$.

Now $(\lambda, h) \mapsto M_{jk}(\lambda, h)$ is analytic near (λ_0, i_Ω) and $M_{jk}(\lambda, i_\Omega) = \int_\Omega \phi_j(\Delta + \lambda)(\phi_k + 0) = (\lambda - \lambda_0)\delta_{jk}$ so det $[M_{jk}(\lambda, i_\Omega)] = (\lambda - \lambda_0)^m$. By Rouché's theorem, there are m roots λ (counting multiplicity) of det $[M_{jk}(\lambda, h)] = 0$ near λ_0 whenever $\|h - i_\Omega\|_{C^2}$ is sufficiently small. (We see later that "most" choices of h split this m-fold root λ_0 into m simple roots, or m simple eigenvalues; see Example 4.4 or 6.1.)

Suppose $h(0, \cdot) = i_\Omega$ and $t \mapsto h(t, \cdot) \in C^2(\Omega, \mathbb{R}^n)$ is real-analytic near $t = 0$; for example $h(t, x) = x + tV(x)$, for some $V \in C^2$. Then $(\lambda, t) \mapsto$ det $[M_{jk}(\lambda, h(t, \cdot))]_{j,k=1}^m$ is analytic near $t = 0$, $\lambda = \lambda_0$, and by Puiseux's theorem the roots $\lambda = \lambda(t)$ may be expressed as analytic functions of some fractional power of t, $t^{1/k}$ for some integer $k \geq 1$. However we know $\lambda(t)$ is real for every (small) real t. If $at^{j/k}$ and $a(e^{i\pi}t)^{j/k}$ are both real with $t > 0$, $j =$ integer, then a is real and eigher $a = 0$ or $j/k =$ integer; thus the "fraction" $1/k = 1$[14] and the eigenvalues $\lambda(t)$ are actually analytic in t, as are the eigenfunctions $u(t, \cdot)$, $F(u(t, \cdot), \lambda(t), h(t, \cdot)) = 0$. We will calculate the derivatives of $\lambda(t)$ at $t = 0$ on such a branch (there will ordinarily be m branches). As usual, it is easier to calculate in the Eulerian mode, and for this it is convenient to assume more smoothness. If Ω is $C^{m,\alpha}$, for example, all the arguments above may be carried out in the $C^{m,\alpha}$ setting and $t \mapsto (\lambda(t, \cdot), (u(t, \cdot)) \in \mathbb{R} \times C^{m,\alpha}(\Omega)_0$ is analytic near $t = 0$ if $t \mapsto h(t, \cdot) \in C^{m,\alpha}(\Omega, \mathbb{R}^n)$ is analytic. We simply assume "sufficient smoothness," and $m = 3$, $0 < \alpha < 1$, appears to be sufficient.

Suppose therefore $\Omega(t) = h(t, \Omega)$, and $\lambda = \lambda(t)$, $v = v(t, x)$ are smooth functions such that

$$\Delta v + \lambda v = 0 \text{ in } \Omega(t), \quad v = 0 \text{ on } \partial\Omega(t),$$

with $v(0, \cdot) \not\equiv 0$, and let $\dot{v} = \partial v / \partial t$; then

$$(\Delta + \lambda)\dot{v} + \dot{\lambda} v = 0 \text{ in } \Omega(t), \quad \dot{v} + V \cdot \nabla v = 0 \text{ on } \partial\Omega(t)$$

where $\partial h / \partial t = V(t, h)$, $h(0, \cdot) = i_\Omega$. When $t = 0$, $v = v(0, \cdot) = \sum_1^m c_j \phi_j$ for some $c_1, \ldots, c_m$, not all zero and

$$\dot{\lambda}(0) c_k = \int_\Omega \dot{\lambda}(0) \phi_k v = -\sum_{j=1}^m c_j \int_{\partial\Omega} V \cdot N \frac{\partial \phi_j}{\partial N} \frac{\partial \phi_k}{\partial N} \,.$$

Let $M_{jk}^0 = \int_{\partial\Omega} V \cdot N_\Omega \frac{\partial \phi_j}{\partial N} \frac{\partial \phi_k}{\partial N}$ at $t = 0$, $c = \mathrm{col}(c_1, \ldots, c_m)$ and it follows $\dot{\lambda}(0) c + M^0 c = 0$, $c \neq 0$, so $\lambda(0)$ is an eigenvalue of $-M^0$. In fact, it is easy to prove that for each simple eigenvalue μ^0 of M^0, there is a corresponding (unique) branch of eigenvalues $t \mapsto \lambda(t)$ of the form $\lambda(t) = \lambda_0 - t\mu^0 + o(t)$ as $t \to 0$. $\lambda(t)$ is a simple eigenvalue of the Dirichlet problem in $\Omega(t)$, for small $t \neq 0$. But the eigenvalue $-\dot{\lambda}(0)$ of interest may not be simple, so we continue.

To compute the second derivative at $t = 0$, we must know $\dot{v}|_{t=0}$. Now there is a unique $\dot{\phi}_j \in H^2(\Omega)$ orthogonal to $\phi_1, \ldots, \phi_m$, such that $(\Delta + \lambda_0)\dot{\phi}_j \in \mathrm{span}\,\{\phi_1, \ldots, \phi_m\}$ and $\dot{\phi}_j = -V \cdot \nabla \phi_j$ on $\partial\Omega$. It follows that $\dot{v}|_{t=0} = \sum_1^m (c_j \dot{\phi}_j + \dot{c}_j \phi_j)$ for some $\dot{c}_t, \ldots, \dot{c}_m$ (the c_j are the same as before). Then at $t = 0$

$$(\Delta + \lambda_0)\ddot{v} + 2\dot{\lambda}(0)\dot{v} + \ddot{\lambda}(0) v = 0 \text{ in } \Omega$$
$$\ddot{v} + 2V \cdot \nabla \dot{v} + (V \cdot \nabla)^2 v + \dot{V} \cdot \nabla v = 0 \text{ on } \partial\Omega$$

and so $\ddot{\lambda}(0) c_k + 2\dot{\lambda}(0)\dot{c}_k = \int_{\partial\Omega} \ddot{V} \frac{\partial \phi_k}{\partial N}$, $1 \leq k \leq m$. This gives the equation in $\mathbb{R}^m$

$$0 = (\ddot{\lambda}(0) I + \dot{M}^0) c + 2(\dot{\lambda}(0) I + M^0)\dot{c}$$

together with $0 = (\dot{\lambda}(0) I + M^0) c$, $c \neq 0$, where $\dot{M}^0 = [\dot{M}_{jk}^0]_{j,k=1}^m$ is

$$\dot{M}_{jk}^0 = \int_{\partial\Omega} \frac{\partial \phi_k}{\partial N} ((V \cdot \nabla)^2 \phi_j + 2V \cdot \nabla \dot{\phi}_j + \dot{V} \cdot \nabla \phi_j).$$

If $\dot{\lambda}(0)$ is a simple eigenvalue of $-M^0$ then $\ddot{\lambda}(0)$ must be chosen so $(\ddot{\lambda}(0) + \dot{M}^0) c$ is the image of $\dot{\lambda}(0) + M^0$, i.e., orthogonal to c, i.e.,

$$\ddot{\lambda}(0) = -c\dot{M}^0 c / c \cdot c$$

Another case of interest is when $M^0 = 0$; then $\dot{\lambda}(0) = 0$ and $\ddot{\lambda}(0)$ is an eigenvalue of $-\dot{M}^0$ with eigenvector c.

Now we study in more detail the case when $\Omega =$ unit disc in $\mathbb{R}^2$, $\Omega(t) =$ the ellipse with semi-axes e^t, e^{-t}. When $t = 0$ we have a double eigenvalue $\lambda > 0$ where $J_p(\sqrt{\lambda}) = 0$ for some integer $p \geq 1$, and ϕ_1, ϕ_2 are given by

$$\phi_1 + i\phi_2 = kJ_p(\sqrt{\lambda}r)e^{ip\theta}$$

in polar coordinates, with $k = \sqrt{\frac{2}{\pi}}/J_p'(\sqrt{\lambda})$, so $\int_\Omega \phi_j\phi_k = \delta_{jk}$. For t near 0 there is a pair of eigenvalues $\lambda_1(t), \lambda_2(t)$ near λ, analytic functions of t. Since $\Omega(-t)$ is geometrically the same as $\Omega(t)$, the pair $\{\lambda_1(t), \lambda_2(t)\}$ is an even function of t, so either $\lambda_1(t), \lambda_2(t)$ are both even or $\lambda_2(t) = \lambda_1(-t)$.

We have $y = h(t, x) = \mathrm{col}(x_1 e^t, x_2 e^{-t})$, $V(t, y) = \mathrm{col}(y_1, -y_2)$, $V \circ \nabla = y_1\frac{\partial}{\partial y_1} - y_2\frac{\partial}{\partial y_2} = \cos 2\theta r\frac{\partial}{\partial r} - \sin 2\theta\frac{\partial}{\partial \theta}$ and the matrix M^0 is

$$M^0 = \frac{2\lambda}{\pi}\int_0^{2\pi} \cos 2\theta \begin{pmatrix} \cos^2 p\theta & \cos p\theta \sin p\theta \\ \sin p\theta & \sin^2 p\theta \end{pmatrix}$$

which is zero when $p \geq 2$, but for $p = 1$

$$M^0 = \begin{pmatrix} \lambda & 0 \\ 0 & -\lambda \end{pmatrix}.$$

Thus for $p = 1$, the double eigenvalue λ $[J_1(\sqrt{\lambda} = 0]$ splits into

$$\lambda_1(t) = \lambda - t\lambda + O(t^2) \quad \text{for } kJ_1(\sqrt{\lambda}r)\cos\theta + O(t),$$
$$\lambda_2(t) = \lambda + t\lambda + O(t^2) \quad \text{for } kJ_1(\sqrt{\lambda}r)\sin\theta + O(t)$$

with

$$\lambda_2(t) = \lambda_1(-t).$$

One may compute the second derivative as well:

$$\lambda_2(t) = \lambda + t\lambda + t^2\frac{\lambda^2}{8} + O(t^3).$$

Now suppose $p \geq 2$ so $M^0 = 0, \dot{\lambda}(0) = 0$. We have $(\Delta + \lambda)\dot{\phi}_j \in \mathrm{sp}[\phi_1, \phi_2]$, $\dot{\phi}_j = -\cos 2\theta\frac{\partial\phi_j}{\partial r}$ so

$$\dot{\phi}_1 + i\dot{\phi}_2|_{r=1} = -\sqrt{\frac{\lambda}{2\pi}}\left(e^{i(p+2)\theta} + e^{i(p-2)\theta}\right)$$

and when $r = 1$

$$\dot{\phi}_1 + i\dot{\phi}_2 = -\sqrt{\frac{\lambda}{2\pi}}\left(\frac{J_{p+2}(\sqrt{\lambda}r)}{J_{p+2}(\sqrt{\lambda})}e^{i(p+2)\theta} + \frac{J_{p-2}(\sqrt{\lambda}r)}{J_{p-2}(\sqrt{\lambda})}e^{i(p-2)\theta}\right),$$

since $p+2 > p > p-2 > -p$ for $p \geq 2$. Then with

$$R_m = \frac{\sqrt{\lambda}J'_m(\sqrt{\lambda})}{J_m(\sqrt{\lambda})},$$

we have

$$2V \cdot \nabla(\dot{\phi}_1 + i\dot{\phi}_2)|_{r=1} = -\sqrt{\frac{\lambda}{2\pi}}e^{ip\theta}(R_{p+2} + R_{p-2} + e^{4i\theta}(R_{p+2} - p - 2) \\ + e^{-4i\theta}(R_{p-2} + p - 2)).$$

$$(V \cdot \nabla)^2(\phi_1 + i\phi_2)|_{r=1} = -\sqrt{\frac{\lambda}{2\pi}}e^{ip\theta}(-2 + e^{4i\theta}(p+1) + e^{-4i\theta}(-p+1)).$$

Thus

$$-\dot{M}^0 = \lambda(R_{p+2} + R_{p-2} + 2)\begin{pmatrix} 1 & 0 \\ 0 & 1 \end{pmatrix} + \delta_{p^2}\lambda(R_{p-2} - 1)\begin{pmatrix} 1 & 0 \\ 0 & -1 \end{pmatrix}.$$

If $p = 2$ we find $J_2(\sqrt{\lambda}) = 0$, $\dot{\lambda}(0) = 0$,

$$\ddot{\lambda}(0) = \begin{cases} -\frac{\lambda}{6}(18 + 5\lambda) & \text{for } kJ_2(\sqrt{\lambda}r)\cos 2\theta + O(t) \\ \frac{\lambda}{6}(\lambda - 6) & \text{for } kJ_2(\sqrt{\lambda}r)\sin 2\theta + O(t) \end{cases}$$

In case $p = 2$, the eigenvalues must be even functions of t.

If $p \geq 3$, both eigenvalues have the form

$$\lambda_{1,2}(t) = \lambda - \frac{\lambda t^2}{2}\left(2 + \frac{\lambda}{p^2 - 1}\right) + O(t^3)$$

but presumably split in higher-order terms (though this is not proved).

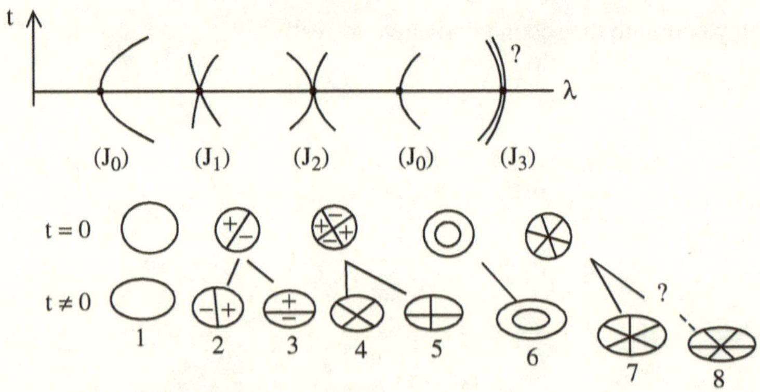

Reyleigh [33] found the nodal-sets for the case $p = 1$ by a variational argument.

Example 4.2. Variation of a Turning Point. Mignon, Murat and Puel [24] studied the variation of the turning point of

$$\Delta u + \lambda e^u = 0 \quad \text{in } \Omega, \quad u = 0 \quad \text{on } \partial\Omega$$

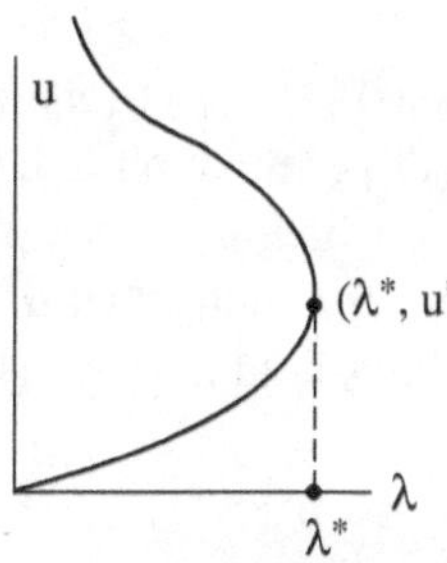

when the domain Ω changes. In a neighborhood of the turning point (λ^*, u^*), the solutions lie on a curve of the form

$$\lambda = \lambda^* + k\tau^2 + O(\tau^3), \quad u = u^* + \tau\phi^* + O(\tau^3)$$

where $k < 0$ and ϕ^* satisfies

$$\Delta\phi^* + \lambda^* e^{u^*}\phi^* = 0 \text{ in } \Omega, \quad \phi^* = 0 \text{ on } \partial\Omega$$

with $\phi^* > 0$ in Ω.

We will treat a more general case: Ω is a bounded C^2-regular region in $\mathbb{R}^n$, $f : \mathbb{R}^n \times \mathbb{R} \to \mathbb{R}$ is C^2 and there is a solution $(\lambda, u) = (\lambda_0, u_0)$ of

$$(*)_\Omega \qquad \Delta u + \lambda f(x, u) = 0 \text{ in } \Omega, \qquad u = 0 \text{ on } \partial\Omega$$

such that the linearization at this point

$$L_0 : \phi \mapsto \Delta\phi + \lambda_0 f'(\cdot, u_0(\cdot))\phi : H^2 \cap H_0^1(\Omega) \to L_2(\Omega)$$

has a one-dimensional kernel, all multiples of ϕ_o. (Here $f'(x, s) = \frac{\partial}{\partial s} f(x, s)$.) We also assume

$$\lambda_0 \neq 0, \qquad \int_\Omega \phi_0 f(\cdot, u_0(\cdot)) \neq 0, \qquad \int_\Omega \phi_0^3 f''(\cdot, u_0(\cdot)) \neq 0.$$

All these hypotheses hold for the example of [24]: the turning point is at the end of the stable branch; so 0 is the principal eigenvalue of the linearization, a simple eigenvalue with a positive eigenfunction.

We prove, for each h C^2-close i_Ω, there is a unique turning point near (λ_0, u_0) and the parameter value λ_h is a differentiable function of h with derivative

$$\frac{d}{dh}\lambda_h|_{h=i_\Omega} : \dot{h} \mapsto -\int_{\partial\Omega} \dot{h} \cdot N_\Omega \frac{\partial u_0}{\partial N}\frac{\partial \phi_0}{\partial N} \Big/ \int_\Omega \phi_0 f(\cdot, u_0(\cdot)).$$

First note, by standard bifurcation arguments (Ω fixed), that the solutions (u, λ) of $(*)_\Omega$ near (u_0, λ_0) lie on a curve

$$\lambda = \lambda_0 + k\tau^2 + o(\tau^2), \qquad u = u_0 + \tau\phi_0 + O(\tau^2), \qquad \tau \text{ small},$$

where $k = -\frac{1}{2}\lambda_0 \int_\Omega \phi_0^3 f''(\cdot, u_0(\cdot)) / \int_\Omega \phi_0 f(\cdot, u_0(\cdot)) \neq 0$. (In the case of [24], it is clear $k < 0$.) For any small perturbation $h(\Omega)$ of Ω, if we find a solution (λ_h, u_h) near (λ_0, u_0) of $(*)_{h(\Omega)}$ which is not simple – so the linearization has eigenvalue zero – it follows by continuity that the kernel is one-dimensional and the corresponding inequalities hold to prove (λ_h, u_h) is a turning point for $(*)_{h(\Omega)}$.

Choose p in $n/2 < p < \infty$, so $W^{2,p}(\Omega) \subset C^0(\Omega)$, and define

$$\begin{aligned} G : (\lambda, v, \phi, h) &\mapsto \left(h^*(\Delta + \lambda f) h^{*-1} u,\ h^*(\Delta + \lambda f') h^{*-1}(u, \phi),\ \int_\Omega \phi^2 \right) \\ &: \mathbb{R} \times W^{2,p} \cap W_0^{1,p}(\Omega) \times H^2 \cap H_0^1(\Omega) \times \mathrm{Diff}^2(\Omega) \\ &\to L_p(\Omega) \times L_2(\Omega) \times \mathbb{R} \end{aligned}$$

on a neighborhood of $(\lambda_0, u_0, \phi_0, i_\Omega)$. Then G is continuously differentiable, $G(\lambda_0, u_0, \phi_0, i_\Omega) = (0, 0, \int_\Omega \phi_0^2)$, and $\frac{\partial G}{\partial(\lambda, v, \phi)}(\lambda_0, u_0, \phi_0, i_\Omega)$ is

$$(\dot\lambda, \dot v, \dot\phi) \mapsto \left(L_0 \dot v + \dot\lambda f(u_0),\ L_0 \dot\phi + \dot\lambda f'(u_0)\phi_0 + \lambda_0 f''(u_0)\phi_0 \dot v,\ 2\int_\Omega \phi_0 \dot\phi \right),$$

which is an isomorphism. Indeed, given $(a, b, c) \in L_p \times L_2 \times \mathbb{R}$, the inverse image $(\dot\lambda, \dot v, \dot\phi)$ is uniquely determined as follows: Choose $\dot\lambda$ so $\dot\lambda \int_\Omega \phi_0 f(u_0) = \int_\Omega a\phi_0$, and then $L_0 \dot v = a - \dot\lambda f(u_0) \perp \phi_0$ determines $\dot v$ to within a multiple of ϕ_0; this multiple is chosen so that $\dot\lambda f'(u_0)\phi_0 + \lambda_0 f''(u_0)\phi_0 \dot v \perp \phi_0$ (recall $\lambda_0 \int_\Omega f''(u_0)\phi_0^3 \neq 0$), hence $\dot\phi$ is determined to within a multiple of ϕ_0; and *this* multiple is chosen by requiring $2\int_\Omega \phi_0 \dot\phi = c$.

By the implicit function theorem, there is a C^1 solution $h \mapsto (\lambda_h, v_h, \phi_h)$ of $G(\lambda, v, \phi, h) = (0, 0, \int_\Omega \phi_0^2)$ with $(\lambda_h, v_h, \phi_h) = (\lambda_0, u_0, \phi_0)$ when $h = i_\Omega$. Also $h^{*-1} v_h \equiv u_h$, $h^{*-1}\phi_h \equiv \psi_h$ satisfy

$$\Delta u_h + \lambda_h f(\cdot, u_h) = 0 \text{ in } h(\Omega), \qquad u_h = 0 \text{ on } \partial h(\Omega)$$

$$(\Delta + \lambda_h f'(\cdot, u_h))\psi_h = 0 \text{ in } h(\Omega), \qquad \psi_h = 0 \text{ on } \partial h(\Omega)$$

and $\psi_h \neq 0$ since $\int_{h(\Omega)} \psi_h^2/(\det h') = \int_\Omega \phi_0^2 > 0$. Thus (λ_h, u_h) is the (unique) turning-point near (λ_0, u_0) for the problem in $h(\Omega)$, h near i_Ω.

The calculation of the derivative, in the Eulerian mode, is simple.

Example 4.3. A Bifurcation Problem with Two-Dimensional Kernel. We seek small solutions u of

$$\Delta u + f(x, u) + \lambda u = \alpha(x) + \beta(x)u \text{ in } \Omega, \qquad u = 0 \text{ on } \partial\Omega$$

where $f(x, u) = O(u^2)$ as $u \to 0$, f and λ are fixed (f is C^3, λ is constant), and (α, β, Ω) are parameters near (0, 0, unit disc of $\mathbb{R}^2$). We suppose $\lambda > 0$ satisfies $J_1(\sqrt{\lambda}) = 0$, so the linearization at $\alpha = 0$, $\beta = 0$, $\Omega = D$ (= unit disc) has two-dimensional kernel with basis $\{\phi_1, \phi_2\}$, $\phi_1 + i\phi_2 = J_1(\sqrt{\lambda}r)e^{i\theta}$ in polar coordinates. We impose a "non-degeneracy" condition on $m(x) \equiv \frac{\partial^2 f}{\partial u^2}(x, 0)$, satisfied for most choices of m, namely: For some complex constants A, B, C

$$\int_D m(x)(\phi_1 + i\phi_2)(\xi_1\phi_1 + \xi_2\phi_2)^2 = A\xi_1^2 + B\xi_1\xi_2 + C\xi_2^2, \qquad \xi \in \mathbb{R}^2$$

and we require $A\xi_1^2 + B\xi_1\xi_2 + C\xi_2^2 \neq 0$ whenever $(\xi_1, \xi_2) \neq (0, 0)$. In this case $\{A\cos^2\theta + B\cos\theta\sin\theta + C\sin^2\theta \mid 0 \le \theta \le 2\pi\}$ is an ellipse in the complex plane which does not pass through 0, although it may have one semi-axis zero. We suppose it does not reduce to a point ($A = C$, $B = 0$) nor does this occur after a non-singular linear change of ξ-variables. Analytically, we suppose either $(A - C)\bar{B}$ is not real, or it is real but $\frac{(A+C)^2}{\{(A-C)^2+B^2\}}$ does not belong to $[1, \infty] \subset \mathbb{C}$. If the ellipse encloses 0, we have an elliptic umbilic i if 0 is outside the ellipse, a hyperbolic umbilic.

We prove that as (α, Ω) vary on a neighborhood of $(0, D)$ with β fixed (small), or as (α, β) vary near (0, 0) with Ω fixed (near D), or various combinations, we produce the universal unfolding of this umbilic, and there may be 0, 2 or 4 solutions, depending on the case.

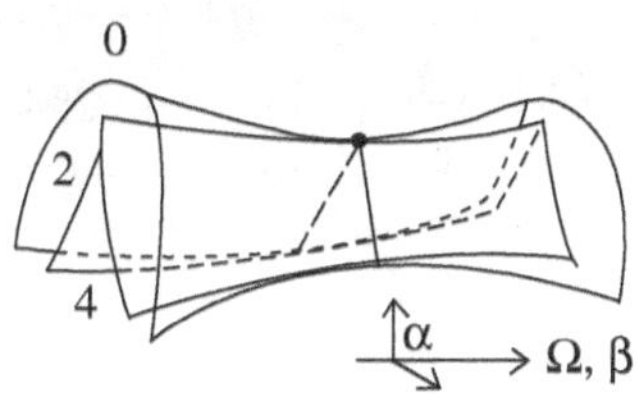

hyperbolic umbilic

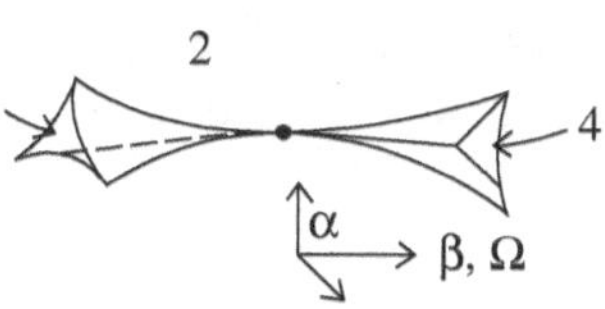

elliptic umbilic

(The case $\alpha \equiv 0$ is quite different, as $u = 0$ is then always a solution.) The solutions u are critical points of

$$u \mapsto \int_\Omega \left(-\frac{1}{2}|\nabla u|^2 + \frac{1}{2}\lambda u^2 + \alpha(x)u + \frac{1}{2}\beta(x)u^2 + \int_0^u f(x, \cdot) \right) dx$$

so it should not be surprising that catastrophe theory applies.

Example:

$$m(x_1, x_2) = m_0 + m_1 x_1 + m_2 x_2 + \sum_0^2 p_j x_1^j x_2^{2-j} + \sum_0^3 q_j x_1^j x_2^{3-j}$$

for some real constants $m_0, m_1, \ldots, q_3$. The corresponding quadratic form does not depend on m_0 or the p_j, and is

$$\begin{aligned} A\xi_1^2 + B\xi_1\xi_2 + C\xi_2^2 = Q_1 \{ & 2(m_1 + im_2)(\xi_1^2 + \xi_2^2) \\ & + (m_1 - im_2)(\xi_1 + i\xi_2)^2 \} \\ & + Q_2 \{ (\xi_1^2 + \xi_2^2)(3q_0 - iq_1 + q_2 - 3iq_3) \\ & + 2(\xi_1^2 - \xi_2^2)(q_0 + iq_3) \\ & - 2i\xi_1\xi_2(2q_0 + iq_1 + q_2 + 2iq_3) \} \end{aligned}$$

where $Q_1 = \frac{\pi}{4} \int_0^1 J_1(\sqrt{\lambda} r)^3 r^2 dr$, $Q_2 = \frac{\pi}{8} \int_0^1 J_1(\sqrt{\lambda} r)^3 r^4 dr$. Suppose Q_1, $Q_2 \neq 0$; this certainly holds if λ is the first (double) eigenvalue or if λ is large ($J_1(\sqrt{\lambda}) = 0$), and numerical integration suggests that it holds in every case. Then depending on the $m_1, m_2, q_0, \ldots, q_3$, we could be in the hyperbolic case (e.g., $m(x) = m_1 x_1 + m_2 x_2 \not\equiv 0$) or the elliptic case (e.g., $m(x) = x_2^3 - 3x_1^2 x_2$) or the degenerate case (e.g., $m(x) = m_0 + \sum_0^2 p_j x_1^j x_2^{2-j}$, when $A = B = C = 0$). Note we do *not* allow $m(x) =$ constant.

Define $F_\Omega(u; \alpha, \beta) = \Delta u + \lambda u + f(\cdot, u) - \alpha(\cdot) - \beta(\cdot)u$, so

$$F_\Omega : H^2 \cap H_0^1(\Omega) \times C^3(\mathbb{R}^2) \times C^3(\mathbb{R}^2) \to L_2(\Omega)$$

is C^3. Also $(u, \alpha, \beta, h) \mapsto h^* F_{h(D)} h^{*-1}(u; \alpha, \beta)$ is C^3 from $H^2 \cap H_0^1(D) \times C^3 \times C^3 \times \text{Diff}^2$ to $L_2(D)$. Using the method of Liapunov-Schmidt (but supressing details) if $h = i_D + g$, $\|g\|_{C^2(D)}$ small $u = \xi_1\phi_1 + \xi_2\phi_2 + \psi = \phi + \psi$ with $\psi \perp \{\phi_1, \phi_2\}$ and ξ_1, ξ_2, ψ small, and $\|\alpha\|_{C^3} + \|\beta\|_{C^3}$ is small, the equation becomes

$$\begin{aligned} 0 &= h^*(\Delta + \lambda)h^{*-1}(\phi + \psi) - \alpha \circ h - \beta \circ h(\phi + \psi) + f(h(\cdot), \phi + \psi) \\ &= (\Delta + \lambda)\psi + [g \cdot \nabla, \Delta + \lambda](\phi + \psi) + \alpha + \beta(\phi + \psi) + \frac{1}{2} m(\phi + \psi)^2 + \cdots \end{aligned}$$

and we solve the component in $\{\phi_1, \phi_2\}^\perp$ for

$$\begin{aligned} \psi = \psi(\xi, \alpha, \beta, g) = O(& \|\alpha\| + \|g\|(|\xi| + \|\alpha\| + \|\beta\|) \\ & + |\xi|\|\beta\| + \|g\|^2 + |\xi|^3). \end{aligned}$$

Substitution in the remainder of the equation gives the bifurcation equation $E(\xi;\alpha,\beta,g)=0$ in $\mathbb{R}^2$,

$$\begin{aligned}(E_1+iE_2)(\xi,\alpha,\beta,g)\\ &=\int_D(\phi_1+i\phi_2)h^*F_{h(D)}(\cdot,\alpha,\beta)h^{*-1}(\xi_1\phi_1+\xi_2\phi_2+\psi(\xi,\alpha,\beta,g))\\ &=\alpha_1+i\alpha_2+\xi_1(\beta_{11}+i\beta_{21})+\xi_2(\beta_{12}+i\beta_{22})\\ &\quad+\xi_1(\gamma_{11}+i\gamma_{21})+\xi_2(\gamma_{12}+i\gamma_{22})\\ &\quad+\frac{1}{2}(A\xi_1^2+B\xi_1\xi_2+C\xi_2^2)+\text{(higher order terms)}\end{aligned}$$

where

$$\begin{pmatrix}\alpha_1\\ \alpha_2\end{pmatrix}=\int_D\alpha(x)\begin{pmatrix}\phi_1\\ \phi_2\end{pmatrix},\qquad \begin{pmatrix}\beta_{11}&\beta_{12}\\ \beta_{21}&\beta_{22}\end{pmatrix}=\int_D\beta(x)\begin{pmatrix}\phi_1^2&\phi_1\phi_2\\ \phi_1\phi_2&\phi_2^2\end{pmatrix},$$

$$\begin{aligned}\begin{pmatrix}\gamma_{11}&\gamma_{12}\\ \gamma_{21}&\gamma_{22}\end{pmatrix}&=\int_{\partial D}g\cdot N\begin{pmatrix}\cos^2\theta&\sin\theta\cos\theta\\ \sin\theta\cos\theta&\sin^2\theta\end{pmatrix}d\theta\\ &=\frac{\pi}{2}(\sqrt{\lambda}J_1'(\sqrt{\lambda}))^2\begin{pmatrix}p_0+p_1&p_2\\ p_2&p_0-p_1\end{pmatrix}\end{aligned}$$

when $g\cdot N|_{\partial D}=p_0+p_1\cos 2\theta+p_2\sin 2\theta+\{\text{terms orthogonal to }1,\cos 2\theta,\sin 2\theta\}$.

To obtain the universal unfolding of the umbilic, $\binom{\alpha_1}{\alpha_2}$ should vary over a full neighborhood of $\binom{0}{0}$ in $\mathbb{R}^2$ and

$$\begin{pmatrix}\beta_{11}&\beta_{12}\\ \beta_{21}&\beta_{22}\end{pmatrix}+\frac{\pi}{2}(\sqrt{\lambda}J_1'(\sqrt{\lambda}))^2\begin{pmatrix}p_0+p_1&p_2\\ p_2&p_0-p_1\end{pmatrix}$$

should include all (small) symmetric 2×2 matrices. The last condition could be achieved with $\beta=0$ (or small), considering only ellipses Ω centered at the origin but with arbitrary orientation and arbitrary semi-axes (near 1). If we restricted attention to ellipses with a fixed area, then $p_0=0$ and we should also allow β to vary (e.g., β an arbitrary small constant).

Example 4.4. Generic Simplicity of Eigenvalues of a Self-Adjoint 2*m*-Order Dirichlet Problem. Following Micheletti [23], grosso modo, we show that for most C^{2m}-regular bounded regions $\Omega\subset\mathbb{R}^n$ (in the sense of Baire category), a self-adjoint strongly elliptic operator with the unique continuation property has only simple eigenvalues for the Dirichlet problem. I am indebted to M. Saut for bringing Micheletti's work to my attention.

Specifically, let Q be an open set in $\mathbb{R}^n$ and suppose $a_{\alpha\beta}=\overline{a_{\beta\alpha}}:Q\to\mathbb{C}$ is of class C^s, $s=\max(|\alpha|,|\beta|)+1$, when the multi-indices α,β have $|\alpha|\le m$,

$|\beta| \leq m$, and for some constant $c_0 > 0$

$$\operatorname{Re} \sum_{|\alpha|=|\beta|=m} a_{\alpha\beta}(x)\xi^{\alpha+\beta} \geq c_0|\xi|^{2m} \qquad \text{for } x \in Q, \xi \in \mathbb{R}^n.$$

Finally assume "unique continuation" holds, in the sense:

If λ is real, $u \in H^{2m}(Q)$ with $\operatorname{support}(u)$ compact in Q and $\sum_{|\alpha|,|\beta|\leq m} D^\alpha(a_{\alpha\beta}D^\beta u) = \lambda u$ as distributions in Q then $u \equiv 0$ in Q. (Here $D^\alpha = D_1^{\alpha_1} \cdots D_n^{\alpha_n}$, $D_j = \frac{1}{i}\frac{\partial}{\partial x_j}$.) This is certainly true if the $a_{\alpha,\beta}$ are analytic (Holmgren's theorem) or if $m = 1$ (see [2] or [12, Thm. 8.9.1]).

Let Ω be a bounded C^{2m}-regular region with closure $\overline{\Omega} \subset Q$. Then for an ample (= residual = category II) set of h in a $C^{2m}(\Omega, \mathbb{R}^n)$-neighborhood of i_Ω, the Dirichlet problem in $h(\Omega)$

$$\sum_{|\alpha|,|\beta|\leq m} D^\alpha(a_{\alpha\beta}D^\beta u) = \lambda u \text{ in } h(\Omega), \qquad u \in H^{2m} \cap H_0^m(h(\Omega)), \quad u \not\equiv 0$$

has only simple eigenvalues.

In fact the argument is more general. Without assuming unique continuation, for most choices of h near i_Ω the eigenvalues will all have minimal multiplicity; and if λ is such an eigenvalue whose minimal multiplicity is $p \geq 2$, there is a basis $\{\phi_1, \ldots, \phi_p\}$ for the eigenfunctions all of which vanish throughout a neighborhood of $\partial h(\Omega)$. Of course such an eigenvalue must remain of multiplicity $p \geq 2$ for any small perturbation of $h(\Omega)$; and, of course, this cannot occur (i.e., the eigenvalues must be simple) when unique continuation holds.

Assume λ is an eigenvalue of multiplicity p for the Dirichlet problem in some Ω, and that it has minimal multiplicity with respect to small perturbations of $\partial\Omega$; that is, for any $\epsilon > 0$, if $\|h - i_\Omega\|_{C^{2m}}$ is sufficiently small, the problem in $h(\Omega)$ has an eigenvalue λ_h of multiplicity $\geq p$ in $|\lambda_h - \lambda| < \epsilon$. (In this case, as we see below, λ_h is the only eigenvalue close to λ, and in fact $\lambda_h = \lambda$.)

It will be sufficient to work with a restricted class of perturbations so let $h(t, x) = x + tV(x)$ for small real t and any given $V \in C^{2m}(\Omega, \mathbb{R}^n)$. Let $\Omega(t) = h(t, \Omega)$ and define

$$A_{\Omega(t)} : v \mapsto \sum_{|\alpha|,|\beta|\leq m} D^\alpha(a_{\alpha\beta}D^\beta v) : H^{2m} \cap H_0^m(\Omega(t) \to L_2(\Omega(t))$$

and

$$B(t) = h(t, \cdot)^* A_{\Omega(t)} h(t, \cdot)^{*-1} : H^{2m} \cap H_0^m(\Omega) \to L_2(\Omega).$$

Then $B(t)$ may be considered a self-adjoint operator with respect to $L_2(\Omega; \det h_x(\cdot, t)dx)$ for each small t. When $t = 0$, $\Omega(0) = \Omega$, $B(0) = A_\Omega$ and there exist $\phi_1, \ldots, \phi_p \in H^{2m} \cap H_0^1(\Omega)$, a basis for $\mathcal{N}(A_\Omega - \lambda)$, such that $\int_\Omega \phi_j\bar{\phi}_k = \delta_{jk}$ $(1 \leq j, k \leq p)$. We apply Gram-Schmidt orthogonalization with respect to the inner product of $L_2(\Omega; \det h_x(\cdot, t)dx)$ to produce another basis $\{\phi_1(t,\cdot), \ldots, \phi_p(t,\cdot)\}$ for $\mathcal{N}(A_\Omega - \lambda)$ with $\int_\Omega \phi_j(t,\cdot)\overline{\phi_k(t,\cdot)}\det h_x(t,\cdot) = \delta_{jk}$ and $\phi_j(0,\cdot) = \phi_j$. Note that $t \mapsto \phi_j(t,\cdot)$ is C^1 (in fact, rational). Define $P(t) = \sum_1^p \phi_j(t) \otimes \phi_j(t)$, the orthogonal projection (in this inner product) onto $\mathcal{N}(A_\Omega - \lambda)$. By the implicit function theorem, there is a C^1 solution $(t, \mu) \mapsto w_j(t, \mu) \in H^{2m} \cap H_0^m(\Omega)$ of

$$(I - P(t))\big(h^* A_{\Omega(t)} h^{*-1} - \lambda_0 - \mu\big)(\phi_j(t) + w_j(t, \mu)) = 0$$
$$P(t) w_j(t, \mu) = 0$$

for small real t, μ, with $w_j(0, \mu) = 0$. In fact $\mu \mapsto w_j(t, \mu)$ is analytic.

By hypothesis, for each t near 0 ($h(t,\cdot)$ near i_Ω) there is an eigenvalue $\lambda_0 + \mu$ near λ_0 of the Dirichlet problem in $\Omega(t)$, having multiplicity $\geq p$. Thus there exists nontrivial ψ in $H^{2m} \cap H_0^m(\Omega)$ with $B(t)\psi = (\lambda_0 + \mu)\psi$, so, for some constants $c_1, \ldots, c_p$, not all zero, $\psi = \sum_j c_j(\phi_j(t,\cdot) + w_j(t, \mu, \cdot))$; note the c_j may depend on t, μ. For each $k = 1, \ldots, p$

$$\mu c_k = \int_\Omega \bar{\phi}_k (B(t)\psi - \lambda_0 \psi) \det h_x(t,\cdot)\, dx = \sum_{j=1}^p c_j p_{jk}(t, \mu)$$

where

$$p_{jk}(t, \mu) = \int_\Omega \bar{\phi}_k (B(t) - \lambda_0)(\phi_j(t,\cdot) + w_j(t, \mu, \cdot)),$$

and

$$\Delta(t, \mu) = \det\,[\mu\delta_{jk} - p_{jk}(t, \mu)]_{jk=1}^p = 0.$$

Note the $p_{jk}(t,\cdot)$ are analytic functions of μ near $(t, \mu) = (0, 0)$, and C^1 in both variables. When $t = 0$, $p_{jk}(0, \mu) = 0$ and $\Delta(0, \mu) = \mu^p$; by Rouché's theorem, there are p roots μ near 0 counting multiplicity, for each small t. But by hypothesis, there is some real root μ near 0 such that $[\mu\delta_{jk} - p_{jk}(t, \mu)]_{j,k=1}^p$ has at least p independent null-vectors, so (since $p_{jk} = \overline{p_{kj}}$)

$$p_{jk}(t, \mu) = \mu\delta_{jk}, \qquad 1 \leq j, k \leq p,$$

for some small $\mu = \mu(t)$. For small $\epsilon > 0$ at t sufficiently small

$$p\mu(t) = \frac{1}{2\pi i} \int_{|\zeta|=\epsilon} \frac{\zeta\, d\zeta}{\Delta(t, \zeta)} \frac{\partial}{\partial \zeta} \Delta(t, \zeta)$$

so $t \mapsto \mu(t)$ is C^1 near $t = 0$.

Now

$$\dot{p}_{jk} \equiv \frac{\partial}{\partial t} p_{jk}(0,0) = \int_\Omega \bar{\phi}_k D_t\{h^*(A_{\Omega(t)} - \lambda)h^{*-1}(\phi_j + w_j)\}_{\substack{t=0\\ \mu=0}}$$
$$+ \int_\Omega \bar{\phi}_k V \cdot \nabla(A_\Omega - \lambda)\phi_j$$
$$= \int_\Omega \bar{\phi}_k (A_\Omega - \lambda) z_j,$$
$$z_j = D_t(\phi_t + w_j(t,0,\cdot))|_{t=0},$$

where $z_j \in H^{2m-1} \cap H_0^{m-1}(\Omega)$ with $(\frac{\partial}{\partial N})^{m-1} z_j = -V \cdot N \frac{\partial^m \phi_j}{\partial N^m}$ on $\partial\Omega$, and we must interpret $(A_\Omega - \lambda)z_j$ as a distribution in $H^{-1}(\Omega)$. Integration by parts gives

$$\dot{p}_{jk} = \int_\Omega \bar{\phi}_k A_\Omega z_j - z_j \overline{A_\Omega \phi_k}$$
$$= -\int_{\partial\Omega} \left(\sum_{|\alpha|=|\beta|=m} a_{\alpha\beta} N^\alpha N^\beta \right) V \cdot N \frac{\partial^m \phi_j}{\partial N^m} \overline{\frac{\partial^m \phi_k}{\partial N^m}}$$
$$= \overline{\dot{p}_{jk}}.$$

Now $\dot{\mu}(0)$ is an eigenvalue of $[\dot{p}_{jk}]$, and in fact the only eigenvalue: otherwise the multiplicity would be reduced for $t \neq 0$. But a Hermitian matrix with only one eigenvalue is a multiple of the identity, so $\dot{p}_{jk} = 0$ and $\dot{p}_{jj} = \dot{p}_{kk}$ whenever $j \neq k$. This must hold for every choice of V, so on $\partial\Omega$

$$\frac{\partial^m \phi_j}{\partial N^m} \overline{\frac{\partial^m \phi_k}{\partial N^m}} = 0, \qquad \left|\frac{\partial^m \phi_j}{\partial N^m}\right|^2 = \left|\frac{\partial^m \phi_k}{\partial N^m}\right|^2 \quad \text{if } j \neq k.$$

Since the multiplicity $p \geq 2$, we conclude $\partial^m \phi_j / \partial N^m = 0$ on $\partial\Omega$, i.e., $\phi_j \in H^{2m} \cap H_0^{m+1}(\Omega)$ for each $j = 1, \ldots, p$. Also we have $\dot{\mu}(0) = 0$.

Summarizing: If $\lambda = \lambda(\Omega)$ is a p-fold eigenvalue of the problem in Ω whose multiplicity cannot be reduced by small perturbations of Ω, then the eigenfunctions belong to $H^{2m} \cap H_0^{m+1}(\Omega)$ and, for every $\tilde{\Omega}$ near Ω, there is a unique eigenvalue $\lambda(\tilde{\Omega})$ near $\lambda(\Omega)$, which also has multiplicity p, depends differentiably on $\tilde{\Omega}$, and has derivative zero when $\tilde{\Omega} = \Omega$. Thus, in fact, $\lambda(\tilde{\Omega}) = \lambda(\Omega) = \lambda$ is constant.

We next show that we may choose a basis for the eigenfunctions which does not depend on small perturbations of Ω.

Returning to the previous notation ($h(t,x) = x + tV(x)$, $\Omega(t) = h(t,\Omega)$) we now know $\mu \equiv 0$, i.e., $p_{jk}(t,0) \equiv 0$ for small t, and $\{\psi_j(t,\cdot)\}_{j=1}^p$, defined by $h(t,\cdot)^*\psi_j(t,\cdot) = \phi_j(t,\cdot) + w_j(t,0,\cdot)$, is a basis for the eigenfunctions of $A_{\Omega(t)}$ for the eigenvalue λ; thus we also have the $\psi_j(t,\cdot) \in H^{2m} \cap H_0^{m+1}(\Omega(t))$. Define $\dot{\psi}_j(t,\cdot) = \frac{\partial}{\partial t}\psi_j(t,\cdot)$; then $h(t,\cdot)^*\dot{\psi}_j(t,\cdot) = D_t(\phi_j(t,\cdot) + h(t,0,\Omega))$ is in $H^{2m-1} \cap H_0^m(\Omega)$ so $\dot{\psi}_j(t,\cdot) \in H^{2m-1} \cap H_0^m(\Omega(t))$. But also $A_{\Omega(t)}\dot{\psi}_j(t,\cdot) = \lambda\dot{\psi}_j(t,\cdot)$ as distributions in $\Omega(t)$, so in fact $\dot{\psi}_j(t,\cdot) \in H^{2m} \cap H_0^m(\Omega(t))$. Since $\{\psi_j(t,\cdot)\}_{j=1}^p$ is a basis for $\mathcal{N}(A_{\Omega(t)} - \lambda)$ there is a continuous matrix $B(t) = [b_{jk}(t)]$ so

$$\dot{\psi}_j(t,\cdot) = \sum_{k=1}^{p} b_{jk}(t)\psi_k(t,\cdot).$$

Define $C(t)$ as the matrix solution of $\dot{C}(t) + C(t)B(t) = 0$, with $C(0) = I$, and let $\theta_j(t,\cdot) = \sum_{k=1}^p C_{jk}(t)\psi_k(t,\cdot)$. This is another basis for the eigenfunctions of $\mathcal{N}(A_{\Omega(t)} - \lambda)$, with the additional property

$$\frac{\partial\theta_j}{\partial t}(t,\cdot) \equiv 0, \qquad 1 \le j \le p.$$

If $y \in \Omega(s)$ for all s between 0 and t, we have $\theta_j(t,y) = \theta_j(0,y) = \phi_j(y)$.

Choose V close to N_Ω and on $\partial\Omega$; then for small $t < 0$, $\Omega(t) \subset \Omega$ and $\cup_{-\delta<t\le 0}\partial\Omega(t)$ is a neighborhood of $\partial\Omega$ (in $\overline{\Omega}$), small $\delta > 0$. For any y in this neighborhood, $y \in \partial\Omega(t)$ for some, $t -\delta < t \le 0$, so $\theta_j(t,y) = 0 = \theta_j(0,y) = \phi_j(y)$. Thus the eigenfunctions ϕ_j must all vanish on a neighborhood of $\partial\Omega$, as was to be proved.

For each $j = 1, 2, \ldots$, there is an open dense set among the C^{2m}-regular bounded regions Ω with closure in Q such that every eigenvalue λ in $-j \le \lambda \le j$ has minimal multiplicity: λ is either simple, or all the eigenfunctions corresponding to λ vanish on a neighborhood of $\partial\Omega$. This condition is obviously open, and the argument above shows it is dense. Taking the intersection over all $j = 1, 2, \ldots$ gives the result claimed at the beginning.

Remark. For the Laplacian, Karen Uhlenbeck [41] proved this result independently, using the transversality theorem; this argument will be our introductory example in Chapter 6.

Note that $t \mapsto h(t,\cdot)^* A_{\Omega(t)} h(t,\cdot)^{*-1}$ is *not* analytic in general, unless the coefficients are analytic. Thus we cannot a priori expect to find a basis for the eigenfunctions continuous as $t \to 0$ [14, p. 111]; hence the careful construction of a basis.

Chapter 5

The Transversality Theorem

The transversality theorem (or transversal density theorem) of Thom and Abraham is a tool useful in many branches of geometry and analysis. The usual form [1, 32, 35, 42] makes unnecessarily strong hypotheses for the case of negative Fredholm index. We will prove some generalizations, aimed at infinite-dimensional applications, with special attention to the case of negative index. (We will see many examples with index $-\infty$ in Ch. 6).

Our terminology is that of Lang [17] and Abraham and Robbin [1] except for "codimension," defined in 5.15, and "σ-proper," defined in 5.4(3). See 5.13–14 for semi-Fredholm operators.

This chapter is logically self-contained except for Sard's theorem. We will use only Theorem 5.4 in the sequel, but other variations on the theme of transversality seemed of sufficient interest to be worth recording.

Theorem 5.1. Sard's Theorem *(Brown-Morse-Sard). Let $A \subset \mathbb{R}^n$ be an open set and $f : A \to \mathbb{R}^m$ a C^k map, where k is a positive integer and $k > n - m$. Then the set of critical values of f has measure zero in $\mathbb{R}^m$ and is meager (=first Baire category).*

Proof. See [1] or [37].

Recall that x is a regular point of f if the derivative $f'(x)$ is surjective; otherwise x is a *critical point*. A *critical value* of f is the image of some critical point; all other points of the codomain are *regular values* of f, including all points outside the image of f. The only modification in infinite dimensions is that, for x to be a regular point of f, we require not only that $f'(x)$ is surjective but also that its kernel splits – which always holds in our applications, as the kernel is finite dimensional.

When $n < m$, the set of critical values of $f : A \subset \mathbb{R}^n \to \mathbb{R}^m$ is simply the image of f, and to treat the image we don't need differentiability.

Theorem 5.2. *Suppose* $1 \le n < m$, *A is any subset in* $\mathbb{R}^n$ *and* $f : A \to \mathbb{R}^m$ *is of class* $C^{\mu+}$, $\mu = n/m$, *that is*

$$\frac{|f(x) - f(y)|}{|x - y|^{\mu}} \to 0 \quad \text{as } x - y \to 0$$

uniformly for $x \neq y$ *in bounded subsets of A. Then* $f(A)$ *has measure zero and is meager in* $\mathbb{R}^m$.

Proof. It suffices to show meas $f(C \cap A) = 0$ for every compact cube $C < \mathbb{R}^n$. Let $\epsilon > 0$ and choose $\delta > 0$ so $|f(x) - f(y) \le \epsilon |x - y|^{\mu}$ when $|x - y| \le \delta$ in $C \cap A$. Divide C into N^n disjoint congruent cubes $\{C'\}$, with N so large that diam $C' = \frac{1}{N}$ diam $C < \delta$. Then for each such C', diam $f(C' \cap A) \le \epsilon$ (diam $C/N)^{\mu}$ so $f(C' \cap A)$ is contained in a cube of edge $\le 2\epsilon$ (diam $C/N)^{\mu}$ and volume $\le (2\epsilon)^m$ (diam $C/N)^n$. Therefore meas $f(C \cap A) \le (2\epsilon)^m$ (diam $C)^n$, with $\epsilon > 0$ arbitrary.

Theorem 5.3

(i) *Let* $n > m \ge 1, \epsilon > 0$; *there is a function* $f : \mathbb{R}^n \to \mathbb{R}^m$ *of class* $C^{n-m+1-\epsilon}$ *with compact support whose critical values include* $[0, 1]^m$. (*The set of critical points includes a set* $\mathbb{R}^{m-1} \times W$ *where* $W \subset \mathbb{R}^{n-m+1}$ *is homeomorphic to* $[0, 1]$ (*but has Hausdorff dimension* $n - m + 1 - \epsilon$).)

(ii) *Let* $1 \le n < m, \epsilon > 0$; *there is a function* $f : \mathbb{R}^n \to \mathbb{R}^m$ *of class* $C^{(n-\epsilon)/m}$ *with compact support whose image contains* $[0, 1]^m$.

Remark. This shows the smoothness conditions of (5.1) for $n \ge m$ and (5.2) for $n < m$, are nearly optimal. If $n = m$, (5.1) only requires continuous differentiability, and we need one derivative to define "critical value."

Proof. (i) Whitney [43] showed how to construct a curve $K_q \subset [0, 1]^q$ (for each $q \ge 2$), homeomorphic to $[0, 1]$, and a function $W_q : K_q \to [0, 1]$ such that $W_q(K_q) = [0, 1]$ and $|W_q(x) - W_q(x')| = 0(|x - x'|^{q-\epsilon})$ for all $x, x' \in K_q$. (To treat arbitrarily small $\epsilon > 0$, we must choose the scaling factor of Whitney's construction close to 1/2; Whitney used the factor 1/3.) Then by Whitney's

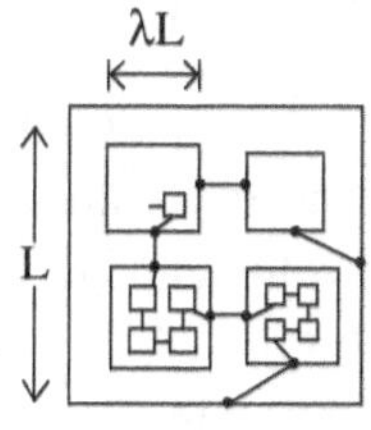

0 < λ = scale factor < 1/2

extension theorem – in the more general version of E. Stein [38] – there is a $C^{q-\epsilon}(\mathbb{R}^n)$ extension, still denoted W_q, such that $(\partial/\partial x)^j W_q(x) = 0$ for all $x \in K_q$ and $1 \le |j| \le q - 1$. If $m = 1$, multiplying W_n by an appropriate "cutoff" gives the required example. In general, let $q = n - m + 1 \ge 2$ and then $x \mapsto (x_1, \ldots, x_{n-1}, W_q(x_m, \ldots, x_n))$ is $C^{q-\epsilon}$ with $\mathbb{R}^{m-1} \times [0, 1]$ among its critical values.

(ii) There is a $C^{1/m}$ Peano curve $P_m : [0, 1]$ onto $[0, 1]^m$ for each $m \ge 2$; the construction generalizes the method of Moore [25] for $m = 2$, by induction on m, as shown in the figure (for $m = 3$).

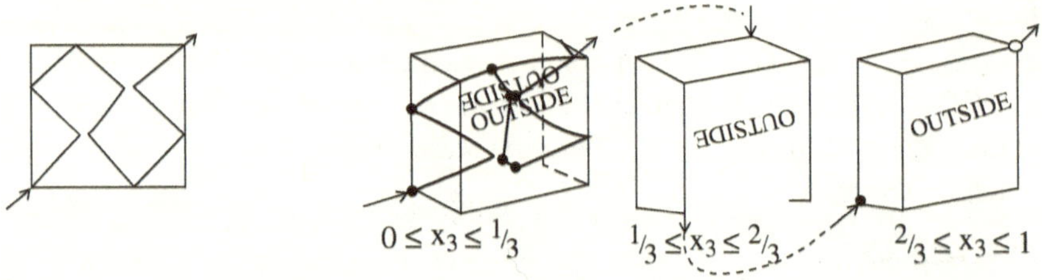

If $1 < n < m$, consider $\phi = P_m \circ W_n | K_n : K_n \subset \mathbb{R}^n \to \mathbb{R}^m$, where W_n is the $C^{n-\epsilon}$ Whitney function. For $x, x' \in K_n$,

$$|\phi(x) - \phi(x')| = O\big(|W_n(x) - W(n(x')|^{1/m}\big) = O\big(|x - x'|^{(n-\epsilon)/m}\big)$$

and there is a $C^{(n-\epsilon)/m}$ extension of ϕ which satisfies the requirements of the theorem – in particular, the image contains $P_m(W_n(K_n)) = P_m[0, 1] = [0, 1]^m$.

Kupka [16] gave an example of a C^∞ map from ℓ_2 to $\mathbb{R}$ with critical values $=$ $[0, 1]$. As modified by Dovady and Bonic, the example is a cubic polynomial $f : \ell_2 \to \mathbb{R}$

$$f(x) = \sum_{n=1}^{\infty} 2^{-n} \lambda(n x_n), \quad x = (x_n)_{n=1}^{\theta} \quad \text{with} \quad \sum_{1}^{\infty} |x_n|^2 < \infty,$$

where $\lambda(t) = 3t^2 - 2t^3$. The critical points are $\{x \in \ell_2 | n x_n = 0 \text{ or } 1, \text{ for all } n\}$ and the critical values $\{\sum_{n=1}^{\infty} 2^{-n} \epsilon_n | \epsilon_n = 0 \text{ or } 1, \text{ for all } n\} = [0, 1]$. Thus smoothness is not enough; but Smale [37] showed we can generalize Sard's theorem to infinite dimensions provided we restrict attention to smooth Fredholm maps (the derivative at each point is a Fredholm linear operator) of separable spaces. Quinn [32] noted that left-Fredholm and σ-proper is sufficient. (Kupka's example is right-Fredholm with index $+\infty$). It is customary [1, 32] to first prove Smale's theorem, and then obtain the transversality theorem from that by a neat geometric argument. I have sinned against geometry in order to allow weaker hypotheses in the case of negative index. We will prove the transversality theorem directly, using a Liapunov-Schmidt argument (as in

[37]) to reduce to finite dimensional problem, and then appealing to Sard's theorem. Smale's theorem is then an immediate corollary. We treat here only the case of regular values – see 5.10 for more general transversality.

Theorem 5.4. Transversality Theorem. *Suppose given positive integers k, m; Banach manifolds X, Y, Z of class C^k; an open set $A \subset X \times Y$; a C^k map of $f : A \to Z$; and a point $\zeta \in Z$.*

Assume for each $(x, y) \in f^{-1}(\zeta)$ that:

(1) $\frac{\partial f}{\partial x}(x, y) : T_x X \to T_\zeta Z$ is semi-Fredholm with index $< k$.
(2) Either

$$(\alpha)\, Df(x, y) = \left(\frac{\partial f}{\partial x}, \frac{\partial f}{\partial y}\right) : T_x X \times T_y Y \to T_\zeta Z \text{ is surjective}$$

or

$$(\beta)\ \dim\left\{\mathcal{R}(Df(x, y))/\mathcal{R}\left(\frac{\partial f}{\partial x}(x, y)\right)\right\} \geq m + \dim \mathcal{N}\left(\frac{\partial f}{\partial x}(x, y)\right).$$

Further assume:

(3) $(x, y) \mapsto y : f^{-1}(\zeta) \to Y$ is σ-proper, that is $f^{-1}(\zeta) = \bigcup_{j=1}^{\infty} M_j$ is a countable union of sets M_j such that $(x, y) \mapsto u : M_j \to Y$ is a proper map, for each j. [Given $(x_\nu, y_\nu) \in M_j$ such that $\{y_\nu\}$ converges in Y, there is a convergent subsequence (or subnet) of $\{(x_\nu, y_\nu)\}$ with limit in M_j.]

We note that (3) holds if $f^{-1}(\zeta)$ is Lindelöf [every open cover has a countable subcover], or more specifically, if $f^{-1}(\zeta)$ is a separable metric space or if X, Y are separable metric spaces.

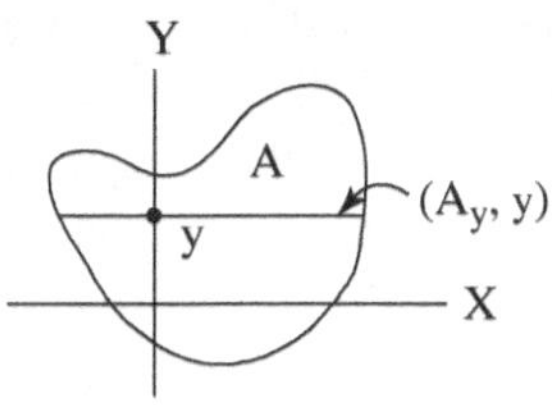

Let $A_y = \{x \mid (x, y) \in A\}$ and

$$\dot{Y}_{\text{crit}} = \{y \mid \zeta \quad \text{is a critical value of} \quad f(\cdot, y) : A_y \to Z\}.$$

Then Y_{crit} is a meager set in Y and, if $(x, y) \mapsto y : f^{-1}(\zeta) \to Y$ is proper, Y_{crit} is also closed. If ind $\partial f/\partial x \leq -m < 0$ on $f^{-1}(\zeta)$, then 2(α) implies 2 (β) and

$$Y_{\text{crit}} = \{y \mid \zeta \in f(A_y, y)\}$$

has codimension $\geq m$ in Y. [See (5.15) for definition of codimension; note Y_{crit} is meager iff codim $Y_{\text{crit}} \geq 1$.]

Remark. The usual hypothesis is that ζ is a regular value of f, so $2(\alpha)$ always holds. If $2(\beta)$ holds at some point then index $\partial f/\partial x \leq -m$ at this point, since

$$\text{codim } \mathcal{R}\left(\frac{\partial f}{\partial x}\right) \geq \dim \left\{\frac{\mathcal{R}(Df)}{\mathcal{R}(\partial f/\partial x)}\right\}.$$

If ind $(\partial f/\partial x) \leq -m$ and $2(\alpha)$ holds. Then $2(\beta)$ also holds. Thus $2(\beta)$ is more general for the case of negative index (and we'll see examples with index $-\infty$).

It is customary to assume $A = X \times Y$. Our generalization $A \subset X \times Y$, while mathematically trivial, is very convenient in some problems.

Terminological quibble. I have deliberately avoided the use of "residual". Category I = "meager" is more or less accepted, and I would suggest "ample" for the complement. "Category II" is merely uninformative; "residual" is misleading.

Lemma 5.5. *Let $k, m = 1, X, Y, Z, A, f, \zeta$ be gives as above, $(x_0, y_0) \in f^{-1}(\zeta)$, and assume hypotheses (1) and (2) – (α) or (β) – hold at (x_0, y_0). Then there exists open neighborhoods U of x_0, V of y_0, and an open dense subset $V^0 \subset V$, such that $\overline{U} \times \overline{V} \subset A$ and ζ is a regular value of $f(\cdot, y)\,|\,U$ whenever $y \in V^0$. (If $\operatorname{ind}(\partial f/\partial x) < 0$, this means $\zeta \notin f(U \times V^0)$.)*

Sublemma 5.6. *Suppose $f(x_0, y_0) = \zeta$, $\frac{\partial f}{\partial x}(x_0, y_0)$ is left-Fredholm and f is continuously differentiable on a neighborhood W of (x_0, y_0). Then there is a neighborhood W of (x_0, y_0) such that $\overline{W} \subset W_0$ and $(x, y) \mapsto y : f^{-1}(\zeta) \cap \overline{W} \to Y$ is proper.*

Proof of Sublemma. The result is local, so we may assume X, Y, Z are Banach spaces. Now $L \equiv \frac{\partial f}{\partial x}(x_0, y_0)$ is left-Fredholm so $X_1 = N(L)$ is finite dimensional and splits $X = X_1 \oplus X_2$, and the restriction of L is an isomorphism from X_2 onto $\mathcal{R}(L)$, with a continuous inverse ($\mathcal{R}(L)$) is closed. There exists $c_0 > 0$ so $|Lx_2| \geq c_0|x_2|$ for all $x_2 \in X_2$. Also, if $B = 1 + \|\frac{\partial f}{\partial y}(x_0, y_0)\|$, there is a bounded neighborhood W of (x_0, y_0), so small that $\overline{W} \subset W_0$ and

$$|f(x, y) - f(x', y') - L(x - x')| \leq \frac{c_0}{2}|x - x'| + B|y - y'|$$

for (x, y) and (x', y') in W. Now suppose $\{(x_n, y_n)\}_{n\geq 1}$ is a sequence in $f^{-1}(\zeta) \cap \overline{W}$ such that $\{y_n\}$ converges. Then x_n^1 (= the components of x_n in X_1) is bounded in a finite-dimensional space and has a convergent subsequence – we suppose, in fact, that $\{x_n^1\}$ converges.

Now

$$c_0\left|x_n^2 - x_m^2\right| \le \left|L\left(x_n^2 - x_m^2\right)\right| = |L(x_n - x_m)| \le \frac{c_0}{2}|x_n - x_m| + B|y_n - y_m|$$

so $|x_n^2 - x_m^2| \le |x_n^1 - x_m^1| + \frac{2B}{c_0}|y_n - y_m| \to 0$ as $n, m \to \infty$. Thus $\{x_n\}$ converges, which proves Sublemma 5.6.

Next we show $f^{-1}(\zeta)$ Lindelöf implies (3). Indeed, by Sublemma 5.6, each point of $f^{-1}(\zeta)$ has an open neighborhood $W \subset \overline{W} \subset A$ such that $(x, y) \mapsto y : f^{-1}(\zeta) \cap \overline{W} \to Y$ is proper. By hypothesis, there is a countable subcover $\{f^{-1}(\zeta) \cap W_j\}_{j=1}^{\infty}$ so (3) holds with $M_j = f^{-1}(\zeta) \cap \overline{W}_j$.

Assuming the Lemma 5.5 and that $(x, y) \mapsto y : f^{-1}(\zeta) \mapsto Y$ is proper, we prove $\{y \in Y \mid \zeta$ is a critical value of $f(\cdot, y)\}$ is closed and without interior. Let $\{y_n\}_{n \ge 1}$ be a sequence in this set which converges in Y; for each n, there exists $x_n \in X$ so $(x_n, y_n) \in f^{-1}(\zeta)$ and x_n is a critical point of $f(\cdot, y_n)$. By hypothesis, we may suppose $(x_n, y_n) \to (x, y)$, so $(x, y) \in f^{-1}(\zeta)$. If $\frac{\partial f}{\partial x}(x, y)$ were onto, then so would be $\frac{\partial f}{\partial x}(x_n, y_n)$ for n large; hence x is a critical point of $f(\cdot, y)$ and closedness is proed. It remains to show that for each $y \in Y$, there exists y' arbitrarily close to y so ζ is a regular value of $f(\cdot, y')$. Let $K_y = \{x | f(x, y) = \zeta\}$; by the properness assumption, this is a compact set.

By Lemma 5.5, for each $x \in K_y$ there are open sts $U_x \ni x$, $V_x \ni y$ and V_x^0 – an open dense subset of V_x – such that $\overline{U}_x \times \overline{V}_x \subset A$ and $f(\cdot, y')|U_x$ has ζ as a regular value for all $y' \in V_x^0$. Choose a finite subcover $U_{x_1}, \ldots, U_{x_N}$ for K_y and let $\tilde{U} = U_{x_1}| \cup \cdots \cup U_{x_N}$, $\tilde{V} = \bigcap_1^N V_{x_j}$, $\tilde{V}^0 = \bigcap_1^N V_{x_1}^0$; $\tilde{V}^0$ is open and dense in $\tilde{V}$, $\tilde{V}$ and $\tilde{U}$ are open, $y \in \tilde{V}$, $K_y \subset \tilde{U}$, and ζ is a regular value of $f(\cdot, y')|\tilde{U}$ when $y' \in \tilde{V}^0$. In fact, ζ is a regular value of $f(\cdot, y')$ for all $y' \in \tilde{V}_0$ sufficiently close to y. Otherwise there would exist $y_n \to y$, $y_n \in \tilde{V}^0$, and critical points x_n of $f(\cdot, y_n)$ with $(x_n, y_n) \in f^{-1}(\zeta)$, such that $\lim_{n \to \infty} x_n = x$ exists; then $(x, y) \in f^{-1}(\zeta)$, $x \in K_y$, and $x_n \in \tilde{U}$, $y_n \in \tilde{V}^0$ for n large, so x_n is not a critical point of $f(\cdot, y_n)$, a contradiction.

If Lemma 5.5 and (3) hold, the same argument shows $\{y \in Y|$ there is a critical point x of $f(\cdot, y)$ with $(x, y) \in M_j \subset f^{-1}(\zeta)\}$ is closed and nowhere dense for each $j = 1, 2, \ldots$; hence the union of these, $\{y \in Y|$ there is a critical point x of $f(\cdot, y)$ with $(x, y) \in f^{-1}(\zeta)\}$ is meager. This completes the first step of the demonstration of the theorem.

Now we show the case $m > 1$ of Theorem 5.4 may be reduced to the case $m = 1$, by change of variables. Suppose therefore $m > 1$, $k = 1$ and 2(β) holds and let $\tilde{X} = X \times S^{m-1}$, $\tilde{Y} = C^1(S^{m-1}, Y)$, $\tilde{A} = \{(x, t; \tilde{y}) \in \tilde{X} \times \tilde{Y} | (x, \tilde{y}(t)) \in A\}$ and $\tilde{f} : \tilde{A} \to Z(= \tilde{Z}) : (x, t; \tilde{y}) \mapsto f(x, \tilde{y}(t))$. Then $\tilde{f}$ is C^1 and the problem "tilde" satisfies the same hypotheses as the original problem, except that "m" is replaced by "1." If 3(β) [or 3(γ)] holds for the original problem, it also holds for "tilde," and we recall $3(\alpha) \Rightarrow 3(\gamma)$. If $f(x, y) = \zeta$, $y = \tilde{y}(t)$, so $\tilde{f}(x, t; \tilde{y}) = \zeta$, we choose a maximal subset $\{\dot{t}_1, \ldots, \dot{t}_q\} \subset T_t(S^{m-1})$ so $\{\frac{\partial f}{\partial y}\tilde{y}'(t)\cdot t_j\}_{j=1}^q$ are independent relative to $R(\partial f/\partial x)$; then $\dim N(\partial f/\partial(x, t)) = \dim N(\partial f/\partial x) + m - 1 - q$. Then we choose $\{\eta_1, \ldots, \eta_p\} \subset T_yY$ so $\{\frac{\partial f}{\partial x}(x, y)\dot{y}k\}_{k=1}^p$ are independent relative to $R(\partial\tilde{f}/\partial(x, t)) = R(\partial f/\partial(x) + R(\frac{\partial f}{\partial y}\tilde{y}'(t))$; by $2(\beta)$, we may assume $p + q \geq m + \dim N(\partial f/\partial x)$, that is

$$\dim\left\{\frac{R(D\tilde{f})}{R(\partial\tilde{f}/\partial(x, t))}\right\} \geq p \geq m - q + \dim N\left(\frac{\partial f}{\partial x}\right) = 1 + \dim N\left(\frac{\partial\tilde{f}}{\partial(x, t)}\right).$$

Thus assuming the theorem is valid for $m = 1$ (the problem "tilde"), we see

$$\{\tilde{y} \in C^1(S^{m-1}, Y) \mid \zeta = f(x, \tilde{y}(t)) \quad \text{for some } x, t \text{ with } (x, \tilde{y}(t) \in A\}$$

is a meager set in $C_1(S^{m-1}, Y)$, which means C^1-codimen $\{y \in Y | \zeta \in f(A_y, y)\} > m - 1$, the C^1-codimension is $\geq m$ and so the codimension is $\geq m$. (See the Definition 5.15 and property (1) of codimension below.)

It remains only to prove Lemma 5.5.

Proof of Lemma 5.5. Since the result is local – near $x_0, y_0 \in X \times Y$ and $\zeta \in Z$ – we may assume X, Y, Z are Banach spaces, $x_0 = 0$, $y_0 = 0$, $\zeta = 0$, f is C^k on a neighborhood of $(0, 0) \in X \times Y$, $f(x, y) = Lx + My + o(|x| + |y|)$, L is semi-Fredholm with index $< k$, and either (α) $\mathcal{R}(L, M) = \{Lx + My|$ for all $x, y\} = Z$ or $(\beta) \dim\{\mathcal{R}(L, M)/\mathcal{R}(L)\} > \dim N(L)$. Since $(\alpha) \Rightarrow (\beta)$ for negative index, it is enough to prove the result in case (α) when $\operatorname{ind} L \geq 0$ and in case (β) when $\operatorname{ind} L < 0$.

Case (α). $\operatorname{ind} L \geq 0$ (or L is Fredholm), $\mathcal{R}(L, M) = Z$.

L is Fredholm so $X = X_1 \oplus X_2$, $Z = Z_1 \oplus Z_2$, $X_1 = N(L)$, $Z_2 = \mathcal{R}(L)$, $L_2 \equiv L|X_2 : X_2 \to Z_2$ is an isomorphism, and $\operatorname{ind} L = \dim X_1 - \dim Z_1$. The complement Z_1 to $\mathcal{R}(L)$ is not unique and we may choose $Z_1 \subset \mathcal{R}(M)$. Then there is a subspace $Y_1 \subset Y$ so $M_1 \equiv M \mid Y_1 : Y_1 \to Z_1$ is an isomorphism; defining $Y_2 = M^{-1}Z_2$, it follows that $Y = Y_1 \oplus Y_2$. Writing f in terms of its components in these spaces,

$$f : (x, y) = (x_1, x_2, y_1, y_2) \mapsto (M_1y_1 + g(x, y), L_2x_2 + h(x, y))$$

where g, h are C^k and g, g_x, g_y, h, h_x all vanish at (0,0), but perhaps $h_y \neq 0$. By the implicit function theorem, we may solve $f(x, y) = (0, 0)$ for $y_1 = \phi(x_1, y_2)$ and $x_2 = \psi(x_1, y_2)$, with ϕ, ψ of class C^k near (0,0). In matrix form, $\frac{\partial f}{\partial x}(x, y)$ is

$$\begin{pmatrix} g_{x_1} & g_{x_2} \\ h_{x_1} & L_2 + h_{x_2} \end{pmatrix} = \begin{pmatrix} 1 & p \\ 0 & 1 \end{pmatrix} \begin{pmatrix} \Delta & 0 \\ 0 & L_2 + h_{x_2} \end{pmatrix} \begin{pmatrix} 1 & 0 \\ q & 1 \end{pmatrix}$$

for appropriate operators p, q, where $\Delta = g_{x_1} - g_{x_2}(L_2 + h_{x_2})^{-1}h_{x_1}$. Then $\partial f/\partial x$ is surjective if and only if $\Delta = X_1 \to Z_1$ is surjective. Now by the definition of ϕ and ψ, we have

$$\{M_1 + g_{y_1} - g_{x_2}(L_2 + h_{x_2})^{-1} h_{y_1}\}\phi_{x_1} + \Delta = 0,$$

and the coefficient of ϕ_{x_1} is an isomorphism when we are close to (0,0). Thus in a neighborhood $\max(|y_1|, |y_2|) < \delta$ of $y = 0$, 0 is a regular value of $f(\cdot, y)|\max(|x_1|, |x_2|) < \epsilon$, with $y = y_1 + y_2$, if and only if y_1 is a regular value of $\phi(\cdot, y_2)|_{\{|x_1|<\epsilon\}}$. Since $\phi(\cdot, y_2) : X_1 \to Y_1$ is C^k near 0 and $k > \operatorname{ind} L = \dim X_1 - \dim Y_1$, Sard's theorem says, for every small y_2 ($|y_2| < \delta$), there is a dense set of y_1 in $|y_1| < \delta$ such that zero is a regular value of $f(\cdot, y_1 + y_2)$ on $\{|x_1| < \epsilon, |x_2| < \epsilon\}$. This proves denseness, while openness follows from the Sublemma 5.6.

Case (β): $\operatorname{ind} L < 0, k = 1, \dim\{\mathcal{R}(L, M)/\mathcal{R}(L)\} > \dim \mathcal{N}(L)$.

Let $n = \dim \mathcal{N}(L)$. There exist $\{f_1, \ldots, f_{n+1}\}$ in $\mathcal{R}(L, M)$ independent relative to $\mathcal{R}(L)$, i.e., $\sum_1^{n+1} c_j f_j \in \mathcal{R}(L)$ implies all $c_j = 0$. Then $f_j = Lx_j + My_j$ where $\{y_1, \ldots, y_{n+1}\}$ are linearly independent in Y, a basis for a subspace $Y_1 \subset Y$ such that $MY_1 \cap \mathcal{R}(L) = \{0\}$ and M is injective on Y_1. Let $Z_1 = MY_1$ and choose $Z_2 \supset \mathcal{R}(L)$ so that $Z_1 \oplus Z_2 = Z$. Let $Y_2 = M^{-1}Z_2$; then $Y = Y_1 \oplus Y_2$. Also let $X_1 = \mathcal{N}(L)$, $X = X_1 \oplus X_2$, so $n = \dim X_1 < \dim Z_1 = \dim Y_1$.

Now f has the form

$$(x, y) = (x_1, x_2, y_1, y_2) \mapsto (M_1 y_1 + g(x, y), L_2 x_2 + h(x, y))$$

where $M_1 = M \mid Y_1 : Y_1 \to Z_1$ is an isomorphism, and $L_2 = L \mid X_2 : X_2 \to Z_2$ is injective with closed image. Further, g, h are C^1 and at (0,0), g, g_x, g_y, h, h_x all vanish. We may solve $M_1 y_2 + g(x, y) = 0$ for $y_1 = \psi(x, y_2)$ a C^1 function with $\psi = 0$ and $\psi_x = 0$ at the origin.

Choose small $\delta > 0$. Fix $y_2 \in Y_2$, $|y_2| < \delta$, and let

$$S(y_2) = \{x \mid x = (x_1, x_2), |x_1| \leq \delta, |x_2| \leq \delta, f(x_1, x_2, \psi(x, y_2), y_2) = 0\}$$

Also let $p_1 : X_1 \times X_2 \to X_1$ be the projection on X_1 and $\pi(y_2) = p_1|S(y_2) : S(y_2) \to X_1$. If δ is small, $\pi(y_2)$ is injective, with Lipschitz inverse; assuming this,

$$\psi(S(y_2), y_2) = \psi(\cdot, y_2) \circ \pi(y_2)^{-1}(\pi(y_2)S(y_2)) \subset Y_1$$

is the Lipschitz image of a set in X_1, and $\dim X_1 < \dim Y_1$ so $\psi(S(y_2), y_2)$ has measure zero in Y_1. Thus given any $|y_1| < \delta$, $|y_2| < \delta$, there exists y_1' arbitrarily close to y_1 but outside $\psi(S(y_2), y_2)$, hence $f(x_1, x_2, y_1', y_2) \neq 0$ for all $|x_1| < \delta$, $|x_2| < \delta$. Openness follows from (5.6), so it only remains to show $\pi(y_2)$ has Lipschitz inverse.

Now $|Lx_2| \geq c_0|x_2|$ for all $x_2 \in X_2$ and some constant $c_0 > 0$. Since $\psi_x(0,0) = 0$, for sufficiently small $\delta > 0$ and $|x| \leq \delta$, $|\bar{x}| \leq \delta$, $|y_2| \leq \delta$

$$|f(x, \psi(x, y_2), y_2) - f(\bar{x}, \psi(\bar{x}, y_2), y_2) - L(x - \bar{x})| \leq \frac{c_0}{2}|x - \bar{x}|.$$

If also $x, \bar{x} \in S(y_2)$ then

$$c_0|x_2 - \bar{x}_2| \leq |L(x - \bar{x})| \leq \frac{c_0}{2}|x - \bar{x}| \leq \frac{c_0}{2}(|x_1 - \bar{x}_1| + |x_2 - \bar{x}_2|)$$

so $|x_2 - \bar{x}_2| \leq |x_1 - \bar{x}_1| = |\pi x - \pi\bar{x}|$, $|x - \bar{x}| \leq 2|\pi x - \pi\bar{x}|$, which completes the proof.

Warning: The remainder of this chapter (except the "Fredholm facts" at the end) will not be used in the sequel. Proceed at your own risk.

Example. As a simple application, suppose E is a Banach space and $M \subset E$ is a C^2 compact submanifold, of dimension $\mathcal{M}$. We show there is an open dense set in $\mathcal{L}(E, \mathbb{R}^{2m+1})$ such that, for each L in this set, $L|M : M \to \mathbb{R}^{2m+1}$ is an embedding: $Lx \neq Lx'$ when $x \neq x'$ in M and $Lv \neq 0$ when $v \neq 0$ in T_xM, $x \in M$.

Openness is easily proved, since M is compact. Let $M^{(2)} = \{(x, x') \in M \times M | x \neq x'\}$ and

$$f : M^{(2)} \times \mathcal{L}(E, \mathbb{R}^{2n+1}) \to \mathbb{R}^{2m+1} : (x, x', L) \mapsto L(x - x').$$

Then f is C^2 and (with $X = M^{(2)}$, $Y = \mathcal{L}(E, \mathbb{R}^{2m+1})$, $\zeta = 0 \in \mathbb{R}^{2m+1}$), hypothesis (1) holds (index $= 2m - (2m+1) < 0$), and also 2(α) holds. In fact, if $f(x, x', L) = 0$, then $x - x' \neq 0$ and $\{\dot{L}(x - x') \mid \text{for all } L\} = \mathbb{R}^{2m+1}$ so $Df(x, x', L)$ is surjective. If E^* is separable, then (3) holds; in fact (3) holds in very case (write $M^{(2)}$ as a countable union of compact sets). Thus $\{L \in \mathcal{L}(E, \mathbb{R}^{2m+1}) \mid L|M \text{ is } \textit{not} \text{ injective}\}$ is a meager set, and also closed. Similarly, by consideration of

$$(x, v, L) \mapsto Lv : (TM)^0 \times \mathcal{L}(E, \mathbb{R}^{2m+1}) \to \mathbb{R}^{2m+1},$$

where $(TM)^0 = \{(x, v) \in TM \subset E \times E | v \neq 0\}$, we see

$$\{L \in \mathcal{L}(E, \mathbb{R}^{2m+1}) \mid L|T_xM \text{ is not injective, for some } x \in M\}$$

is a meager, hence, closed without interior. In the open dense complement of these sets, $L|M$ is an embedding.

Professor W. M. Oliva asked me if there is a corresponding result for C^1 manifolds. The short answer is No!, but a more extended response is in Corollary 5.12 below. If $1 > \mu > 0$, a $C^{1+\mu}$ m-dimensional submanifold of E may be embedded by "linear projection" in $\mathbb{R}^N$ provided N is sufficiently large; but we must allow $N \to \infty$ as $\mu \to 0$.

So far, we have considered only regular values, a special case of transversality. To justify the title of this chapter, we treat briefly the general case, transversality to submanifold which need not be a single point.

Definition 5.7.

(1) Let F be a Banach space and F_1 a closed linear subspace of F; we say F_1 *splits in* F (or F_1 has a closed complementary space) if there is a closed subspace $F_2 \subset F$ such that $F_1 \oplus F_2 = F$, i.e., $F_1 + F_2 = F$ and $F_1 \cap F_2 = \{0\}$.

(2) Let E, F be Banach spaces, F_1 a closed subspace of F which splits, and $L \in \mathcal{L}(E, F)$ a continuous linear map. Then L *is transverse to* F_1 (in F), $L \pitchfork F_1$, if
   (i) $L^{-1}(F_1)$ splits in E and
   (ii) $R(L) + F_1 = F$.

(3) Let X, Y be C^1 Banach manifolds and $Z \subset Y$ a C^1 submanifold [so $q \in Z$ implies $T_q Z$ splits in $T_q Y$], and suppose $f : X \to Y$ is C^1. Then f *is transverse to* Z *at* $p \in X$, $f \pitchfork Z$ if either $F(p) \notin Z$ or $f(p) = q \in Z$ and $f'(p) \pitchfork T_q Z$. We say *f is transverse to Z*, $f \pitchfork Z$, if $f \pitchfork Z$ for every $p \in X$.

Remark. This definition is equivalent to that of Lang and, what is less obvious, to that of Abraham and Robbin [1] and the notes of R. Palais. These authors require – instead of (2)(ii) for $L \pitchfork F_1$ – that the image of L contain a closed complement of F_1. This certainly implies 2(ii), and the following lemma (part (c)) shows that such a complement must exist under our definition.

Lemma 5.8. *(Eine kleine Linearwissenschaft)*

(1) Let F_1 be a closed subspace of a Banach space F, if F_1 is finite dimensional or has finite codimension, then F_1 splits in F. If F is a Hilbert space, every closed linear subspace splits in F.

(2) If every closed linear subspace of F splits in F, then F is a Hilbert space, i.e., there is an inner product in F which defines an equivalent norm.

(3) Suppose F_1 splits in F and F_2, F_2' are closed complements: $F = F_1 \oplus F_2 = F_1 \oplus F_2'$. Then there is an isomorphism $L : F \to F$ such that $L|F_1$ is the identity in F_1 and $L(F_2') = F_2$.

(4) Let E, F be Banach spaces and $F_1 \subset F$ a closed splitting subspace. Then $L \in \mathcal{L}(E, F)$ is transverse to F_1 if and only if there exist splittings $E = E_1 \oplus E_2$, $F = F_1 \oplus F_2$, such that the corresponding matrix form of L is

$$\begin{pmatrix} L_{11} & L_{12} \\ 0 & L_{22} \end{pmatrix} : \begin{pmatrix} E_1 \\ E_2 \end{pmatrix} \to \begin{pmatrix} F_1 \\ F_2 \end{pmatrix}$$

where $L_{22} : E_2 \to F_2$ *is an isomorphism. In this case,* $E_1 = L^{-1}(F_1)$ *and* $L(E_2)$ *is a closed complement to* F_1. *If we choose* $F_2 = L(E_2)$ *then in the matrix form* $L_{12} = 0$. *The set* $\{L \in \mathcal{L}(E, F) | L \pitchfork F_1\}$ *is open in* $\mathcal{L}(E, F)$.

Proof. (1) is straightforward and (2) is a theorem of Lindenstrauss and Tzafiri [19]. (3) Let $P, P' \in \mathcal{L}(E)$ be the projections onto F_1 whose kernels are F_2, F_2' respectively. Then for real t, $P(t) = (1-t)P + tP'$ is a projection onto F_1. Choose an integer $N > 2\|P - P'\|(1 + \|P\| + \|P'\|)$ and define $P_j = P(j/N), W_j = P_j P_{j+1} + (I - P_j)(I - P_{j+1})$. Then $\|W_j - I\| < 1$, W_j is an isomorphism, $W_j | F_1$ is the identity in F_1, $P_{j+1} = W_j^{-1} P_j W_j$, and $L = W_0 W_1 \cdots W_{N-1}$ is the desired isomorphism.

In (4), suppose L has the indicated matrix form; then $L^{-1}(F_1) = E_1$ splits and $R(L) + F_1 = F_1 + L(E_2) = F$ so $L \pitchfork F_1$. Conversely suppose $L \pitchfork F_1$, F_2 is a closed complement to F_1, $E_1 = L^{-1}(F_1)$ and E_2 is a closed complement to E_1. In the matrix form for L, $L_{21} : E_1 \to F_2$ is zero since $LE_1 \subset F_1$; L_{22} is injective, since $E_1 = L^{-1}(F_1)$; and L_{22} is surjective, since $F_1 + R(L) = F$. It is easy to see that $F_1 \oplus L(E_2) = F$, and that $L(E_2)$ is closed (a graph over $F_2 = L_{22}E_2$). Openness of transverality also follows easily in the matrix form.

Remark. Transversality of a map to a submanifold, as in (5.7(3)) seems to be the most useful form, but it is enlightening to contemplate the more geometric version, transversality of two submanifolds. We cite only the linear case. If E, F are closed splitting subspaces of a Banach space G, then E, F have transversal intersection ($E \pitchfork F$ or $F \pitchfork E$) when one of the following equivalent conditions holds:

(i) the inclusion $E \subset G$ is transverse to F;
(ii) the inclusion $F \subset G$ is transverse to E;
(iii) $E + F = G$ and $E \cap F$ splits in E or in F or in G, in which case it splits in all three.
(iv) There are closed subspaces $E_1 \subset E$, $F_2 \subset F$ so that

$$E = E_1 \oplus (E \cap F), \qquad F = (E \cap F) \oplus F_2$$

and

$$G = E_1 \oplus (E \cap F) \oplus F_2.$$

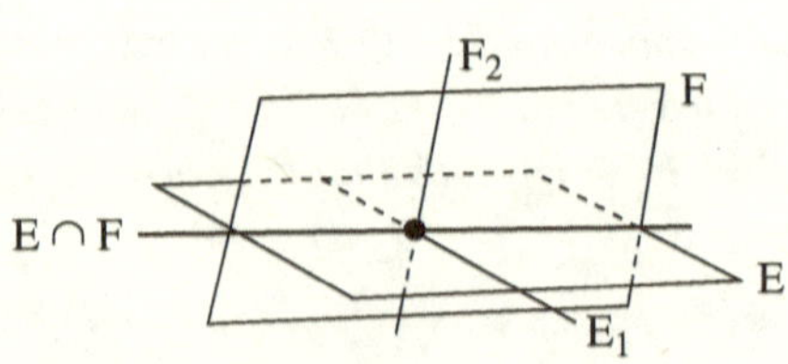

When $\dim G < \infty$, it is easy to see $E + F = G$ implies $(x + E) \cap (y + F)$ is nonempty with the same dimension for all x, y, while $E + F \neq G$ implies $(x + E) \cap (y + F)$ is empty for almost all x, y.

Lemma 5.9. *Let X, Y be C^k Banach manifolds ($k \geq 1$) and let Z be a C^k submanifold of Y, and let $f : X \to Y$ be C^k. Given $q \in Z$ there exists an open neighborhood V of q in Y, a Banach space E and a C^k map $g : V \to E$ such that 0 is regular value of g and $g^{-1}(0) = V \cap Z$. If U is an open set in X such that $F(U) \subset V$, then the restriction $f|U : U \to Y$ is transverse to Z if and only if 0 is a regular value of $g \circ f|U : U \to E$.*

Proof. This follows immediately from the definitions. We only recall that a submanifold of Z (following [1, 17]) is what is sometimes called a split-embedded submanifold, and at a regular point, the kernel of the derivative splits.

The above lemma shows that transversality conditions reduce (locally) to conditions on regular values. The following theorem is, then, merely a translation of Theorem 5.4 to more general language.

Theorem 5.10. *Suppose given positive integers k, m; Banach manifolds X, Y, Z, W of class C^k, with W a C^k submanifold of Z_1; A an open set in $X \times Y$; and a C^k map $f : A \to Z$. Assume the following:*

(1) Either

(α) W is finite dimensional and σ-compact, and for $(x, y) \in f^{-1}(W)$, $\frac{\partial f}{\partial x}(x, y)$ is semi-Freholdm with $\operatorname{ind} \frac{\partial f}{\partial x}(x, y) + \dim W < k$; *or*

(β) X is finite dimensional and σ-compact, W is σ-closed, and $\dim X - \operatorname{codim} W < k$. [*$W$ is σ-closed if Z is metrizable.*]

(2) For each $(x, y) \in f^{-1}(W)$, with $q = f(x, y)$, either

(α) $R(Df(x, y)) + T_qW = T_qZ_1$ or

(β) $\dim\{(R(Df(x, y)) + T_qW)/(R(\partial f/\partial x) + T_qW)\} \geq m + \dim \mathcal{N}(\frac{\partial f}{\partial x}(x, y)) + \dim (T_qW \cap R(\frac{\partial f}{\partial x}(x, y)))$.

(3) $(x, y) \mapsto (f(x, y), y) : f^{-1}(W) \to W \times Y$ is σ-proper which is true, for example, if X is σ-compact and W σ-closed, or if $f^{-1}(W)$ is Lindelöf.

Then $Y_{\text{crit}} \equiv \{y \in Y \,|\, f(\cdot, y)$ is not transverse to $W\}$ is meager in Y. Y_{crit} is closed if X is compact and W and $f^{-1}(W)$ are closed, or if W is compact and $(x, y) \mapsto (f(x, y), y) : f^{-1}(W) \to W \times Y$ is proper. If $\dim W < \infty$ *and* $\operatorname{ind}(\partial f/\partial x) + \dim W \leq -m$ *on $f^{-1}(W)$, or if* $\dim X < \infty$ *and* $\dim X - \operatorname{codimen} W \leq -m$, *then 2($\alpha$) implies 2($\beta$) and $Y_{\text{crit}} = \{y \in Y \mid f(A_y, y)$ meets $W\}$ has codimension $\geq m$.*

Remark. Theorem 5.4 is the case $W = \{\zeta\}$. To obtain the semi-Fredholm property, it seems necessary to suppose at least one of W, X is finite dimensional.

If both X and Z have finite dimension,

$$\text{ind}\left(\frac{\partial f}{\partial x}\right) + \dim W = \dim X - \dim Z + \dim W = \dim X - \text{codim W}.$$

Proof. The "σ-compactness" in (1) and "σ-properness" in (3) permit reduction to a local problem: given $(x_0, y_0) \in f^{-1}(W)$, there are open neighborhoods U of x_0, V of y_0 and an open dense $V^0 \subset V$ such that $\overline{U} \times \overline{V} \subset A$ and $f(\cdot, y)|U \pitchfork W$ for $y \in V^0$. By Lemma 5.9, this local problem reduces to a problem on regular values to which Theorem 5.4 applies. The details are omitted.

Another variation on this theme, using the method rather than the result of Theorem 5.4, gives a "non-differentiable" transversality theorem.

Theorem 5.11. *Let Y, Z, W be C^1 Banach manifolds, W a σ-closed submanifold of Z, and let (X, d) be a metric space, σ-compact, with finite Hausdorff dimension* $\dim X$. *Let A be an open set in $X \times Y$ and assume $f : A \to Z$ is locally of class $C^\mu (0 < \mu < 1)$. Suppose each point of $f^{-1}(W)$ has a neighborhood where $y \mapsto f(x, y)$ is differentiable and $(x, y) \mapsto \frac{\partial f}{\partial y}(x, y)$ is continuous. Further suppose $f(x, y) = z \in W$ implies*

$$\dim\left\{\frac{(\mathcal{R}(\frac{\partial f}{\partial y}(x, y)) + T_z W)}{T_z W}\right\} > \frac{1}{\mu}\dim X.$$

This holds if $\mu > \dim X/\text{codim } W$ and $f(x, \cdot) \pitchfork W$ for each x.

Then $\{y \mid f(X \times \{y\} \cap A)$ meets $W\}$ is a meager set in Y, and it is closed if X is compact and $f^{-1}(W)$ is closed, or if $(x, y) \mapsto y : f^{-1}(W) \to Y$ is proper.

Remark. Both $\dim X$ and the Hölder exponent μ depend on the metric chosen for X.

Proof. We reduce to a local problem as before: each $(x_0, y_0) \in f(W)$ has an open neighborhood $U \times V \subset X \times Y$, with $\overline{U} \times \overline{V} \subset A$, and an open dense $V^0 \subset V$, such that $f(U, V^0)$ does not meet W.

For the local problem with X compact, we may suppose Y, Z, W are Banach spaces, $y_0 = 0$, $f(x_0, 0) = 0$ and there is a subspace $Y_1 \subset Y$ such that $MY_1 \cap W = \{0\}$ $[M = \frac{\partial f}{\partial y}(x_0, 0)]$, $M|Y_1$ is injective and $\infty > \dim Y_1 > \frac{1}{\mu}\dim X$. There is a projection P of Z onto MY_1 and a complement Y_2, $Y = Y_1 \oplus Y_2$. If $f(x, y) \in W$ then $Pf(x, y) = 0$ and near $(x_0, 0)$ we may solve the last equation for $y_1 = \psi(x, y_2)$, $y = y_1 + y_2$, where ψ is C^μ. It follows that $\dim \psi(X, y_2) \le \frac{1}{\mu}\dim X < \dim Y_1$, so $\psi(X, y_2)$ has measure zero in Y_1. For any (small) $y = y_1 + y_2$ we may choose y_1' arbitrarily near y_1 so $y_1' \notin \psi(X, y_2)$, hence $f(X, y_1' + y_2)$ does not meet W.

Example. Let $P_m : [0,1] \xrightarrow{\text{onto}} [0,1]^m$ be a $C^{1/m}$ Peano curve and let $X = [0,1]$, $Y = Z = \mathbb{R}^m$, $W = \{0\}$ and $f(x,y) = P_m(x) - y$. Then f is C^μ with $\mu = \frac{1}{m} = \dim X/\operatorname{codim} W$, but all other hypotheses are satisfied – in particular $\frac{\partial f}{\partial y}(x,y)$ is surjective. And the conclusion fails: $0 \in f([0,1], y)$ for all $y \in [0,1]^m$.

Corollary 5.12. *Let E be a finite or infinite-dimensional Banach space, $0 < \mu < 1$, and $M \subset E$ a compact $C^{1+\mu}$ submanifold of dimension m. (We could allow M to have a boundary or corners.) If $N \geq 2m+1$ and $N > m(1+1/\mu) - 1$, then there is an open dense set of $L \in \mathcal{L}(E, \mathbb{R}^N)$ such that the restriction $L|M : M \to \mathbb{R}^N$ is an embedding.*

For each $d = 2, 3, 4, \ldots$ there exists a curve $K_d \subset \mathbb{R}^{d+1}$, which is a submanifold of class $C^{1+1/d}$ and dimension $m = 1$, such that; $L|K_d : K_d \to \mathbb{R}^d$ is never a C^1 embedding for any $L \in \mathcal{L}(\mathbb{R}^{d+1}, \mathbb{R}^d)$, although it is a topological embedding for most L if $d \geq 3$. Note $d = m(1+1/\mu) - 1$ when $m = 1$, $\mu = 1/d$.

Proof. Openness is clear, and denseness reduces to a local problem: for each $p \in M$, there is a neighborhood U of p and an open dense $S_p \subset \mathcal{L}(E, \mathbb{R}^N)$ so $L|M \cap U$ is an embedding for $L \in S_p$. Since $N \geq 2m+1$, $L|M$ is injective for most L, as noted earlier. For the local problem, suppose $p = 0$, $E = \mathbb{R}^m \times E_2$ and

$$M \cap (\text{neighborhood of } 0) = \{(\xi, \psi(\xi)) \mid \xi \in \mathbb{R}^m \text{ near } 0\}$$

where $\psi : \mathbb{R}^m \to E_2$ is $C^{1+\mu}$ near 0, $\psi(0) = 0$, $\psi'(0) = 0$. A linear map from $\mathbb{R}^m \times E_2$ to $\mathbb{R}^N$ has the form $(\xi, \eta) \mapsto P\xi + Q\eta$ $(P \in \mathbb{R}^{N\times m}, Q \in \mathcal{L}(E_2, \mathbb{R}^N))$ and we want $P + Q\psi'(\xi)$ injective (rank m) for all (small) ξ, for most choices of P, Q. (Apply the theorem above to $f : (\xi, P, Q) \mapsto P + Q\psi'(\xi)$ with $X = \{\text{neighborhood of } 0 \text{ in } \mathbb{R}^m\}$, $Y = \mathbb{R}^{N\times m} \times \mathcal{L}(E_2, \mathbb{R}^N)$, $Z = \mathbb{R}^{N\times m}$ and $W = W_r$ the set of $N \times m$ matrices of rank r (for each $0 \leq r \leq m-1$). Note codim $W_r = (m-r)(N-r) \geq N - m + 1 > m/\mu$.

To construct the curves K_d, first choose a C^1 map $q : [0,1]^d \xrightarrow{\text{onto}} S^d$ and let $q_d = q \circ P_d : [0,1] \xrightarrow{\text{onto}} S^d$, of class $C^{1/d}$. For any ϕ in $C^1([0,1], \mathbb{R})$, let $V_\phi(t) = \int_0^t \phi(s) q_d(s) ds$, $0 \leq t \leq 1$. Applying Theorem 5.4 to $(t, t', \phi) \mapsto V_\phi(t) - V_\phi(t')(t \neq t')$ we see $V_\phi(\cdot)$ is injective for most choices of ϕ. Choose such ϕ with $\varphi(t) > 0$ for all t. Then $t \mapsto V_\phi(t) : [0,1] \to \mathbb{R}^{d+1}$ is a $C^{1+1/d}$ embedding, whose image is K_d; and it is easy to show for each $L : \mathbb{R}^{d+1} \to \mathbb{R}^d$, there exists $t \in (0,1)$ with $L \frac{d}{dt} V_\phi(t) = 0$, so $L|K_d$ is not a C^1 embedding.

Review of Fredholm and Semi-Fredholm Operators

As a general reference for this topic, see Kato [14].

Definition 5.13. Let X, Y be Banach spaces and $T : X \to Y$ a continuous linear operator. Then T is *Fredholm* [or *semi-Fredholm*] if $\mathcal{R}(T)$ is closed and both [or, at least one] of $\dim \mathcal{N}(T)$, codim $\mathcal{R}(T)$ are finite; in this case, the *index* of T is

$$\operatorname{ind}(T) = \dim \mathcal{N}(T) - \operatorname{codim}\mathcal{R}(T).$$

which is an integer or $+\infty$ or $-\infty$. A semi-Fredholm operator with finite index is Fredholm. A semi-Fredholm operator is *left-Fredholm* if its index is not $+\infty$ ($\dim \mathcal{N}(T)$ finite), *right-Fredholm* if its index is not $-\infty$ (codim $\mathcal{R}(T)$ finite).

Examples

(1) If X, Y are finite dimensional, every linear operator $T : X \to Y$ is Fredholm with index $= \dim X - \dim Y$. In fact, suppose T has rank r; then $\dim \mathcal{N}(T) = \dim X - r$, codim $\mathcal{R}(T) = \dim Y - r$ and $\operatorname{ind}(T)$ does not depend on r or T. If at least one of X, Y is finite-dimensional, every $T \in \mathcal{L}(X, Y)$ is semi-Fredholm and $\operatorname{ind}(T) = \dim X - \dim Y$. If both X, Y are infinite dimensional the zero operator is not semi-Fredholm, and any compact operator from X to Y is not semi-Fredholm.

(2) Let $X = Y = \ell_2$, the usual sequence space. The identity in ℓ_2 (or any isomorphism) is Fredholm with index 0. The shift $S : (x_1, x_2, \dots) \mapsto (x_2, x_3, \dots)$ is Fredholm with index 1 and S^m is Fredholm with index m ($m = 1, 2, 3, \dots$). The adjoint $S^* : (x_1, x_2, \dots) \mapsto (0, x_2, x_2, \dots)$ is Fredholm with index -1 and $(S^*)^m$ has index $-m(m = 1, 2, 3, \dots)$. $S^n(S^*)^m$ and $(S^*)^m S^n$ have index $n - m$.

$$L : (x_1, x_2, \dots) \mapsto (x_1, 0, x_2, 0, x_3, 0, \dots)$$

is left-Fredholm with index $-\infty$, and its adjoint $L^* : (x_1, x_2, \dots) \mapsto (x_1, x_3, x_5, \dots)$ is right-Fredholm with index $+\infty$. Note L^*L is the indentity but LL^* is not semi-Fredholm.

(3) If there exists a Fredholm operator from X to Y with (finite) index m, then $X \times \mathbb{R}^m$ is isomorphic to Y (if $m \leq 0$) or X is isomorphic to $Y \times \mathbb{R}^m$ (if $m \geq 0$), so the spaces are "nearly the same." Our familiar ∞-dimensional Banach spaces E have $E \times \mathbb{R}$ isomorphic to E, but it is an open problem whether this holds generally. In any case, many properties (separable, reflexive, uniformly convex, ...) must be the same in both X and Y.

5.14. Fredholm Facts

(1) $T \in \mathcal{L}(X, Y)$ is semi-Fredholm iff $T^* \in \mathcal{L}(Y^*, X^*)$ is semi-Fredholm, and in this case ind $T^* = -$ind T (see Example 2 above).

(2) If $T \in \mathcal{L}(X, Y)$ is semi-Fredholm, there exists $\delta > 0$ such that, whenever $\|S - T\|_{\mathcal{L}(X,Y)} < \delta$, S is semi-Fredholm and ind $S =$ ind T. The set of semi-Fredholm operators with given index is open in $\mathcal{L}(X, Y)$.

(3) If $T \in \mathcal{L}(X, Y)$ is semi-Fredholm and $K \in \mathcal{L}(X, Y)$ is compact, then $T + K$ is semi-Fredholm and ind $(T + K) =$ ind T. (If T is an isomorphism, this result contains a substantial fraction of the Riesz-Schauder theory of compact operators.)

(4) If $T \in \mathcal{L}(X, Y)$ is semi-Fredholm, there exists $F \in \mathcal{L}(X, Y)$ of finite rank such that, for all real $t \neq 0$, $T + tF$ is surjective (if ind $T \geq 0$) or injective (if ind $T \leq 0$), or an isomorphism (if ind $T = 0$).

(5) If $T \in \mathcal{L}(X, Y)$, $S \in \mathcal{L}(Y, Z)$ are both semi-Fredholm and their indices may be summed (excluding only the case $\infty - \infty$), then $ST \in \mathcal{L}(X, Z)$ is semi-Fredholm and ind$(ST) =$ ind $S +$ ind T. In the case "$\infty - \infty$," ST need not be semi-Fredholm, and if it is, the index may take any value in $Z \cup \{+\infty, -\infty\}$ (see Example 2 above).

(6) For each finite i and integer $j \geq 0$, the set

$$\left\{\begin{array}{l} T \in \mathcal{L}(X, Y) \mid T \text{ Fredholm, ind } T = 1 \text{ and} \\ \dim \mathcal{N}(T) = j \ (\text{if } i \leq 0) \\ \text{or} \\ \text{codim } R(T) = j \ (\text{if } i \geq 0). \end{array}\right\}$$

is either empty or an analytic submanifold of $\mathcal{L}(X, Y)$ with codimension $j(j + |i|)$, and is an open set when $j = 0$.

(7) If H_1, H_2 are Hilbert spaces, the set of semi-Fredholm operators is open and dense in $\mathcal{L}(H_1, H_2)$. In fact

$$\begin{aligned} &\{T \in \mathcal{L}(H_1, H_2) \mid T \text{ is injective with closed image or } T \text{ is surjective}\} \\ &= \{T \in \mathcal{L}(H_1, H_2) \mid T \text{ has a left-inverse or a right-inverse}\} \end{aligned}$$

is open and dense. (See Halmos, *A Hilbert Space Problem Book*, #109.) The set of Fredholm operators is not dense unless both H_1 and H_2 are finite dimensional.

(8) Let $L_j \in \mathcal{L}(X, Y_j)\,(j = 1, \ldots, n)$ and suppose L_1 is left Fredholm. Then $x \mapsto (L_1 x, \ldots, L_n x)$ is left-Fredholm, and it has index $-\infty$ if some $Y_j (j > 1)$ has infinite dimension. This simple fact is used frequently in Chapter 6.

Codimension of Sets

Definition 5.15. Let E be a topological Baire space. For any closed or σ-closed (= countable union of closed sets = F_σ) subset $F \subset E$ and integer $m \geq 0$, *codim* $F > m$ means $\{\phi \in C(I^m, E) | \phi(I^m) \text{ meets } F\}$ is meager in $C(I^m, E)$.

Here $I^m = [0, 1]^m$, $I^0 = \{0\}$, and $C(I^m, E)$ has the obvious compact-open topology. In place of I^m, we could use (for example) $\overline{B}^m$, the closed unit ball in $\mathbb{R}^m$, or $S^m = \partial \overline{B}^{m+1}$, which give equivalent conditions.

Remark. If E is a Banach space or C^k Banach manifold, we may also define *C^k-codim $F > m$* to mean $\{\phi \in C^k(S^m, E) \mid \phi(S^m) \text{ meets } F\}$ is meager in $C^k(S^m; E)$. It is clear that C^k-codim $F > m$ implies C^j-codim $F > m$ for $0 \leq j \leq k$, and C^0-codim $F > m$ is equivalent to codim $F > m$. Most of the properties below hold also for C^k-codimension (in particular (7)), but we will use only C^0-codim = codim. Note that, if F is closed, codim $F > m$ is equivalent to saying $\{\phi \in C^k(S^m, E) \mid \phi(S^m) \text{ meets } F\}$ is nowhere dense in the C^0-topology; thus we need not worry about Peano curves.

It is easy to prove

$$\operatorname{codim}(F_1 \cup F_2 \cup \ldots) > m \text{ iff } \operatorname{codim} F_j > m \text{ for every } j = 1, 2, \ldots$$

and $m > 0$, codim $F > m$ imply codim $F > m - 1$. By the first, we need only consider closed sets F; and by the second, we may define codimension (not only the phrase "codim $F > m$").

Definition. (Continued) *codim $F = m$* means codim $F \not> m$ and either $m = 0$ or codim $F > m - 1$. *codim $F = \infty$* means codim $F > m$ for every $m \geq 0$. Then for every closed or σ-closed $F \subset E$ there is a well-defined $m \in \{0, 1, 2, \ldots, \infty\}$ such that codim $F = m$, and we call m the *codimension* of F. Obviously the codimension of F depends on E, so we sometimes specify "codimension of F in E."

5.16. Simple Properties of Codimension

(1) $F_1 \subset F_2$ implies codim $F_1 \leq$ codim F_2.

(2) codim $F > 0$ if and only if F is meager in E. (If E were not a Baire space, one could have codim $E > 0$.)

(3) codim $(F_1 \cup F_2 \cup \ldots) = \min\{\operatorname{codim} F_j \mid j = 1, 2, \ldots, \}$

(4) codim $\phi = \infty$. If $E = \mathbb{R}^n$, any nonempty $F \subset E$ has codim $F \leq n$. Antoine's necklace [Hocking and Young, *Topology*, p. 177] is a compact totally-disconnected set in $\mathbb{R}^3$, so it has topological dimension zero. But its codimension is *not* 3; $\mathbb{R}^3-$ (Antoine's necklace) is not simply connected, so the codimension is ≤ 2, by (5) below. In fact the codimension equals 2 : it is greater than 1, since any continuous curve in $\mathbb{R}^3$ may be slightly perturbed so it misses the necklace, since it misses the (sufficiently small) approximating tori. (One may conjecture that generally dim $F +$ codim $F \leq$ dim E, using topological dimension. What is the Hausdorff dimension? For a particular (economical) construction it is no more than $1+ \log^2/\log^3$, which *might* be exact. If dim $A > 1$ we would

have the amusing result top-dim $A +$ codim $A <$ dim $E < H -$ dim $A +$ codim A for $A =$ Antoine's necklace $CE = \mathbb{R}^3$

(5) Suppose E is a Banach space, $F \subset E$ is closed; then

$$\text{codim } F > 1 \text{ implies } E \backslash F \text{ is connected,}$$
$$\text{codim } F > 2 \text{ implies } E \backslash F \text{ is simply connected,}$$

codim $F > m + 1$ implies every map in $C^0(S^m, E\backslash E)$ is homotopic to a constant.

(6) If $h : E \to E_1$ is homeomorphism and $F_1 = h(F)$, then codim F (in E) = codim F_1 (in E_1).

(7) Suppose E is a C^1 Banach manifold and F is C^1 submanifold with finite codimension m in the usual sense, and F has a countable base. Then codim $F = m$ according to our definition.

The proofs are obvious, except possibly for (7). In (7), since F has a countable basis for its topology, we may work locally and (applying a C^1 diffeomorphism) reduce to the case: $E = \mathbb{R}^m \times E_2$, $F = \{0\} \times F_2$ where F_2 is the closure of a neighborhood of 0 in E_2. Given any C^1 $\phi : S^{m-1} \to E$, we have $\phi = (\phi_1, \phi_2)$, $\phi_j : S^{m-1} \to E_j$, and $\phi(S^{m-1})$ meets F only if $0 \in \phi_1(S^{m-1}) \subset \mathbb{R}^m$. But a C^1-small perturbation $\tilde{\phi}_1$ of ϕ_1 makes $0 \notin \tilde{\phi}_1(S^{m-1})$ [apply transversality to $(t, \psi) \mapsto \psi(t)$] hence C^1-codim $F > m - 1$, so also codim $F > m - 1$ (see remark above). To prove the codimension is not greater than m, let $\phi(t) = (t, 0)$, $\phi : \overline{B^m} \to E = \mathbb{R}^m \times E_2$. Any $\tilde{\phi} = (\tilde{\phi}_1, \tilde{\phi}_2) \in C(\overline{B^m}, E)$ uniformly close to ϕ has $\tilde{\phi}(\overline{B^m}) \cap F$ nonempty, i.e., $\tilde{\phi}_1(t) = 0$ for some t; otherwise we could construct a retraction of $\overline{B^m}$ on ∂B^m.

Interlude

When condition (2.β) of Theorem 5.4 fails, we typically conclude that a certain complicated operator has finite rank. Chapters 7 and 8 provide different ways to obtain a more concrete condition ($G(x, y) = 0$, in certain Banach spaces) from the "finite-rank" hypothesis. Both methods are difficult, so we pause to show the importance of these results in a fairly general context.

We start with some smooth $f : X + Y \to Z$ such that $f_x = \partial f / \partial x$ is Fredholm, and hope to show $0 \in Z$ is a regular value of $f(x, y)$ for most choices of $y \in Y$ – at least, a dense set. We argue by contradiction: assume for every y near $y_0 \in Y$, these exists $x \in X$ with $f(x, y) = 0$ and $f_x(x, y)$ not onto. Then in fact $Df(x, y) = (f_x, f_y)$ is not onto, by the transversality theorem, and we derive from this an additional infinite-dimensional condition $\partial(x, y) = 0 \in W_1$ (dim $W_1 = \infty$), hoping for a contradiction. But perhaps it is not contradictory, or not obviously so; we continue.

For each y near y_0, there exists x so (x, y) is a critical point of f and $(f(x, y), g(x, y)) = (0, 0) \in Z \times W_1$. Since dim $W_1 = -\infty$, the x-derivative of (f, g) is semi-Fredholm with index $-\infty$, and condition (2.β) of Theorem 5.4 must fail. It follows that, for each $\dot{y} \in T_yY$, there exists $\dot{x} \in T_xX$ so that $(f_x\dot{x} + f_y\dot{y}, g_x\dot{x} + g_y\dot{y}) \in E$, for a fixed finite-dimensional subspace $E \subset T_0Z \times T_0W_1$. Since f_x is Fredholm, it has an "inverse" $[f_x^{-1}]$, module finite rank operators. Then we can "solve" for $\dot{x} = -[f_x^{-1}]f_y\dot{y}$ (module finite rank) and conclude $g_y - g_x[f_x^{-1}]f_y : T_yy \to T_0W_1$ has finite rank. The methods of Chapters 7 and 8 show that this (in typical problems) implies yet another infinite-dimensional condition : $h(x, y) = 0 \in W_2$ (dim $W_2 = \infty$).

Thus for each y near y_0 there is a solution X of the even-more-overdetermined problem:

$$(f(x, y), g(x, y), h(x, y)) = (0, 0, 0) \in Z \times W_1 \times W_2, DF(x, y) \text{ not onto,}$$

If this is not yet clearly contradictory, we may repeat the process. Or we may get tired, return to the beginning and modify our hypotheses so that we do have a contradiction.

Essential steps in this cyclic argument come from condition (2.β) of Theorem 5.4 and the methods of Chapters 7 and 8.

Chapter 6
Generic Perturbation of the Boundary

One of the compensations for the difficulty of PDEs, compared to ordinary differential equations, is that the regions in which we work can have almost any shape. This may seem only an added difficulty, but it turns into an advantage if we restrict attention to properties which are generic with respect to perturbation of the boundary. This is quite a strong condition, as the class of perturbations is infinite-dimensional (but it is finite-dimensional for ODEs). In many problems it is also reasonable to require "genericity" with respect to certain coefficients, etc. (See Uhlenbeck [42] and Saut and Teman [34] for some results of this kind.) But some problems are very rigid in this regard (e.g., the Navier-Stokes equation). Perturbation of the boundary is almost always reasonable, if we allow for occasional symmetry constraints as in Example 6.2 below. We will in any event perturb *only* the boundary – as a way of testing the strength of our tools, as a first step in more general "genericity" arguments, and for the intrinsic interest of the problems.

In this chapter, "generic" properties are those that hold for most domains Ω, in the sense of Baire category, or for most domains in a certain stated class $\mathcal{C}$ (C^m-regular, bounded, perhaps connected, perhaps contained in a certain open set or the boundary is contained in a certain open set, or the boundary meets a certain open set, or ...). More precisely, given any region Ω_0 of the class $\mathcal{C}$, if $h \in \mathrm{Diff}^m(\Omega_0)$ is outside a certain meager subset $F \subset \mathrm{Diff}^m(\Omega_0)$ and if $h(\Omega_0)$ is in the class $\mathcal{C}$, then the property (supposed generic) holds. In all our applications, the excluded set of embeddings F is defined by properties of the *image* of the embedding, so it is invariant under composition with C^m-diffeomorphisms $\Omega_0 \to \Omega_0$. This implies (though we don't prove it here) that the image $F(\Omega_0)$ in Micheletti's metric space [22] of regions C^m-diffeomorphisms to Ω_0 is also meager, and in fact has the same codimension as F (in the sense of Definition 5.15).

Frank Morgan ("Measures on spaces of surfaces," *Arch. Rat. Mech. Anal.* **78** (1982)) has defined a class of measures of smooth embeddings – generalizing Wiener measure for continuous curves – which induces a notion of "measure-zero" for surfaces (images of the embeddings). He obtained interesting results holding "almost-everywhere" in this sense.

Some work has been done on generic geometry without direct mention of PDEs – though many of the results are motivated by applications to PDEs. See, for example, "Generic caustics by reflection," *Amer. Math. Mo.*, V. I. Arnold, "Singularities of systems of rays," *Russ. Math. Surveys* **38** (1983), V. M. Petkov and L. Stojanov, "Periodic geodesics of generic nonconvex domains," Univ. Fed. Pernambuco, preprint (1985). Some of our results (as the lemma of Example 6.8) might also be classed as generic geometry.

The main tool in this chapter is the transversality Theorem 5.4. Our first example was proved by Micheletti [21] and Uhlenbeck [41] and is also contained in Example 4.4 above; but it is a convenient example to introduce the use of transversality.

Example 6.1. Generic Simplicity of Eigenvalues of the Dirichlet Problem for Δ. We prove, for most C^2 regular bounded regions $\Omega \subset \mathbb{R}^n$, that all eigenvalues λ of

$$\Delta u + \lambda u = 0 \text{ in } \Omega; \quad u = 0 \text{ on } \partial\Omega, \quad u \not\equiv 0$$

are simple.

First suppose Ω_0 is connected, as well as bounded and C^2. Consider the map

$$\begin{aligned} F &: (u, \lambda, h) \mapsto h^*(\Delta + \lambda)h^{*-1}u \\ &: \left(H^2 \cap H_0^1(\Omega)\backslash\{0\}\right) \times \mathbb{R} \times \mathrm{Diff}^2(\Omega_0) \to L_2(\Omega_0). \end{aligned}$$

Suppose, for some such h, $(u, \lambda) \mapsto F(u, \lambda, h)$ has zero as a regular value; then with $\Omega = h(\Omega_0)$, $0 \in L_2(\Omega)$ is a regular value of

$$(v, \lambda) \mapsto \Delta v + \lambda v : \left(H^2 \cap H_0^1(\Omega)\backslash\{0\}\right) \times \mathbb{R} \to L_2(\Omega).$$

This says, whenever $\Delta v + \lambda v = 0$, $v \neq 0$, then

$$(\dot{v}, \dot{\lambda}) \mapsto (\Delta + \lambda)\dot{v} + \dot{\lambda} v : H^2 \cap H_0^1(\Omega) \times \mathbb{R} \to L_2(\Omega)$$

is surjective, hence $(\Delta + \lambda) \mid H^2 \cap H_0^1(\Omega)$ has (closed) image of codimension ≤ 1, and in fact codimension $= 1$, since λ is an eigenvalue; therefore λ is a simple eigenvalue. Thus all eigenvalues of the Dirichlet problem in $h(\Omega_0)$ are simple. Therefore it is sufficient to show zero is a regular value of $F(\cdot, \cdot, h)$

for most h, or (since F is analytic, $F(\cdot, \cdot, h)$ is Fredholm for each h and the spaces are separable) to show zero is a regular value of F, verifying 2(α) of Theorem 5.4.

This is possible, but raises some technical questions of smoothness which we may avoid by a slight reformulation of the problem.

For each integer j (= 1, 2, 3, ...), the set of $h \in \mathrm{Diff}^2(\Omega_0)$ such that all eigenvalues $\lambda \leq j$ are simple (for the problem in $h(\Omega_0)$) is an open set in $\mathrm{Diff}^2(\Omega_0)$: there are only finitely many such eigenvalues and each depends continuously (analytically!) on h, according to Example 3.2. We show the set is also dense, for each j, and then taking the intersection over all j gives the result desired.

Thus proof of denseness is the key step, and for this we may use stronger norms. By Theorem 1.8, we may approximate Ω_0 (in the C^2 sense) by a C^3 (or C^∞ or analytic) region and then show denseness in the stronger C^3 topology.

Suppose therefore Ω_0 is in fact C^3; we show zero is a regular value of $(u, \lambda, h) \mapsto F(u, \lambda, h)$ for $h \in \mathrm{Diff}^3(\Omega_0)$. If that were false, there would exist a critical point (u_0, λ_0, h_0) such that $F(u_0, \lambda_0, h_0) = 0$. We transfer the "origin" from Ω_0 to $\Omega = h_0(\Omega_0)$ defining $v_0 = h_0^{*-1} u_0 \in H^2 \cap H_0^1(\Omega) \backslash \{0\}$ and

$$\begin{aligned} \tilde{F} &: (v, \lambda, h) \mapsto h^*(\Delta + \lambda) h^{*-1} v \\ &: \left(H^2 \cap H_0^1(\Omega) \backslash \{0\}\right) \times \mathbb{R} \times \mathrm{Diff}^3(\Omega) \to L_2(\Omega) \end{aligned}$$

so $\tilde{F}(v, \lambda, h) = h_0^{*-1} F(h_0^* v, \lambda, h \circ h_0)$ has a critical point $(v_0, \lambda_0, i_\Omega)$ with $\tilde{F}(v_0, \lambda_0, i_\Omega) = 0$ (recall h_0^* is an isomorphism). (Such "changes of origin" may be made later without comment.)

By hypothesis, the derivative $D\tilde{F}(v_0, \lambda_0, i_\Omega)$

$$\begin{aligned} (\dot{v}, \dot{\lambda}, \dot{h}) &\mapsto (\Delta + \lambda_0)\dot{v} + [\dot{h} \cdot \nabla, \Delta + \lambda_0] v_0 + \dot{\lambda} v_0 \\ &= (\Delta + \lambda_0)(\dot{v} - \dot{h} \cdot \nabla v_0) + \dot{\lambda} v_0 \end{aligned}$$

is not surjective. The image is closed in $L_2(\Omega)$, since the image of $\{\dot{v} \in H^2 \cap H_0^1(\Omega), \dot{\lambda} = 0, \dot{h} = 0\}$ is closed with finite codimension, so there exists $\psi \neq 0$ in $L_2(\Omega)$ orthogonal to the image. Keeping $\dot{v} = 0, \dot{h} = 0$, we see $\int_\Omega \psi v_0 = 0$; keeping $\dot{h} = 0$, $\int_\Omega \psi (\Delta + \lambda_0)\dot{v} = 0$ for all $\dot{v} \in H^2 \cap H_0^1(\Omega)$ says ψ is a weak (hence, strong) solution of $(\Delta + \lambda_0)\psi = 0$ in Ω, $\psi = 0$ on $\partial\Omega$, so $\psi \in H^2 \cap H_0^1(\Omega)$. In fact, since $\partial\Omega$ is C^3, both v_0 and ψ are in $C^{2,\alpha}(\Omega) \cap W^{3,p}(\Omega)$ for any $\alpha < 1, p < \infty$. Now allowing $\dot{h}$ to vary in $C^3(\Omega, \mathbb{R}^n)$

we have

$$0 = \int_\Omega \psi(\Delta + \lambda_0)\dot{h} \cdot \nabla v_0 = \int_\Omega \{\psi(\Delta + \lambda_0)\dot{h} \cdot \nabla v_0 - (\Delta + \lambda_0)\psi \dot{h} \cdot \nabla v_0\}$$

$$= \int_{\partial\Omega} \left\{\psi \frac{\partial}{\partial N}\dot{h} \cdot \nabla v_0 - \frac{\partial \psi}{\partial N}\dot{h} \cdot \nabla v_0\right\} = -\int_{\partial\Omega} \dot{h} \cdot N \frac{\partial \psi}{\partial N}\frac{\partial v_0}{\partial N}$$

for all $\dot{h} \in C^3(\Omega, \mathbb{R}^n)$. [Note $v_0 \in W^{3,p}(\Omega)$ so $\dot{h} \cdot \nabla v_0 \in W^{2,p}(\Omega)$ – which is why we chose to work in a C^3 domain.] Therefore our hypothesis (which we hope to contradict) implies the existence of a non-zero C^2 solution ψ of $(\Delta + \lambda_0)\psi = 0$ in Ω, $\psi = 0$ on $\partial\Omega$, such that $\frac{\partial \psi}{\partial N}\frac{\partial v_0}{\partial N} = 0$ everywhere on $\partial\Omega$.

Suppose at some $x_0 \in \partial\Omega$, that $\frac{\partial v_0}{\partial N} \neq 0$; then $\frac{\partial v_0}{\partial N} \neq 0$ on a neighborhood of this point so $\frac{\partial \psi}{\partial N} = 0$ and $\psi = 0$ on a neighborhood (in $\partial\Omega$) of x_0, while $\Delta\psi + \lambda\psi = 0$ in Ω. By uniqueness in the Cauchy problem for second-order elliptic equations (see [2] or [12, Th. 8.9.1], or the statement of the result at the end of this section), it follows $\psi \equiv 0$ in Ω, since Ω is connected; but this is false. Therefore $\frac{\partial v_0}{\partial N} = 0$ at every point of $\partial\Omega$ – but this implies $v_0 \equiv 0$ in Ω, which is also false. Thus we have proved that zero is a regular value of F (restricted to C^3 deformations), so there is a C^3-dense set of $h \in \mathrm{Diff}^3(\Omega_0)$ such that all eigenvalues of the Dirichlet problem in $h(\Omega_0)$ are simple, and the argument is complete for connected (bounded, C^2) regions.

Further, given a simple eigenvalue λ and corresponding eigenfunction u for the problem in Ω (connected), if $V \in C^2(\Omega, \mathbb{R}^n)$ and $\Omega(t) = \{x + tV(x) \mid x \in \Omega\}$ for t near 0, there is unique eigenvalue $\lambda(t)$ near λ of the problem in $\Omega(t)$, $\lambda(0) = \lambda$, and

$$\frac{d}{dt}\lambda(t)\bigg|_{t=0} = -\frac{\int_{\partial\Omega} V \cdot N_\Omega\left(\frac{\partial u}{\partial N_\Omega}\right)^2}{\int_\Omega u^2},$$

by Example 3.2. By uniqueness in the Cauchy problem, $\frac{\partial u}{\partial N_\Omega}|_{\partial\Omega}$ does not vanish identically ($\int_{Q\cap\partial\Omega}(\frac{\partial u}{\partial N_\Omega})^2 > 0$ for any open $Q \subset \mathbb{R}^n$ which meets $\partial\Omega$) so we can move the eigenvalue λ by perturbing Ω.

In the general case, suppose Ω_0 has several components $\Omega_1, \ldots, \Omega_N$ [finitely many, since Ω_0 is bounded and C^2-regular]. We may first perturb each Ω_i ($1 \leq i \leq N$) separately, to ensure all eigenvalues of the problem restricted to any Ω_i and with modulus $\leq j$, are simple; a further slight perturbation will separate any eigenvalues which occur simultaneously for two or more components. For an open dense set of embedding $h \in \mathrm{Diff}^2(\Omega_0)$, all eigenvalues $\lambda \leq j$ are simple ($j = 1, 2, \ldots$); taking the intersection, we see *all* eigenvalues are simple in most bounded C^2 regions.

Remark. We will often have occassion to cite the theorem on uniqueness in the Cauchy problem, so we state the result in more detail here. The proof is in Hörmander [12, Theorem 8.9.1] and – more accessible, but assuming a_{ij} are C^2 – in Agmon [2].

Uniqueness in the Cauchy Problem for Second-Order Elliptic Equations

Suppose Q is a connected open set in $\mathbb{R}^n$, B is a ball which meets ∂Q in a C^2 hypersurface $B \cap \partial Q$, $a_{ij} = a_{ji} : Q \to \mathbb{R}$ are C^1 functions with $\sum_{i,j} a_{ij}(x)\xi_i\xi_j \geq c_0|\xi|^2$ for $x \in Q$, $\xi \in \mathbb{R}^n$ and some constant $c_0 > 0$. Assume $u \in H^2(Q)$and for some constant K

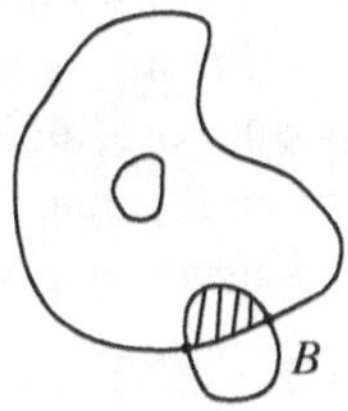

$$\left| \sum a_{ij}(x) \frac{\partial^2 u}{\partial x_i \partial x_j} \right| \leq K(|\nabla u(x)| + |u(x)|) \text{ a.e. in } Q$$

and a.e. on $\partial Q \cap B$ we have $u = 0$, $\frac{\partial u}{\partial N} = 0$. [More exactly,

$$\int_{\partial Q \cap B} |u|^2 + \left| \frac{\partial u}{\partial N} \right|^2 = 0$$

so, if we set $u = 0$ in $B \backslash Q$, then $u \in H^2(B \cup Q)$.] Then $u = 0$ a.e. in Q.

Example 6.2. Simplicity of Eigenvalues with a Reflection-Symmetry Constraint. Sometimes one imposes geometrical symmetry conditions on the regions Ω, of the form: $\gamma(\Omega) = \Omega$ for all $\gamma \in \Gamma$, where Γ is a given subgroup of the orthogonal group $O(n)$. We treat here the simplest example, when $\Gamma = \{I, J\}$, $J \neq I$, $J^2 = I$ so $J = J^{-1} = J$. We may rotate coordinates so J has the form $J = \left(\begin{smallmatrix} I_p & 0 \\ 0 & -I_q \end{smallmatrix}\right)(p + q = n, q > 0)$, a reflection. (If q is even, J may also be considered a rotation.) We will prove, for most bounded C^2 regions $\Omega \subset \mathbb{R}^n$ such that $J\Omega = \Omega$, all eigenvalues of the Dirichlet problem for the Laplacian in Ω are simple, and so every eigenfunction ϕ is either even ($\phi \circ J = \phi$) or odd ($\phi \circ J = -\phi$).

As was pointed out to me by A. L. Pereira, for a third-order symmetry ($J^3 = I$, $J \neq I$), one may have double eigenvalues which cannot be split by perturbing the boundary unless the symmetry is violated.

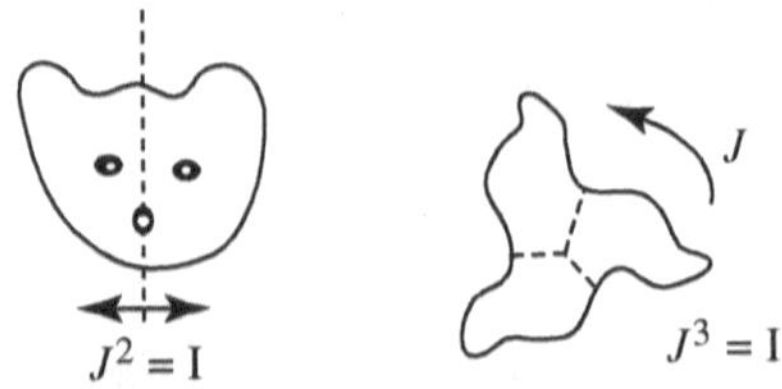

To simplify, we assume Ω is connected. For the involution $J(J^2 = I,$ $J \neq I)$ we may write any function f on Ω as $f = f_{\text{even}} + f_{\text{odd}}$ where $f_{\text{even}} = \frac{1}{2}(f + f \circ J) = f_{\text{even}} \circ J$, $f_{\text{odd}} = \frac{1}{2}(f - f \circ J) = -f_{\text{odd}} \circ J$, and observe $\int_\Omega$ (even). (odd) $= 0$ so $\int_\Omega f^2 = \int_\Omega f_{\text{even}}^2 + \int_\Omega f_{\text{odd}}^2$ when $f \in L_2(\Omega)$. Thus $L_2(\Omega) = L_2^{\text{even}}(\Omega) \oplus L_2^{\text{odd}}(\Omega)$ as an orthogonal sum, and the Laplacian preserves eveness or oddness. We restrict attention to one of these (even or odd) subspaces, and prove, within this subspace, all eigenvalues are generically simple; then we show how to split an eigenvalue which happens to occur simultaneously in both spaces (has both an even and odd eigenfunction).

Lemma. *Let $J^2 = I$, $J \neq I$, $J(\Omega) = \Omega$ and let $\phi : \partial\Omega \to \mathbb{R}$ be a continuous even function ($\phi \circ J = \phi$). Then there exists $V : \mathbb{R}^n \to \mathbb{R}^n$ of class C^∞ with $V(Jx) = JV(x)$ everywhere and such that $V \cdot N_\Omega \mid \partial\Omega$ is uniformly arbitrarily close to ϕ· (If $\phi \in L_p(\partial\Omega)$, $1 \leq p < \infty$, and is even, we may approximate in $L_p(\partial\Omega)$.)*

Proof. Choose $W : \mathbb{R}^n \to \mathbb{R}^n, C^\infty$, uniformly close to ϕN_Ω on $\partial\Omega$, and define $V(x) = \frac{1}{2}(W(x) + J^{-1}W(Jx))$. Then V is $C^\infty, V(Jx) = JVx$ and on $\partial\Omega$, $V(x)$ is close to $\frac{1}{2}\phi(x)(N_\Omega(x) + JN_\Omega(Jx)) = \phi(x)N_\Omega(x))$.

Remark. The same "symmetrizing" argument shows any J-invariant region C^2-close to Ω is of the form $h(\Omega)$ for some h near i_Ω with $h(Jx) = Jh(x)$ for all $x \in \Omega$. Let $\text{Diff}_J^2 \Omega = \{h \in \text{Diff}^2(\Omega) \mid h(Jx) = Jh(x) \text{ for all } x \in \Omega\}$ so, for every such h, $h(\Omega)$ is also J-invariant. By the same calculation as in Example 6.1, above applied to

$$(u, \lambda, h) \mapsto h^*(\Delta + \lambda)h^{*-1}u : \left(H^2 \cap H_0^1(\Omega)\right)^{\text{even}} \times \text{Diff}_J^2(\Omega) \to L_2(\Omega)^{\text{even}},$$

we must show the impossibility of $\int_{\partial\Omega} \dot{h} \cdot N_\Omega \frac{\partial u}{\partial N}\frac{\partial \psi}{\partial N} = 0$ for all $\dot{h} \in C^2(\Omega, \mathbb{R}^n)$ with $\dot{h}(Jx) = J\dot{h}(x)$, when u, ψ are both even eigenfunctions. Similarly in the odd subspace. In either case the product $\frac{\partial u}{N}\frac{\partial \psi}{\partial N}$ is even on $\partial\Omega$ and, by the lemma, we may find C^∞ V so that $V \circ J = J \cdot V$ and $V \cdot N_\Omega$ is close to sgn $(\frac{\partial u}{\partial N}\frac{\partial \psi}{\partial N})$ (in $L_1(\partial\Omega)$). But if $h(t, x) = x + tV(x)$ for small real t, then $h(t, \cdot) \in \text{Diff}_J^2 \Omega$ so $V \cdot N_\Omega = \dot{h} \cdot N_\Omega \mid_{t=0}$ is orthogonal to $\frac{\partial u}{\partial N}\frac{\partial \psi}{\partial N}$. It follows that $\frac{\partial u}{\partial N}\frac{\partial \psi}{\partial N} = 0$ on

$\partial\Omega$, violating uniqueness in the Cauchy problem. Thus, in each subspace, the eigenvalues are generically simple.

Now suppose λ_0 is an eigenvalue in Ω for both even and odd subspaces, which is simple in each space separately. There are eigenfunctions u_0, v_0 corresponding to λ_0, with $u_0 \circ J = u_0$, $v_0 \circ J = -v_0$, $\int_\Omega u_0^2 = 1 = \int_\Omega v_0^2$. A small perturbation $h(t, x) = x + tV(x)$ (with $V \circ J = J \cdot V$) moves the eigenvalues to

$$\lambda_{\text{even}}(t) = \lambda_0 - t \int_{\partial\Omega} V \circ N_\Omega \left(\frac{\partial u_0}{\partial N}\right)^2 + O(t^2)$$
$$\lambda_{\text{odd}}(t) = \lambda_0 - t \int_{\partial\Omega} V \circ N_\Omega \left(\frac{\partial v_0}{\partial N}\right)^2 + O(t^2)$$

Thus we can split λ_0 into two simple eigenvalues by (symmetric) perturbation of Ω, unless perhaps

$$\left(\frac{\partial u_0}{\partial N}\right)^2 = \left(\frac{\partial v_0}{\partial N}\right)^2 \quad \text{on } \partial\Omega.$$

Let $w_\pm = u_0 \pm v_0$; w_+, w_- are eigenfunctions with $\int_\Omega w_\pm^2 = \int_\Omega u_0^2 + \int_\Omega v_0^2 = 2$, $w_+ = w_- = 0$ on $\partial\Omega$, and $\frac{\partial w_+}{\partial N}\frac{\partial w_-}{\partial N} = (\frac{\partial u_0}{\partial N})^2 - (\frac{\partial v_0}{\partial N})^2$, which cannot vanish a.e. in $\partial\Omega$ by uniqueness in the Cauchy problem. Thus, the eigenvalues may always be split into simple eigenvalues, by small (symmetric) perturbations of the boundary. For an open dense set of $h \in \text{Diff}_J^2(\Omega)$, all eigenvalues with modulus $\leq k$ for $h(\Omega)$ are simple ($k = 1, 2, 3, \ldots$); so for an ample set of such h, all eigenvalues for $h(\Omega)$ are simple.

Remark. According to Theorem 1.7, a J-invariant C^m-regular region Ω may be represented in the form $\{x : \phi(x) > 0\}$ where ϕ is C^m, $\phi(x) = 0 \Rightarrow |\nabla\phi(x)| \geq 1$ and $\phi(Jx) = \phi(x)$ for all x. To approximate Ω by C^∞ (or analytic) J-invariant regions, it suffices to approximate ϕ by appropriately smooth ψ, define $\tilde{\psi}(x) = \frac{1}{2}(\psi(x) + \psi(Jx))$, and then $\{x : \tilde{\psi}(x) > 0\}$ is a smooth J-invariant region C^m near Ω. (Theorem 1.6). Thus, we could carry out the calculation above under the hypothesis that Ω is C^m, for any large m; then conclude all eigenvalues with modulus $\leq k$ are simple for an open dense set of J-invariant C^2-regular regions ($k = 1, 2, \ldots$).

Example 6.3. Simplicity of Real Eigenvalues of a General Second-Order Dirichlet Problem. We study elliptic operators of the form

$$A(\lambda) = \sum_{i,j=1}^{n} a_{ij}(x) \frac{\partial^2}{\partial x_i \partial x_j} + \sum_{j=1}^{n} b_j(x, \lambda) \frac{\partial}{\partial x_j} + c(x, \lambda)$$

where $a_{ij}(x) = a_{ji}(x)$ is C^2, and $b_j(x, \lambda), c(x, \lambda)$ are polynomials in λ with C^2 coefficients, all real-valued when λ is real. We also assume that $\sum_{i,j} a_{ij}(x)\xi_i\xi_j \geq c_0|\xi|^2$ for all x, ξ in $\mathbb{R}^n$and some constant $c_0 > 0$. Given bounded C^2-regular $\Omega \subset \mathbb{R}^n$, λ is an *eigenvalue* if there is a solution $u \not\equiv 0$ of

$$A(\lambda)u = 0 \text{ in } \Omega, \quad u = 0 \text{ on } \partial\Omega;$$

λ is a *simple eigenvalue* if the kernal of $A(\lambda) \mid H^2 \cap H_0^1(\Omega)$ is one-dimensional (so the image has codimension one) and the image does not contain $(\frac{d}{d\lambda}A(\lambda))u$, for $u \neq 0$ in the kernel. This definition ensures that simple eigenvalues may be studied without trouble using the implicit function theorem – as in the usual case $b(x, \lambda) = b(x)$, $c(x, \lambda) = c(x) + \lambda$. Such polynomials occur, for example, when we seek exponential solutions $u(t, x) = e^{\lambda t}v(x) \not\equiv 0$ of the wave equation

$$u_{tt} + r(x)u_t = \Delta u \text{ in } \Omega, \quad u = 0 \text{ on } \partial\Omega,$$

so $\Delta v - (\lambda^2 + \lambda r(x))v = 0$ in Ω, $v = 0$ on $\partial\Omega$, $v \not\equiv 0$. It is important to know also the complex eigenvalues, but these are more difficult to handle; they will be discussed in Example 6.8 below. The important question is whether the *coefficients* are real or complex – thus in the example, if $r(x) \equiv 0$, we could treat λ on the imaginary axis by the methods of this section. (Example 4.4 treats self-adjoint problems, possibly complex-valued, with the more usual form of eigenvalue parameter.)

Returning to the general problem above, let $A_\Omega(\lambda)$ be the restriction of $A(\lambda)$ to a map on $H^2 \cap H_0^1(\Omega)$ into $L_2(\Omega)$ (the real spaces, when λ is real). It is then easy to show zero is a regular value of $(\lambda, u) \mapsto A_\Omega(\lambda)u : \mathbb{R} \times (H^2 \cap H_0^1(\Omega)\backslash\{0\}) \to L^2(\Omega)$ if and only if all real eigenvalues λ are simple. We will prove, for most bounded C^2 regions $\Omega \subset \mathbb{R}^n$, that all real eigenvalues are simple, and that all simple real eigenvalues can be moved by small perturbations of Ω. Our hypotheses allow the case when $A(\lambda)$ does not depend on λ – in that case, there can be no "simple eigenvalues," so generically A_Ω is an isomorphism.

Suppose Ω is connected. Consider the map

$$\begin{aligned} F &: (u, \lambda, h) \mapsto h^* A_{h(\Omega)}(\lambda) h^{*-1}(u) \\ &: (H^2 \cap H_0^1(\Omega)\backslash\{0\}) \times \mathbb{R} \times \mathrm{Diff}^2(\Omega) \to L_2(\Omega), \end{aligned}$$

which is of class C^2. For fixed h, $(u, \lambda) \mapsto F(u, \lambda, h)$ is Fredholm with index 1 (index 0, if we keep λ fixed). The spaces are separable, and we only need to prove zero is a regular value of F.

Suppose to the contrary that F has a critical point (u, λ, h) with $F(u, \lambda, h) = 0$; we may suppose $h = i_\Omega$. Then $A_\Omega(\lambda)u = 0$ in Ω, $u = 0$ on $\partial\Omega$, $u \not\equiv 0$, and the derivative of F at (u, λ, i_Ω)

$$(\dot{u}, \dot{\lambda}, \dot{h}) \mapsto A_\Omega(\lambda)(\dot{u} - \dot{h} \cdot \nabla u) + A'_\Omega(\lambda) u \dot{\lambda} \quad \left[A'_\Omega(\lambda) = \frac{d}{d\lambda} A_\Omega(\lambda) \right]$$

is not surjective, so there exists $\psi \neq 0$ in $L_2(\Omega)$ orthogonal to the image. It follows that $\psi \in H^2 \cap H^1_0(\Omega)$, $A^*_\Omega(\lambda)\psi = 0$ in Ω, $\int_\Omega \psi A'_\Omega(\lambda) u = 0$, and for all $\dot{h} \in C^2(\Omega, \mathbb{R}^n)$

$$\begin{aligned} 0 = \int_\Omega \psi A_\Omega(\lambda) \dot{h} \cdot \nabla u &= \int_\Omega \{\psi A_\Omega(\lambda) \dot{h} \cdot \nabla u - A^*_\Omega(\lambda) \psi \dot{h} \cdot \nabla u\} \\ &= \int_{\partial\Omega} -\frac{\partial \psi}{\partial N} \frac{\partial u}{\partial N} \sum_{i,j} a_{ij} N_i N_j \dot{h} \cdot N \end{aligned}$$

hence $\frac{\partial \psi}{\partial N} \frac{\partial u}{\partial N} = 0$ on $\partial\Omega$. But this violates uniqueness in the Cauchy problem, so zero is not a critical value of F. Thus, generically in Ω, all real eigenvalues are simple.

Now suppose λ is a simple eigenvalue for the problem in Ω. Let $\phi, \psi \in H^2 \cap H^1_0(\Omega) \backslash \{0\}$ with $A_\Omega(\lambda)\phi = 0$, $A^*_\Omega(\lambda)\psi = 0$; ϕ, ψ are unique (up to constant real multiples) and $\int_\Omega \psi A'_\Omega(\lambda)\phi \neq 0$, by definition of "simplicity." For any C^2 $V : \mathbb{R}^n \to \mathbb{R}^n$, let $h(t, x) = x + tV(x)$, $\Omega(t) = h(t, \Omega)$ for small real t. A slight variant of the argument in Example 3.2 (or 3.6) shows there is a unique eigenvalue $\lambda(t)$ near λ for the problem in $\Omega(t)$, $\lambda(t)$ is a simple eigenvalue and differentiable in t and

$$\left. \frac{d}{dt} \lambda(t) \right|_{t=0} = \frac{-\int_{\partial\Omega} V \cdot N_\Omega (\sum a_{ij} N_i N_j) \frac{\partial \phi}{\partial N} \frac{\partial \psi}{\partial N}}{\int_\Omega \psi A'_\Omega(\lambda)\phi}.$$

Since $\frac{\partial \phi}{\partial N} \frac{\partial \psi}{\partial N} |_{\partial\Omega} \not\equiv 0$ (in fact, it is nonzero a.e. in $\partial\Omega$) we may choose V so the derivative is nonzero. Thus every simple eigenvalue can be moved by small perturbations of $\partial\Omega$, and indeed by small perturbations restricted to a small neighborhood of a given point of $\partial\Omega$.

So what's wrong with complex coefficients?

The problem with complex coefficients is that we cannot work entirely in the complex field: the embeddings h are necessarily real. If we take λ, u in complex spaces, the image of the derivative of F will be a real-linear subspace of $L_2(\Omega, \mathbb{C})$, but need not be linear over the complex field. Following the argument above, there exists $\psi \neq 0$ so that $\operatorname{Re} \int_\Omega \psi DF(u, \lambda, i_\Omega)(\dot{u}, \dot{\lambda}, \dot{h}) = 0$ for all $(\dot{u}, \dot{\lambda}, \dot{h})$ and we wind up with the condition: $\operatorname{Re}\left(\frac{\partial \psi}{\partial N} \frac{\partial u}{\partial N}\right) = 0$ on $\partial\Omega$. It is not at all clear that this is impossible. We return to the complex case later (Example 6.8).

Although complex coefficients cause trouble, we can easily treat more general *real* dependence on λ. Examining the argument above, we see that when Λ

is a C^{m+1} m-dimensional manifold and $(x, \lambda) \mapsto b_j(x, \lambda), c(x, \lambda) : \mathbb{R}^n \times \Lambda \to \mathbb{R}$ are C^{m+1}, then for most bounded C^{m+1}-regular regions $\Omega \subset \mathbb{R}^n$, $(u, \lambda) \mapsto A_\Omega(\lambda)u : (H^2 \cap H_0^1(\Omega)\backslash\{0\}) \times \Lambda \to L_2(\Omega)$ has zero as a regular value. Thus whenever $A_\Omega(\lambda)u = 0$ for some $u \neq 0$, it follows that $A_\Omega(\lambda)(H^2 \cap H_0^1(\Omega)) + (\frac{d}{d\lambda}A_\Omega(\lambda)u)T_\lambda\Lambda = L_2(\Omega)$. This implies dim $\mathcal{N}(A_\Omega(\lambda)) \leq m = \dim \Lambda$, and provides at least a first step in the study of multi-parameter eigenvalue problems.

Example 6.4. Generic Simplicity of Eigenvalues for the Neumann Problem for Δ. The Neumann eigenvalue problem in $\Omega \subset \mathbb{R}^n$ is

$$\Delta u + \lambda u = 0 \text{ in } \Omega, \qquad \frac{\partial u}{\partial N} = 0 \text{ on } \partial\Omega, \qquad u \not\equiv 0$$

and always has the eigenvalue $\lambda = 0$. The corresponding eigenfunctions are constant in each component of Ω, and the multiplicity of the zero eigenvalue is the number of components of Ω. We prove that, for most bounded C^2-regular regions $\Omega \subset \mathbb{R}^n$, all non-zero eigenvalues are simple, and may be moved by small perturbations of Ω. As usual, it is enough to do this for connected regions Ω.

Given a C^2-regular bounded region Ω, the Neumann problem in Ω has all non-zero eigenvalues simple if and only if (0, 0) is a regular value of

$$F_\Omega : (u, \lambda) \mapsto \left(\Delta u + \lambda u, \left.\frac{\partial u}{\partial N}\right|_{\partial\Omega}\right)$$
$$: (H^2(\Omega)\backslash\{0\}) \times (\mathbb{R}\backslash\{0\}) \to L_2(\Omega) \times H^{1/2}(\partial\Omega).$$

Suppose in fact Ω is C^3-regular connected and bounded and consider the map $G : (u, \lambda, h) \mapsto h^* F_{h(\Omega)} h^{*-1}(u, \lambda)$ with F as above and $h \in \text{Diff}^3(\Omega)$, G is an analytic map of separable spaces, $(u, \lambda) \mapsto G(u, \lambda, h)$ is Fredholm (index 1) for each H, and we wish to prove (0, 0) is a regular value of G. Suppose to the contrary there is a critical point (u, λ, h) of G with $G(u, \lambda, h) = (0, 0)$; we may suppose $h = i_\Omega$. Then $\Delta u + \lambda u = 0$ in Ω, $\partial u/\partial N = 0$ on $\partial\Omega$, $u \not\equiv 0$, $\lambda \neq 0$ and there exists non-zero $(\psi, \theta) \in L_2(\Omega) \times H^{-1/2}(\partial\Omega)$ orthogonal to the image of the derivative.

$$0 = \int_\Omega \psi\{(\Delta + \lambda)(\dot{u} - \dot{h} \cdot \nabla_u) + \dot{\lambda} u\}$$
$$+ \left\langle \theta, \frac{\partial}{\partial N}(\dot{u} - \dot{h} \cdot \nabla u) + \dot{N} \cdot \nabla u + \dot{h} \cdot \nabla\left(\frac{\partial u}{\partial N}\right)\right\rangle_{\partial\Omega}$$

where $\langle \cdot, \cdot \rangle_{\partial\Omega}$ indicates the duality of $H^{-1/2}(\partial\Omega) \times H^{1/2}(\partial\Omega)$. Keeping $\dot{h} = 0$, $\dot{\lambda} = 0$, it follows that $\psi \in H^2(\Omega)$, $\Delta\psi + \lambda\psi = 0$ in Ω, $\partial\psi/\partial N = 0$ on $\partial\Omega$

and $\langle \theta, g \rangle_{\partial\Omega} = -\int_{\partial\Omega} \psi g$ for all $g \in H^{1/2}(\partial\Omega)$. As $\dot\lambda$ varies, we see $\int_\Omega \psi u = 0$; and when $\dot h$ varies, we find (supposing, as we may, $\partial N/\partial N = 0$ on $\partial\Omega$)

$$0 = \int_{\partial\Omega} \psi\left(-\nabla_{\partial\Omega}\sigma \cdot \nabla u + \sigma\frac{\partial^2 u}{\partial N^2}\right) \quad \text{for all} \quad \sigma = \dot h \cdot N \in C^3(\partial\Omega)$$

since $\dot N = -\nabla_{\partial\Omega}\sigma$ (2.3).

Now $\partial\Omega$ is C^3 so, u, ψ are in the fact $C^{2,\alpha}$ functions for any $\alpha < 1$, and $-\psi\nabla_{\partial\Omega}\sigma \cdot \nabla u = -\text{div}_{\partial\Omega}(\sigma\psi\nabla_{\partial\Omega}u) + \sigma(\nabla_{\partial\Omega}\psi \cdot \nabla_{\partial\Omega}u + \psi\Delta_{\partial\Omega}u)$ so $0 = \nabla_{\partial\Omega}\psi \cdot \nabla_{\partial\Omega}u + \psi(\Delta_{\partial\Omega}u + \partial^2 u/\partial N^2)$ on $\partial\Omega$. Since $\partial u/\partial N = 0$ on $\partial\Omega$ and $\Delta u = -\lambda u$, this (with 1.12) says $\nabla_{\partial\Omega}\psi \cdot \nabla_{\partial\Omega}u = \lambda u\psi$ on $\partial\Omega$, which we show is impossible.

By uniqueness in the Cauchy problem $u\psi \neq 0$ on a dense open set of $\partial\Omega$; choose some $x_0 \in \partial\Omega$ where $u\psi \neq 0$. Define a curve $t \mapsto x(t) \in \partial\Omega$ by $x(0) = x_0$, $dx/dt = \nabla_{\partial\Omega}u(x)$; $\nabla_{\partial\Omega}u(\cdot)$ is a C^1 tangent-vector field on $\partial\Omega$, and $\partial\Omega$ is compact, so this curve is defined on $-\infty < t < \infty$. Further, $\frac{d}{dt}u(x(t)) = |\nabla_{\partial\Omega}u(x)|^2 \geq 0$ and $\psi(x(t)) = \psi(x_0)\exp(\lambda\int_0^t u(x(s))ds)$ for all t. If $u(x_0) > 0$ then $u(x(t)) \geq u(x_0) > 0$ for all $t \geq 0$, and $\lambda > 0$, so $\lambda\int_0^t u(x(s))ds \to +\infty$ as $t \to +\infty$. This is impossible, since ψ is bounded. But if $u(x_0) < 0$, the same argument applies at $t \to -\infty$; in any case, we have a contradiction. Uhlenbeck [41] gives a similar argument.

Thus (0, 0) is a regular value of G, and in most C^3 (bounded, connected) regions $\Omega \subset \mathbb{R}^n$, all eigenvalues of the Neumann problem are simple. The set of (bounded, connected) C^2-regions Ω, such that the eigenvalues of modulus $\leq k$ are simple, is open (by Example 3.5) and dense (by the C^3 case). This holds for each $k = 1, 2, 3, \ldots$, and taking the intersection gives the general (connected) case.

If λ is a non-zero simple eigenvalue with eigenfunction u, $\int_\Omega u^2 = 1$, then according to Example 3.5 we can move λ by small perturbations of Ω, except possibly in the case where $|\nabla_{\partial\Omega}u|^2 = \lambda u^2$ everywhere on $\partial\Omega$. In this case, let $u(x_0) = \max u \mid \partial\Omega$; then $\nabla_{\partial\Omega}u(x_0) = 0$ so $u(x_0) = 0$ and similarly $\min\, u \mid \partial\Omega = 0$, so $u \equiv 0$ on $\partial\Omega$, $u \equiv 0$ in Ω, which false. (Since $\partial\Omega$ is C^2, $u \in W^{2,p}(\Omega)$ for any $p < \infty$ so u is at least C^1.)

Example 6.5. Generic Simplicity of Solutions of $\Delta u + f(x, y, \nabla u) = 0$ in Ω, $u \geq 0$ on $\partial\Omega$. Let Q be an open set in $\mathbb{R}^n \times \mathbb{R} \times \mathbb{R}^n$ and let $f : Q \to \mathbb{R}$ be locally C^2. We prove, for most bounded C^2-regular regions $\Omega \subset \mathbb{R}^n$, all solutions u of

$$\Delta u + f(x, u, \text{grad}\, u) = 0 \text{ in } \Omega, \quad u = 0 \text{ on } \partial\Omega$$

are simple, i.e., the linearization

$$\phi \mapsto \Delta\phi + \frac{\partial f}{\partial p}(\cdot, u, \nabla u) \cdot \nabla\phi + \frac{\partial f}{\partial u}(\cdot, u, \nabla u)\phi$$
$$: \ H^2 \cap H_0^1(\Omega) \to L_2(\Omega)$$

is an isomorphism. (One of the requirements for a C^1 function u to be a solution is that $(x, u(x), \operatorname{grad} u(x)) \in Q$ for all $x \in \bar{\Omega}$). It follows that the set of solutions is (generically) discrete, and if f is bounded (for example), this set is finite. Thus $\Delta u + \sin u = 0$, in Ω, $u = 0$ on $\partial\Omega$, has only finitely many solutions for most Ω.

The same result holds with any uniformly elliptic second-order operator in place of the Laplacian.

Saut and Teman [34] proves this for $\Delta u + f(x, u) = 0$, under the hypothesis – inadvertantly omitted from [34] – that $f(x, 0) \equiv 0$. (The author had independently proved this case). The general form stated above seems to require the extension of the transversality theorem given in Chapter 5 (Theorem 5.4.). Indeed this is one of the examples which motivated study of this generalization.

We prove the result first under the hypothesis that f is everywhere defined and bounded (as well as locally C^2), and Ω is C^3 regular and connected.

Consider the map

$$F : (u, h) \mapsto h^*(\Delta + f)h^{*-1}(u) : W^{2,p} \cap W_0^{1,p}(\Omega) \times \operatorname{Diff}^3(\Omega) \to L_p(\Omega)$$

for some p in $n < p < \infty$ (so $W^{2,p}(\Omega) \subset C^1(\Omega)$). If zero is a regular value of $u \mapsto F(u, h)$, it is easily shown that all solutions of the problem in $h(\Omega)$ are simple. As usual, the only difficulty is to know whether zero is a regular value of F. We will try to prove this by contradiction – and fail. For certain f and Ω, zero may be a critical value – but then the critical point has very special properties. Using conditions $2(\beta)$ of Theorem 5.4, we show there is an open dense subset $\mathcal{D}_0 \subset \operatorname{Diff}^3(\Omega)$ such that the restriction of F to $\mathcal{D}_0$ has zero as a regular value, and then there is an open dense $\mathcal{D}_1 \subset \mathcal{D}_0$ such that, if $h \in \mathcal{D}_1$, the problem in $h(\Omega)$ has only finitely-many simple solutions. The usual tricks then give the general result.

Suppose now (u, i_Ω) is a critical point of F with $F(u, i_\Omega) = 0$. There exists $\psi \neq 0$ in $L_{p'}(\Omega)\,(1/p + 1/p' = 1)$ orthogonal to the image of the derivative,

$$\int_\Omega \psi L(\dot{u} - \dot{h} \cdot \nabla u) = 0 \quad \text{for all } \dot{u} \in W^{2,p} \cap W_0^{1,p}(\Omega), \dot{h} \in C^3(\Omega)$$

where $L = \Delta + \sum_1^n \frac{\partial f}{\partial p_k}(\cdot, u(\cdot), \nabla u(\cdot))\frac{\partial}{\partial x_k} + \frac{\partial f}{\partial u}(\cdot, u(\cdot), \nabla u(\cdot))$. Since $\partial\Omega$ is C^3, u is $C^{2,\alpha}$ for any $\alpha < 1$, so the coefficients of L are C^1. Fixing $\dot{h} = 0$, we see $L^*\psi = 0$ in Ω, $\psi = 0$ on $\partial\Omega$ and $\psi \in W^{2,q}(\Omega)$ for any $q < \infty$. Allowing

$\dot{h}$ to vary, we see $0 = \int_{\partial\Omega} \dot{h} \cdot N \frac{\partial\psi}{\partial N} \frac{\partial u}{\partial N}$ for all $\dot{h}$, so $\frac{\partial\psi}{\partial N} \frac{\partial u}{\partial N} \equiv 0$ on $\partial\Omega$. Since $\psi \not\equiv 0$ it follows that $\frac{\partial u}{\partial N} \equiv 0$ on $\partial\Omega$.

If $f(x, 0, 0) \equiv 0$, it follows by uniqueness in the Cauchy problem that $u \equiv 0$. But the zero solution may be studied separately; by Example 6.3 (in the case where the coefficients don't depend on λ), for most regions Ω, the zero solution is simple. Applying the above argument for $u \in W^{2,p} \cap W_0^{1,p}(\Omega)\backslash\{0\}$, we see th nonzero solutions are also generically simple. (This is the case treated in [34].)

More generally, suppose $f(x, 0, 0)$ vanishes on some neighborhood of $\partial\Omega$; then also u vanishes near $\partial\Omega$, so we have a special solution $u = u_0$ which does not depend on Ω (for small perturbations), and the same argument applies. u_0 is generically simple, by Example 6.3, and solutions in $W^{2,p} \cap W_0^{1,p}(\Omega)\backslash\{u_0\}$ are generically simple.

It is easy to find examples u, f, Ω, such that $f(x, 0, 0) \neq 0$ on a dense set and u is a solution with $u = 0$, $\partial u/\partial N = 0$ on $\partial\Omega$. (Hint: First choose u.) But we want to show this is an unusual property of Ω, given f.

For an open dense set of (bounded, connected, C^3-regular) regions Ω, either $f(x, 0, 0) \equiv 0$ on a neighborhood of $\partial\Omega$, or $f(x, 0, 0) \neq 0$ at some point of $\partial\Omega$. The first case is treated above, and in second case we show there is an open dense subset of a C^3-neighborhood of i_Ω such that, for any h in this set, there is no solution u of the problem in $h(\Omega)$ with both $u = 0$ and $\partial u/\partial N = 0$ on $\partial h(\Omega)$. (We assume here $n \geq 2$; the case $n = 1$ is easy, but requires separate treatment.) When $n \geq 2$, the excluded set of embeddings has infinite codimension.

Choose $\delta_0 > 0$ so $\|h - i_\Omega\|_{C^3} < \delta_0$ implies h is an embedding and $f(x, 0, 0) \neq 0$ somewhere on $\partial h(\Omega)$. Define

$$G : (h, u) \mapsto h^*(\Delta + f)h^{*-1}(u) : B_{\delta_0}^{C^3}(i_\Omega) \times W_0^{2,p}(\Omega) \to L_p(\Omega).$$

(Note $u \in W^{2,p}(\Omega)$ is in $W_0^{2,p}(\Omega)$ if and only if $u = 0$ and $\partial u/\partial N = 0$ on $\partial\Omega$.) Now G is of class C^1 and $G(\cdot, h)$ is semi-Fredholm with index $-\infty$ for each h. We verify condition 2(β) of Theorem 5.4 at (u, i_Ω), supposing $G(u, i_\Omega) = 0$. Observe that $\frac{\partial}{\partial u} G(u, i_\Omega) = L|W_0^{2,p}(\Omega)$ is injective, by uniqueness in the Cauchy problem, though we don't use this fact. Let us suppose

$$m = \dim\left\{\mathcal{R}(DG(u, i_\Omega))/\mathcal{R}\left(\frac{\partial}{\partial u}G(u, i_\Omega)\right)\right\} < \infty$$

Then there exist $f_1, \dots, f_m$ in $L_p(\Omega)$ such that $f_j = L(\dot{u}_j - \dot{h}_j \cdot \nabla u)$ for some $\dot{u}_j \in W_0^{2,p}(\Omega)$, $\dot{h}_j \in C^3(\Omega)$, and such that, for every $(\dot{u}, \dot{h})$, there exists $\dot{v} \in W_0^{2,p}(\Omega)$ and unique $c_1, \dots, c_m \in \mathbb{R}$ so

$$\sum_1^m c_j f_j + L\dot{v} = L(\dot{u} - \dot{h} \cdot \nabla u)$$

or

$$\left(L(\dot{u} - \dot{v} - \sum_1^m c_j \dot{u}_j - \left(\dot{h} - \sum_1^m c_j \dot{h}_j\right) \cdot \nabla u\right) = 0$$

Let $\{\phi_1, \ldots, \phi_k\}$ be a basis for the kernal of $L|W^{(2,p)} \cap W_0^{1,p}(\Omega)$. Since $u \in W_0^{2,p}(\Omega) \cap W^{3,p}(\Omega)$, it follows $\frac{\partial u}{\partial x_j} \in W^{2,p} \cap W_0^{1,p}(\Omega)$ for each j and so

$$\dot{u} - \dot{v} - \sum_1^m c_j \dot{u}_j - \left(\dot{h} - \sum_1^m c_j \dot{h}_j\right) \cdot \nabla u = \sum_1^k b_i \phi_i \in \Omega$$

for some constants $b_1, \ldots, b_k$. Compute the normal derivative

$$-\left(\dot{h} - \sum_1^m c_j \dot{h}_j\right) \cdot N \frac{\partial^2 u}{\partial N^2} = \sum_1^k b_i \frac{\partial \phi_i}{\partial N} \text{ on } \partial\Omega$$

and $\frac{\partial^2 u}{\partial N^2}|_{\partial\Omega} = \Delta u|_{\partial\Omega} = -f(x, 0, 0)$ so

$$\dot{h} \cdot N(x) f(x, 0, 0) = \sum_1^m c_j \dot{h}_j \cdot N f(x, 0, 0) + \sum_1^k b_i \frac{\partial \phi_i}{\partial N}(x).$$

For *every* $\dot{h} \in C^3(\Omega, \mathbb{R}^n)$, there exist $c_1, \ldots, c_m, b_1, \ldots b_k$ such that this equation holds on $\partial\Omega$, that is

$$\dot{h} \mapsto \dot{h} \cdot N f(\cdot, 0, 0)|_{\partial\Omega}$$

has finite rank. But this is false, since $f(\cdot, 0, 0)|_{\partial\Omega} \not\equiv 0$ and $n \geq 2$ (so dim $\partial\Omega > 0$).

Thus $\mathcal{R}(DG(u, i_\Omega))/\mathcal{R}(\frac{\partial}{\partial u} G(u, i_\Omega))$ is infinite-dimensional and condition $2(\beta)$ holds. (It holds at any point $(u, h) \in G^{-1}(0)$, by usual change of origin.) It is also easy to verify that $(u, h) \mapsto h : G^{-1}(0) \to B_{\delta_0}$ is proper (recall f is bounded). Thus excluding only a closed subset of infinite codimension, for any h with $\|h - i_\Omega\|_{C^3} < \delta_0$, $G(u, h) \neq 0$ for all $u \in W_0^{2,p}(\Omega)$.

Thus there is an open dense set of $h \in \text{Diff}^3(\Omega)$such that: either $f(x, 0, 0) \equiv 0$ on a neighborhood of $\partial h(\Omega)$ and the solution (if any) in $W_0^{2,p}(h(\Omega))$ is simple, or $f(x, 0, 0) \neq 0$ somewhere on $\partial h(\Omega)$ and the there are no solutions in $W_0^{2,p}(h(\Omega))$. This is the set $\mathcal{D}_0$ claimed above, so the proof (for $n \geq 2$, Ω of class C^3 and f bounded) is complete. With f bounded, the set of C^3-regular regions having all solutions simple, is open (by the implicit function theorem: there are finitely many solutions) and dense, since we may approximate by C^3 regions.

Now suppose $n = 1$:

$$u'' + f(x, u, u') = 0 \quad \text{in} \quad a < x < b, \quad u(a) = u(b) = 0.$$

Perturbation of $\Omega = (a, b)$ means merely changing a and b, but the argument above applies until the last step. We must still show there are no solutions u with $u'(a) = 0$, $u'(b) = 0$, for most choices a, b. Let $U(x; a)$ be the solution of $U_{xx} + f(x, U, U_x) = 0$ with $U = 0$, $U_x = 0$ when $x = a$. Assume for some $a < b$ that $f(b, 0, 0) \neq 0$ and $U(b; a) = 0$, $U_x(b; a) = 0$. Since $U_{xx}(b; a) = -f(b, 0, 0) \neq 0$ we may solve $U_x(x; \alpha) = 0$ for $x = \beta(\alpha)$ near b when α is near a. Then there are no solutions of our problem in $W_0^{2,p}(\alpha, \beta)$ (α near a, β near b) unless $\beta = \beta(\alpha)$ and $U(\beta(\alpha); \alpha) = 0$, a condition at least of codimension 1.

Thus we have the result when f is bounded and defined everywhere. For the general case, choose open sets $Q_1 \subset Q_2 \subset \cdots \subset Q$ whose union is Q with $\overline{Q}_j$compact in Q_{j+1}, and choose $f_j \in C^2(\mathbb{R}^n \times \mathbb{R} \times \mathbb{R}^n)$ such that $f_i = f$ on Q_j. For each f_j there is an open-dense set $\mathcal{D}_j \subset \mathrm{Diff}^2(\Omega)$ of "good" embeddings. Let $h \in \bigcap_1^\infty \mathcal{D}_j$ and let u be any solution of our nonlinear Dirichlet problem in $h(\Omega)$; $\{(x, u(x), \operatorname{grad} u(x)) \mid x \in \overline{\Omega}\}$ is a compact subset of Q, so it is contained in some Q_j. Thus u also satisfies the problem with f_j in place of f, and the linearization L_j is an isomorphism ($h \in \mathcal{D}_j$). But $L_j = L$, and the proof is complete.

Interlude: On the Need for Chapter 7

The last example is the simplest case of a problem we encounter with full force in the succeeding examples. We wish to apply the transversality theorem to some $f(x, y)$, with $\frac{\partial f}{\partial x}(x, y)$ Fredholm, by showing 0 is a regular value of f. We suppose, for contradiction, that there exists (x, y) with $f(x, y) = 0$ and $Df(x, y)$ not surjective, and deduce some other infinite dimensional condition $g(x, y) = 0$, hoping for a contradiction. But it is not contradictory, or at least not obviously so; we continue. Hoping to show it is unusual, if not contradictory, we apply transversality again to $(x, y) \to (f(x, y), g(x, y))$. Now we have index $-\infty$ (see 5.14, #8). Trying to verify 2(β) of Theorem 5.4 by contradiction, we suppose the quotient space has finite dimension n_0, so for appropriate $\{\theta_1, \ldots, \theta_{n_0}\}$:

For any $\dot{y}$ there exist $\dot{x}$ and constants $c_1, \ldots, c_{n_0}$ such that

$$(f_x \dot{x}, g_x \dot{x}) + (f_y \dot{y}, g_y \dot{y}) = \sum_1^{n_0} c_j \theta_j.$$

The operator f_x is Fredholm, so we can almost solve the first component

$$\left[f_x \dot{x} = -f_y \dot{y} + \sum_1^{n_0} c_j \theta_j^1, \theta_j = (\theta_j^1, \theta_j^2) \right]$$

for $\dot{x}$, modulo operators of finite rank; then substitute in the second component to conclude

$$\Gamma : \dot{y} \mapsto g_y \dot{y} - g_x [f_x^{-1}] f_y \dot{y}$$

is of finite rank. (The "inverse" should not be taken too seriously.) In the last example, we could reduce this to a simple form; but in general Γ is a complicated abstract gadget which might (who knows?) be of finite rank. To show it is not of finite rank to obtain the longed-for contradiction, we need quite concrete representation of Γ (especially, $[f_x^{-1}]$), and this is the purpose of Chapter 7. In the remainder of this chapter, we merely quote the results needed from Chapter 7, hopefully fortifying our spirits for the trials to come.

Example 6.6. Generic Simplicity of Eigenvalues of Robin's Problem. Given real-valued $c(x)$, $\beta(x)$ of class C^2, C^3 respectively on $\mathbb{R}^n$, consider the eigenvalue problem

$$(*)_\Omega \qquad \Delta u + (\lambda + c(x))u = 0 \text{ in } (\Omega), \quad \frac{\partial u}{\partial N} + \beta(x)u = 0 \text{ on } \partial\Omega, \quad u \not\equiv 0$$

for bounded C^2-regions $\Omega \subset \mathbb{R}^n$.

In the special case $\beta \equiv 0$, $c \equiv$ constant (Neumann), $\lambda = -c$ is an eigenvalue for any choice of Ω, with multiplicity equal to the number of components of Ω, and locally-constant eigenfunctions. All other eigenvalues are simple for most choices of Ω, and move when we perturb Ω: Example 6.4.

We will prove, for some bounded C^2-regions Ω, all eigenvalues of $(*)_\Omega$ are simple, and may be moved by perturbation of Ω, except possibly in the case; $\beta \equiv 0$ on a neighborhood of Ω and $c \equiv$ constant on a neighborhood of $\partial\Omega_i$, where Ω_i is some component of Ω. In the exceptional case, let c_0 be the constant value of c; then $\lambda = -c_0$ *may* be an eigenvalue which does not move when Ω is slightly perturbed; whose eigenfunctions are locally constant on a neighborhood of $\partial\Omega_i$ (if Ω_i is a component of Ω such that $c(x) \equiv c_0$ near $\partial\Omega_i$) and vanish in any component Ω_k such that $c(x) \not\equiv c_0$ on a neighborhood of $\partial\Omega_k$; and the multiplicity of $\lambda = -c_0$ is at most the number of components of the first kind. Any eigenvalues not of this special type are simple (generically) and move when we perturb $\partial\Omega$.

We will assume Ω is connected; then (generically) *all* eigenvalues are simple, and all may be moved by small perturbation of Ω expect perhaps $\lambda = -c_0$ in case $c \equiv c_0$ and $\beta \equiv 0$ on a neighborhood of $\partial\Omega$ with an eigenfunction locally constant near $\partial\Omega$.

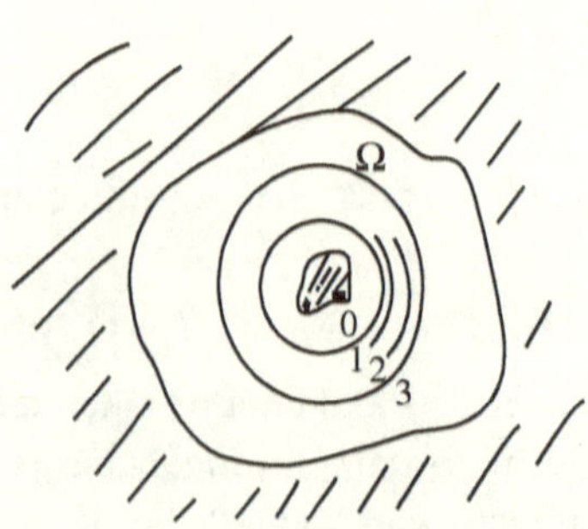

Example. Choose $0 < R_0 < R_1 < R_2 < R_3$ with $J_0(R_1) > 0 > J_0(R_2)$, J_0 is the zero-order Bessel function. Choose $\phi(t)$ of class C^∞ so that $\phi(t) = J_0(t)$ on $R_1 \leq t \leq R_2$, $\phi(t) \neq 0$ outside $[R_1, R_2]$, $\phi(t) = 1$ for $t \leq R_0$, $\phi(t) = -1$ for $t \geq R_3$. Let $u(x) = \phi(|x|)$, $\beta(x) \equiv 0$, $c(x) = -\Delta u(x)/u(x)$ for $x \in \mathbb{R}^2$, so u, c are C^∞ and $c(x)$ has support in $\{R_0 \leq |x| \leq R_1\} \cup \{R_2 \leq |x| \leq R_3\}$. For any annular region $\Omega \subset \mathbb{R}^2$ with inner boundary in $\{|x| < R_0\}$ and outer boundary in $\{|x| > R_3\}$, zero is an eigenvalue with eigenfunction $u(x)$ which is locally constant (but not constant) near $\partial\Omega$, and also not of constant sign. For most such Ω, all eigenvalues (including 0) are simple. If J_0 has k roots in (R_1, R_2), then zero is at least the $(k+1)^{st}$ eigenvalues.

Each of the following conditions is open with respect to perturbation of Ω:

(i) $\beta(x) \neq 0$ somewhere on $\partial\Omega$;
(ii) $\beta(x) \equiv 0$ on a neighborhood of $\partial\Omega$, and $c \mid \partial\Omega$ is not constant;
(iii) $\beta(x) \equiv 0$ and $c(x) \equiv$ constant on a neighborhood of $\partial\Omega$.

[We assume Ω is connected, but $\partial\Omega$ may not be connected.] Further, given any (bounded, connected, C^2-regular) region $\Omega_0 \subset \mathbb{R}^n$, some arbitrarily small C^2-perturbation Ω of Ω_0 will satisfy at least one of these conditions. We may thus assume we are in one of these cases, and restrict ourselves to small perturbations so we always remain in that case.

In this, as in most genericity arguments, the hard part is to prove denseness. Assume for contradiction that there is a bounded connected C^2 region Ω_0, satisfying one of (i), (ii) or (iii) above, such that, for every Ω C^2-close to Ω_0, $(*)_\Omega$ has a multiple eigenvalue. We will show we must be in Case (iii), and then examine this case more closely.

Choose such Ω (close to Ω_0) and define

$$\begin{aligned} F(u, \lambda, h) = \big(&h^*(\Delta + \lambda + c)h^{*-1}u, \\ &h^*(N_{h(\Omega)} \cdot \nabla + \beta)h^{*-1}u\big) \in L_2(\Omega) \times H^{1/2}(\partial\Omega) \end{aligned}$$

for $u \neq 0$ in $H^2(\Omega)$, $\lambda \in \mathbb{R}$, and h near i_Ω in $C^2(\Omega, \mathbb{R}^n)$. If $(u, \lambda) \mapsto F(u, \lambda, h)$ has $(0, 0)$ as a regular value, all eigenvalues λ of $(*)_{h(\Omega)}$ would be simple – which is false when h is close to i_Ω. By the transversality theorem, $(0, 0)$ must be a critical value of F and in fact there is a solution (u, λ) of $(*)_\Omega$ with $u \neq 0$ so (u, λ, i_Ω) is a critical point of F. It follows – as in Example 6.4 – that there exists $\psi \neq 0$ in $H^2(\Omega)$ which also satisfies $(*)_\Omega$ for the same λ and further

$$(**)_\Omega \qquad \nabla_{\partial\Omega}\psi \cdot \nabla_{\partial\Omega}u = \psi u(\lambda + c + \beta^2 - \beta \operatorname{div} N_\Omega - \partial\beta/\partial N_\Omega) \text{ on } \partial\Omega$$

and $\int_\Omega \psi u = 0$. [We will not use this last condition; the argument below only says (λ, u) and (λ, ψ) satisfy $(*)_\Omega$ with $u \not\equiv 0$, $\psi \not\equiv 0$, and (λ, u, ψ) satisfies $(**)_\Omega$.]

If the coefficient of ψu in $(**)_\Omega$ is always positive on $\partial\Omega$ – for example, if λ is large – the argument of Example 6.4 applies to give a contradiction. But in general, $(**)_\Omega$ is not contradictory.

Example. Ω is the unit ball of $\mathbb{R}^n$, $n \geq 2$, $u(x) = x_1$, $\psi(x) = x_2$, $\lambda = 0$, $c \equiv 0$, and on $\partial\Omega$, $\beta = -1$ and $\partial\beta/\partial N = n + 1$; then $(**)_\Omega$ holds. [If $n = 1$, the eigenvalue are always simple.]

It is, however, unusual. Unless we are in Case (iii) – with $\beta \equiv 0$, $c(x) \equiv$ constant near $\partial\Omega$ – there is an open dense set of embeddings h in a C^2-neighborhood of i_Ω such that, if u, ψ are both non-trivial solutions of $(*)_{h(\Omega)}$ for some λ, then $(**)_{h(\Omega)}$ fails. Even in Case (iii), the condition $(**)_{h(\Omega)}$ will fail for most h near i_Ω unless $\lambda + c \equiv 0$ and $|\nabla u||\nabla\psi| \equiv 0$ on a neighborhood of $\partial\Omega$. (This shows that, in Cases (i) and (ii), all eigenvalues are generically simple; and in Case (iii), all are simple except perhaps $\lambda = -c|\partial\Omega$.)

To prove this, consider

$$G(u, \psi, \lambda, h) = F(u, \lambda, h), F(\psi, \lambda, h), h^* K_{h(\Omega)} h^{*-1}(u, \psi, \lambda)|_{\partial\Omega})$$

with F as above, and

$$\begin{aligned} K_\Omega(u, \psi, \lambda) &= \nabla u \cdot \nabla\psi - (\lambda + Q)u\psi, \\ Q &= c + 2\beta^2 - \beta \operatorname{div} N_\Omega - N_\Omega \cdot \nabla\beta, \end{aligned}$$

with values in (say) $L_1(\partial\Omega)$. Assume for every h near i_Ω, there exist u, ψ in $H^2(\Omega)\backslash\{0\}$ and $\lambda \in \mathbb{R}$ such that $G(u, \psi, \lambda, h) = (0, 0, 0, 0, 0) \in (L_2(\Omega) \times H^{1/2}(\partial\Omega))^2 \times L_1(\partial\Omega)$. By Theorem 5.4 [with $x = (u, \psi, \lambda)$, $y = h$, $\zeta = 0$], this can only happen if $2(\beta)$ fails, and in particular, for the solution of $G(u, \psi, \lambda, i_\Omega) = 0$, the quotient space $\mathcal{R}(DG(u, \psi, \lambda, i_\Omega))/\mathcal{R}(\partial G/\partial(u, \psi, \lambda))$ must have finite dimension – say, dimension n_0. We may choose $\theta_1, \ldots, \theta_{n_0}$ in $(L_2(\Omega) \times H^{1/2}(\partial(\Omega))^2 \times L_1(\partial\Omega)$ such that: given $\dot{h} \in C^2(\Omega)$ there exist $(\dot{u}, \dot{\psi}, \dot{\lambda}) \in H^2 \times H^2 \times \mathbb{R}$ and real numbers $c_1, \ldots, c_{n_0}$ so

$$DG(u, \psi, \lambda, i_\Omega)(\dot{u}, \dot{\psi}, \dot{\lambda}, \dot{h}) = \sum_1^{n_0} c_j\theta_j.$$

This is a collection of five equations. Consider the first pair.

$$(\Delta + \lambda + c)/(\dot{u} - \dot{h} \cdot \nabla u) + \dot{\lambda} u = \sum_1^{n_0} c_j\theta_j^1 \in \Omega,$$

$$\left(\frac{\partial}{\partial N} + \beta\right)(\dot{u} - \dot{h} \cdot \nabla u) + \dot{N} \cdot \nabla u + \dot{h} \cdot \nabla\left(\frac{\partial u}{\partial N} + \beta u\right) = \sum_1^{n_0} c_j\theta_j^2 \in \partial\Omega,$$

where $\theta_j = (\theta_j^1, \ldots, \theta_j^5)$. We can almost solve for $\dot{u} - \dot{h} \cdot \nabla u$, modulo finite dimensions. Specifically, let $\{\phi_1, \ldots, \phi_p\}$ be a basis for the eigenfunctions of

$(*)_\Omega$, with eigenvalue λ; there are continuous linear operators $\mathcal{C} : H^{1/2}(\partial\Omega) \to H^2(\Omega)$, $\mathcal{A} : L_2(\Omega) \to H^2(\Omega)$, such that $v \equiv \mathcal{A}f + \mathcal{C}g$ is the solution of

$$\begin{aligned}(\Delta + \lambda + c)v - f|_\Omega &\in \operatorname{span}\{\phi_1, \dots, \phi_p\}\\ \frac{\partial v}{\partial N} + \beta v &= g \text{ on } \partial\Omega\\ \int_\Omega v\phi_j &= 0 \quad \text{for } 1 \le j \le p.\end{aligned}$$

Then $\dot{u} - \dot{h} \cdot \nabla u = -\mathcal{C}(\dot{N} \cdot \nabla u + \dot{h} \cdot \nabla(\frac{\partial u}{\partial N} + \beta u) +$ (finite) where "(finite)" denotes some unspecified element of a specific finite-dimensional space – in this case, having dimension $\le n_0 + p + 1$, as we vary $\dot{\lambda}, c_1, \dots, c_{n_0}$ and $\int_\Omega \phi_j(\dot{u} - \dot{h} \cdot \nabla u)$ for $1 \le j \le p$.

Similarly from the second pair of equations, we have

$$\dot{\psi} - \dot{h} \cdot \nabla\psi = -\mathcal{C}\left(\dot{N} \cdot \nabla\psi + \dot{h} \cdot \nabla\left(\frac{\partial\psi}{\partial N} + \beta\psi\right)\right) + \text{(finite)}.$$

Substitution in the final equation

$$\begin{aligned}\sum_1^{n_0} c_j\theta_j^5 = \dot{h} \cdot \nabla K_\Omega(u, \psi, \lambda) &+ u\psi(\beta \operatorname{div} \dot{N}_\Omega + \dot{N}_\Omega \cdot \nabla\beta)\\ &+ (\nabla u \cdot \nabla - (\lambda + Q)u)(\dot{\psi} - \dot{h} \cdot \nabla\psi)\\ &+ (\nabla\psi \cdot \nabla - (\lambda + Q)\psi)(\dot{u} - \dot{h} \cdot \nabla u)\end{aligned}$$

shows the operator

$$\begin{aligned}\Gamma : \sigma \mapsto \sigma\frac{\partial}{\partial N}K_\Omega &+ u\psi(-\beta\Delta_{\partial\Omega}\sigma - \nabla_{\partial\Omega}\sigma \cdot \nabla\beta)\\ &+ (\nabla u \cdot \nabla - (\lambda + Q)u)\mathcal{C}\left(\nabla_{\partial\Omega}\sigma \cdot \nabla\psi - \sigma\frac{\partial}{\partial N}\left(\frac{\partial\psi}{\partial N} + \beta\psi\right)\right)\\ &+ (\nabla\psi \cdot \nabla - (\lambda + Q)\psi)\mathcal{C}\left(\nabla_{\partial\Omega}\sigma \cdot \nabla u - \sigma\frac{\partial}{\partial N}\left(\frac{\partial u}{\partial N} + \beta u\right)\right)\end{aligned}$$

has finite rank. Here $\sigma = \dot{h} \cdot N$, and we have assumed $\partial N_\Omega/\partial N_\Omega = 0$ on $\partial\Omega$, so $\dot{N}_\Omega = -\nabla_{\partial\Omega}\sigma$.

Case (i): $\beta(x) \neq 0$ somewhere on $\partial\Omega$.

The operator Γ is of order 2 and $\sigma \mapsto \Gamma\sigma + \beta u\psi\Delta_{\partial\Omega}\sigma$ is continuous from (say) $H^{3/2}(\partial\Omega)$ to $L_1(\partial\Omega)$. Since $\beta u\psi \not\equiv 0$ on $\partial\Omega$, by uniqueness in the Cauchy problem, it follows Γ cannot be of finite rank, so we cannot be in Case (i). In more detail, suppose $x_0 \in \partial\Omega$ and $\beta u\psi \neq 0$ at x_0 hence, on a neighborhood of x_0. Choose a smooth function σ_0 supported in a small neighborhood of x_0, $\sigma_0(x_0) \neq 0$, and a smooth $\theta : \partial\Omega \to \mathbb{R}$ with $\nabla_{\partial\Omega}\theta \neq 0$ on supp σ_0. For large real k, let $\sigma_k(x) = k^{-2}\sigma_0(x)\exp(ik\theta(x))$; then $\|\sigma_k\|_{H^{3/2}(\partial\Omega)} \to 0$ as $k \to \infty$ and $\Gamma\sigma_k(x) = \beta u\psi\sigma_0(x) \cdot |\nabla_{\partial\Omega}\theta(x)|^2 \exp(ik\theta(x)) + o(1)$ so Γ is not of finite rank.

Case (ii) or (iii): $\beta \equiv 0$ near $\partial\Omega$.

The condition $(**)_\Omega$ becomes $\nabla u \cdot \nabla\psi = (\lambda + c)u\psi$ on $\partial\Omega$, and $Q = c$ in the definition of K_Ω.

Let $S(x, y)$ be a fundamental solution for $\Delta + \lambda + c$. Then $\mathcal{C}g(x) = -2\int_{\partial\Omega} S(x, y)\tilde{g}(y)dAy$ where $\tilde{g}$ solves

$$g(x) = \tilde{g}(x) - 2\int_{\partial\Omega} \frac{\partial S}{\partial N_x}(x, y)\tilde{g}(y)dAy, \qquad x \in \partial\Omega.$$

(We disregard operator of finite rank.) If $E_n(x - y)$ is the fundamental solution of the Laplacian, it follows (see 7.6) that

$$\mathcal{C}_g(x) = -2\int_{\partial\Omega} E_n(x - y)g(y) + \int_{\partial\Omega} R(x, y)g(y), \qquad x \in \partial\Omega,$$

where the "remainder" R is a kernel of class $C^1_*(2)$ (in the notation of Chapter 7), defining an integral operator of order -2 on $\partial\Omega$, while $\mathcal{C}$ is of order -1. We use this representation of $\mathcal{C}$ to find the principal part of Γ, namely the operator order 1.

$$\sigma \mapsto \frac{2}{w_n} \text{P.V.} \int_{\partial\Omega} dA_y \frac{(y - x) \otimes \nabla_{\partial\Omega}\sigma(y)}{|y - x|^n} : M(x)$$

with $M(x) = \nabla_{\partial\Omega}u(x) \otimes \nabla_{\partial\Omega}\psi(x) + \nabla_{\partial\Omega}\psi(x) \otimes \nabla_{\partial\Omega}u(x)$, $w_n = |S^{n-1}|$ and we use the notation

$$(a \otimes b)_{ij} = a_i b_j, \qquad 1 \le i, j \le n$$
$$P : Q = \sum_{i,j} P_{ij} Q_{ij}.$$

The difference between Γ and the operator above is continuous (for example) from $H^{1/2}(\partial\Omega)$ to $L_1(\partial\Omega)$. Since Γ has finite rank, it is necessary that

$$(\tau_1 \otimes \tau_2) : M(x) = 0 \text{ for all } x \in \partial\Omega \text{ and } \tau_1, \tau_2 \in T_x(\partial\Omega)$$

or $\frac{\partial u}{\partial\tau}(x)\frac{\partial\psi}{\partial\tau}(x) = 0$ for $x \in \partial\Omega$, $\tau \perp N_\Omega(x)$ (see 7.4).

Suppose $\nabla_{\partial\Omega}u(x) \neq 0$ at some point $x \in \partial\Omega$; then for almost all tangent directions τ, $\frac{\partial u}{\partial\tau}(x) \neq 0$ so $\frac{\partial\psi}{\partial\tau}(x) = 0$, so $\nabla_{\partial\Omega}\psi(x) = 0$. Similarly on interchanging u, ψ; thus $|\nabla_{\partial\Omega}u||\nabla_{\partial\Omega}\psi| \equiv 0$ on $\partial\Omega$. By $(**)_\Omega$, $(\lambda + c)u\psi \equiv 0$ on $\partial\Omega$ so in fact $\lambda + c \equiv 0$ on $\partial\Omega$. Thus we cannot be in Case (ii), so we must be in Case (iii), with $\beta \equiv 0$, $c \equiv$ constant $= -\lambda$ near $\partial\Omega$, and also $|\nabla u||\nabla\psi| = 0$ on $\partial\Omega$. On a neighborhood of $\partial\Omega$, u and ψ are harmonic functions; on a neighborhood of a given component of $\partial\Omega$, at least one of u, ψ must be constant, by uniqueness in the Cauchy problem, and so $|\nabla u||\nabla\psi| = 0$ near (as well as on) $\partial\Omega$.

Now suppose we are in Case (iii), $\beta \equiv 0$ and $c(x) \equiv$ constant near $\partial\Omega$ – we may suppose $c(x) \equiv 0$ near $\partial\Omega$. We want to prove 0 is either not an eigenvalue, or is a simple eigenvalue, for most regions near Ω. Suppose 0 is a p-fold eigenvalue ($p \geq 1$) for Ω and $\{\phi_1, \ldots, \phi_p\}$ is an orthonormal basis for the eigenfunctions. Applying the Liapunov-Schmidt method, for h near i_Ω, 0 is an eigenvalue for $h(\Omega)$ only if there is a solution $q \in \mathbb{R}^p \backslash \{0\}$ of $M(h)q = 0$ (and then $h^{*-1}(\sum_{j=1}^p q_j(\phi_j + S_j(h)))$ is an eigenfunction) where

$S_j(h)$ is the solution of

$$0 = (\Delta + c)v + (h^*(\Delta + c)h^{*-1} - \Delta - c)(\phi_j + v) \sum_k (\text{const.})\, \phi_k$$

$$0 = \frac{\partial v}{\partial N} + \left(h^* \frac{\partial}{\partial N} h^* - \frac{\partial}{\partial N}\right)(\phi_j + v) \text{ on } \partial\Omega$$

$\int_\Omega v\phi_k = 0$ for $1 \leq k \leq p$,

and $M(h) = [M_{jk}(h)]_{j,k=1}^p$ is

$$\begin{aligned} M_{jk}(h) = &\int_\Omega \phi_k (h^*(\Delta + c)h^{*-1} - \Delta - c)(\phi_j + S_j(h)) \\ &- \int_{\partial\Omega} \phi_k \left(h^* \frac{\partial}{\partial N} h^{*-1} - \frac{\partial}{\partial N}\right)(\phi_j + S_j(h)). \end{aligned}$$

For any $V \in C^2(\Omega, \mathbb{R}^n)$,

$$\frac{d}{dt} M(i_\Omega + tV)|_{t=0} = \left[- \int_{\partial\Omega} V \cdot N \nabla\phi_j \cdot \nabla\phi_k \right]_{j,k=1}^p$$

We could reduce the multiplicity of the zero eigenvalue by small perturbations of Ω unless $\nabla\phi_j \cdot \nabla\phi_k = 0$ on $\partial\Omega$ for all j, k (If $p = 1$, "reduce multiplicity" means 0 is not an eigenvalue.) Thus if the zero eigenvalue is of minimal multiplicity $p \geq 1$, then in fact $p = 1$. For the eigenfunctions ϕ_j are locally-constant on $\partial\Omega$ (and also near $\partial\Omega$, being harmonic there), so they would be linearly dependent if we had $p > 1$.

This completes the proof of generic simplicity, and we now consider moving eigenvalues by perturbing Ω. In the case just discussed, the eigenvalue (and corresponding eigenfunction) are independent of small perturbations of Ω; but this is the only exception. In general, let λ be a simple eigenvalue with eigenfunction ϕ, $\int_\Omega \phi^2 = 1$. By Example 3.5, there is a unique eigenvalue $\lambda(t)$ near λ for the problem in $\Omega(t) = \{x + tV(x) \,|\, x \in \Omega\}$, t small, and

$$\frac{d}{dt}\lambda(t)|_{t=0} = \int_{\partial\Omega} V \cdot N_\Omega \{|\nabla_{\partial\Omega}\phi|^2 - (\lambda + c + \beta^2 - \beta \operatorname{div} N_\Omega - \partial\beta/\partial N_\Omega)\phi^2\}.$$

Thus the derivative is nonzero for some V unless $K_\Omega(\phi, \phi, \lambda) = 0$ on $\partial\Omega$, in the previous notation. But our argument above shows – for most choices of

Ω – there are no solutions (u, ψ, λ) of

$$\begin{aligned} \Delta u + (\lambda + c)u = 0 = \Delta\psi + (\lambda + c)\psi \text{ in } \Omega, &\qquad u \not\equiv 0 \\ \frac{\partial u}{\partial N} + \beta u = 0 = \frac{\partial \psi}{\partial N} + \beta\psi \text{ on } \partial\Omega, &\qquad \psi \not\equiv 0 \\ K_\Omega(u, \psi, \lambda) = 0 \text{ on } \partial\Omega & \end{aligned}$$

except in the special Case (iii) just discussed.

The argument above assumed $n \geq 2$, in the "finite-rank" discussion; but for $n = 1$ all eigenvalues are always simple. Suppose some eigenvalue λ of

$$u'' + (\lambda + c) \cdot u = 0 \quad \text{in}(a, b), \quad -u + \beta_1 u = 0 \quad \text{at } a, \quad u' + \beta_2 u = 0 \quad \text{at } b$$

cannot be moved by perturbing a, b_i then $\beta_1' + \beta_1^2 + \lambda + c = 0$ near a, $-\beta_2' + \beta_2^2 + \lambda c \equiv 0$ near b. We may change variables by $v = e^{-\varphi}\, u$ where $\varphi(x)$ is chosen so $\varphi^1 = \beta_1$ near a, $\varphi^1 = -\beta_2$ near b_i the problem becomes

$$v'' + 2\varphi'(x)v' + (\lambda + \tilde{c}(x))v = 0 \text{ in } (a, b)$$

$$v' = 0 \text{ at } x = a, x = b.$$

In this form, λ cannot be moved by perturbing a, b only if $\lambda + \tilde{c} \equiv 0$ near a, b, and the eigenfunction v is locally constant near $\partial(a, b)$. At least after changing variables, we get a result similar to the case $n \geq 2$.

Example 6.7. Generic Simplicity of Solutions of $\Delta u + f(u, x) = 0$ in Ω, $\partial u/\partial N = g(x, u)$ on $\partial\Omega$. We wish to prove $(0, 0)$ is a regular value of

$$F_\Omega : u \mapsto \left(\Delta u + f(\cdot, u), \frac{\partial u}{\partial N_\Omega} - g(\cdot, u)|_{\partial\Omega}\right)$$

for most choices of $\Omega \subset \mathbb{R}^n$ (in appropriate spaces), which will follow – under mild smoothness hypotheses – if $(0, 0)$ is a regular value of $(h, u) \mapsto h^* F_{h(\Omega)} h^{*-1}(u)$. Under some plausible "special hypotheses" for f, g, described below, we show that – restricting h to an open dense subset – $(0, 0)$ will be a regular value of $(h, u) \mapsto h^* F_{h(\Omega)} h^{*-1}(u)$, so the solutions will all be simple for most choices of Ω.

As an example (suggested by population genetics theory) suppose $r : \mathbb{R} \to \mathbb{R}$ is C^2 and has zero as a regular value ($r(c) = 0 \Rightarrow r'(c) \neq 0$) and $S : \mathbb{R}^n \to \mathbb{R}$ is C^2 such that and $s : \mathbb{R}^n \to \mathbb{R}$ is ξ^2 with $S(x) \neq 0$ on a dense set. Then for most bounded connected C^2 regions $\Omega \subset \mathbb{R}^n$, all solutions u of

$$\Delta u + r(u)S(x) = 0 \text{ in } \Omega, \qquad \frac{\partial u}{\partial N} = 0 \text{ on } \partial\Omega$$

are simple. Note that, if $c \in r^{-1}(0)$, $u(x) \equiv c$ is a solution for any choice of Ω; it is simple for most choices of Ω. Another case of interest is when

f, g are independent of x, we require f to be C^2, g to be C^3 and $f(c) = g(c) = 0 \Rightarrow |f'(c)| + |g'(c)| \neq 0$. Then the solutions of $\Delta u + f(u) = 0$ in Ω, $\partial u/\partial N = g(u)$ on $\partial\Omega$, are all simple for most choices of Ω. We are assuming $n \geq 2$; some analogous results for $n = 1$ are given in Theorem 11 of [D. Henry, Some infinite-dimensional Morse-Smale systems, *J. Diff. Eq.* **58** (1985)] but the case $n = 1$ is still essentially unsolved. [Indeed the case $n \geq 2$ is not completely satisfactory.]

If we impose the boundary condition $\partial u/\partial N = g(x, u)$ on certain components of $\partial\Omega$ while $u = 0$ on other components (at least one!) a mild variant of the argument in Example 6.5 – restricting our perturbations to the "Dirichlet" components of $\partial\Omega$ shows the solutions are generically simple.

We will work in an anti-logical fashion, starting only with smoothness assumptions (f is C^2, g is C^3) and discovering appropriate additional hypotheses in the course of the calculation. This is the way the assumptions for this and other examples were found. But if you prefer a logical order, you may turn directly to the statement of the result at the end of this section. we always suppose $n \geq 2$ and $\partial\Omega$ is connected, bounded and (in the initial calculation) C^3.

Suppose for contradiction that there exists a solution u of $F_\Omega(u) = 0$ such that (u, i_Ω) is a critical point of $(v, h) \mapsto h^* F_{h(\Omega)} h^{*-1}(v)$, and further, that this holds for every Ω in some (C^3) neighborhood of a given bounded, C^3, connected region Ω_0. It follows (as in Examples 6.4, 6.6) that there exists $\psi \neq 0$ in the kernel of the linearization with $K_\Omega(u, \psi) = 0$ on $\partial\Omega$,

$$K_\Omega(u, \psi) = \nabla_{\partial\Omega}\psi \cdot \nabla_{\partial\Omega} u - \psi(f(\cdot, u) + g(\cdot, u)\,\mathrm{div}\, N_\Omega + N_\Omega \cdot g_x(\cdot, u) + gg')$$

where $g' = \partial g/\partial u$, $g_x = \partial g/\partial x$, evaluated at $(x, u(x))$, $x \in \partial\Omega$. Thus for every small perturbation Ω of Ω_0, there is a solution (u, ψ) of the over-determined problem

$$\begin{cases} \Delta u + f(\cdot, u) = 0 \text{ in } \Omega, & \dfrac{\partial u}{\partial N} = g(\cdot, u) \text{ on } \partial\Omega \\ \Delta\psi + f'(\cdot, u)\psi = 0 \text{ in } \Omega, & \dfrac{\partial \psi}{\partial N} = g'(\cdot, u)\psi \text{ on } \partial\Omega, \quad \psi \not\equiv 0, \\ K_\Omega(u, \psi) = 0 \text{ on } \partial\Omega. & \end{cases}$$

The first four equations (with $\psi \not\equiv 0$) define a Fredholm map for (u, ψ) of index zero; including the fifth, we have a semi-Fredholm map of index $-\infty$. Let $(u, \psi) \mapsto G_\Omega(u, \psi)$ denote the resulting map from $W^{2,p}(\Omega) \times (W^{2,p}(\Omega)\backslash\{0\})$ to $(L_p(\Omega) \times W^{1/p,p}(\partial\Omega))^2 \times L_1(\partial\Omega)$, for some $n < p < \infty$ (so $W^{2,p} \subset C^1$). By hypothesis, $h^* G_{h(\Omega)} h^{*-1}(u, \psi) = (0, 0, 0, 0, 0)$ has a solution (u, ψ) for every h in a C^3-neighborhood of i_Ω, so assumption 2(β) of Theorem 5.4 must fail at (u, ψ, i_Ω). In particular, the quotient space has finite dimension (say, dimension n_0) and for some $\theta_1, \ldots, \theta_{n_0}$ the following holds: given any

$\dot{h} \in C^3(\Omega, \mathbb{R}^n)$ there exist $\dot{u}$, $\dot{\psi}$ in $W^{2,p}(\Omega)$ and real numbers $c_1, \ldots, c_{n_0}$ so

$$DG_\Omega(u,\psi)(\dot{u} - \dot{h}\cdot\nabla u, \dot{\psi} - \dot{h}\cdot\nabla\psi) + \dot{h}\cdot\nabla G_\Omega(u,\psi) + \frac{\partial G_\Omega}{\partial N}(u,\psi)\dot{N}_\Omega = \sum_1^{n_0} c_j\theta_j.$$

Let $\theta_j = (\theta_j^1, \ldots, \theta_j^5)$, and the first two components of this equation are:

$$L_u(\dot{u} - \dot{h}\cdot\nabla u) = \sum_1^{n_0} c_j\theta_j^1 \text{ in } \Omega,$$

$$M_u(\dot{u} - \dot{h}\cdot\nabla u) = \sum_1^{n_0} c_j\theta_j^2 - \dot{N}\cdot\nabla u - \dot{h}\cdot N\frac{\partial}{\partial N}\left(\frac{\partial u}{\partial N} - g(\cdot,u)\right) \text{ on } \partial\Omega$$

where $L_u = \Delta + f'(\cdot,u)$, $M_u = \partial/\partial N_\Omega - g'(\cdot,u)$. Then

$$\dot{u} - \dot{h}\cdot\nabla u = -\mathcal{C}_{L,M}\left(\dot{N}\cdot\nabla u + \dot{h}\cdot N\frac{\partial}{\partial N}\left(\frac{\partial u}{\partial N} - g(\cdot,u)\right)\right) + (\text{finite}),$$

where $\mathcal{C}_{L,M}$ is the appropriate solution operator (studied in 7.6). Similarly the second pair of equations yields

$$\begin{aligned}\dot{\psi} - \dot{h}\cdot\nabla\psi &= -\mathcal{C}_{L,M}\left(\dot{N}\cdot\nabla\psi + \dot{h}\cdot N\frac{\partial}{\partial N}(M_u\psi) - g''(\cdot,u)\psi(\dot{u} - \dot{h}\cdot\nabla u)\right)\\ &= -\mathcal{A}_{L,M}(f''(\cdot,u)\psi(\dot{u} - \dot{h})\cdot\nabla u)) + (\text{finite}),\end{aligned}$$

and we substitute the earlier expression for $\dot{u} - \dot{h}\cdot\nabla u$ on the right-hand side. Then we substitute these in the fifth component equation, to conclude: $\sigma \mapsto \Gamma\sigma$ is an operator of finite rank, where $\sigma = \dot{h}\cdot N_\Omega$, $\dot{N}_\Omega = -\nabla_{\partial\Omega}\sigma$ (i.e., we suppose $\partial N_\Omega/\partial N_\Omega = 0$) and

$$\begin{aligned}\Gamma\sigma = {}& \psi g\Delta_{\partial\Omega}\sigma + \psi\nabla_{\partial\Omega}\sigma\cdot g_x + \sigma\frac{\partial}{\partial N}K_\Omega(u,\psi)\\ &+ (\nabla u\cdot\nabla - Q)\mathcal{C}_{L,M}\left(\nabla_{\partial\Omega}\sigma\cdot\nabla\psi - \sigma\frac{\partial}{\partial N}(M_u\psi)\right)\\ &+ \begin{pmatrix}\nabla\psi\cdot\nabla - \psi Q' + (\nabla u\cdot\nabla - Q)\mathcal{C}_{L,M}g''\psi\\ -(\nabla u\cdot\nabla - Q)\mathcal{A}_{L,M}f''\psi\end{pmatrix}\mathcal{C}_{L,M}\Bigg((\nabla_{\partial\Omega}\sigma\cdot\nabla u\\ &- \sigma\frac{\partial}{\partial N}\left(\frac{\partial u}{\partial N} - g\right)\Bigg)\Bigg)\end{aligned}$$

$$\begin{aligned}K_\Omega(u,\psi) &= \nabla u\cdot\nabla\psi - \psi Q(u) \qquad (= 0 \text{ on } \partial\Omega)\\ Q(u) &= f(\cdot,u) + g(\cdot,u)\operatorname{div}N + 2g'g + N\cdot g_x,\\ Q'(u) &= \frac{\partial}{\partial u}Q(u).\end{aligned}$$

The principal part of Γ is $\sigma \mapsto \psi g\Delta_{\partial\Omega}\sigma$, of second order, while the remainder is first order ($\mathcal{C}_{L,M}$ is order -1, $\mathcal{A}_{L,M}$ is order -2). Thus Γ will not be of finite rank unless $\psi g(\cdot, u) \equiv 0$ on $\partial\Omega$. Since $\psi \not\equiv 0$, this says $g(\cdot, u) \equiv 0$ on $\partial\Omega$.

Thus u satisfies the over-determined problem

$$\Delta u + f(\cdot, u) = 0 \text{ in } \Omega, \qquad \frac{\partial u}{\partial N} = 0 \text{ and } g(\cdot, u) = 0 \text{ on } \partial\Omega.$$

We study such over-determined problems in the lemma below. By hypothesis, a solution u exists for every Ω near Ω_0, hence by the lemma, there is a solution u which also satisfies $g_x(\cdot, u) = 0$ and $g'(\cdot, u)\nabla_{\partial\Omega} u = 0$ on $\partial\Omega$. But this information may be useless – for example, if $g \equiv 0$ (the Neumann problem) – so we continue.

Knowing that $g(\cdot, u) = 0$, $g_x(\cdot, u) = 0$ and $g'(\cdot, u)\nabla_{\partial\Omega} u = 0$ on $\partial\Omega$ we see $K_\Omega(u, \psi) = \nabla u \cdot \nabla\psi - \psi f(\cdot, u) = 0$ on $\partial\Omega$ and the principal part of Γ is of order 1:

$$\begin{aligned}\sigma \mapsto\ & \nabla u \cdot \nabla C_{L,M}(\nabla_{\partial\Omega}\sigma \cdot \nabla\psi) + \nabla\psi \cdot \nabla C_{L,M}(\nabla_{\partial\Omega}\sigma \cdot \nabla u)\\ =\ & \nabla_{\partial\Omega} u \cdot \nabla_{\partial\Omega} C_{L,M}(\nabla_{\partial\Omega}\sigma \cdot \nabla_{\partial\Omega}\psi) + \nabla_{\partial\Omega}\psi \cdot \nabla_{\partial\Omega} C_{L,M}(\nabla_{\partial\Omega}\sigma \cdot \nabla_{\partial\Omega} u)\end{aligned}$$

or, more explicitly,

$$\Gamma\sigma(x) = \frac{2}{w_n}\,\text{P.V.} \int_{\partial\Omega} \frac{(y - x) \otimes \nabla_{\partial\Omega}\sigma(y)}{|x - y|^n} : M(x) + (\text{order zero})$$

with $M(x) = \nabla_{\partial\Omega} u \otimes \nabla_{\partial\Omega}\psi + \nabla_{\partial\Omega}\psi \otimes \nabla_{\partial\Omega} u$ at $x \in \partial\Omega$. For this to be of finite rank, it is necessary that $\frac{\partial u(x)}{\partial\tau}\frac{\partial\psi(x)}{\partial\tau} = 0$ for every $x \in \partial\Omega$, $\tau \perp N_\Omega(x)$. This implies $|\nabla_{\partial\Omega} u||\nabla_{\partial\Omega}\psi| = 0$ on $\partial\Omega$, and so (from $K_\Omega(u, \psi) = 0$, $\psi \not\equiv 0$) it follows $f(\cdot, u) \equiv 0$ on $\partial\Omega$.

Applying the lemma to f, we have also that $f_x(\cdot, u) = 0$ and

$$f'(\cdot, u)\nabla_{\partial\Omega} u = 0 \text{ on } \partial\Omega.$$

Applying the lemma again to f_x, g_x, and then to g_{xx} (since g is C^3) we conclude:

For every Ω C^3-close to Ω_0, there is a solution (u, ψ) of

$$\begin{aligned}&\Delta u + f(\cdot, u) = 0 \text{ in } \Omega, \quad && \Delta\psi + f'(\cdot, u)\psi = 0 \text{ in } \Omega, \quad && \psi \not\equiv 0,\\ &\frac{\partial u}{\partial N} = 0 \text{ on } \partial\Omega, && \frac{\partial\psi}{\partial N} = g'(\cdot, u)\psi \text{ on } \partial\Omega,\end{aligned}$$

and on $\partial\Omega$,

$$\begin{gathered} g(\cdot, u) = 0, \qquad g_x(\cdot, u) = 0, \qquad g_{xx}(\cdot, u) = 0, \qquad g_{xxx}(\cdot, u) = 0,\\ f(\cdot, u) = 0, \qquad f_x(\cdot, u) = 0, \qquad f_{xx}(\cdot, u) = 0,\\ g'(\cdot, u)\nabla_{\partial\Omega} u = 0, \qquad g'_x(\cdot, u)\nabla_{\partial\Omega} u = 0, \qquad g'_{xx}(\cdot, u)\nabla_{\partial\Omega} u = 0,\\ f'(\cdot, u)\nabla_{\partial\Omega} u = 0, \qquad f'_x(\cdot, u)\nabla_{\partial\Omega} u = 0, \qquad \text{and} \qquad |\nabla_{\partial\Omega} u||\nabla_{\partial\Omega}\psi| = 0.\end{gathered}$$

With all these conditions, we should be able to find a contradiction under mild additional assumptions on f, g.

But first, we prove the lemma – in a more general case then we need here.

Lemma. *Suppose* Λ *is an open set in* $\mathbb{R}^q$, V *is open in* $\mathbb{R}^n$ *which meets* $\partial\Omega_0$, W_0 *is open in* $W^{2,p}(\Omega_0)$, $f(x,u,p,\lambda)$ *and* $g(x,u,\lambda)$ *are* C^2 *and* $\phi(x,u,\lambda)$ *is* C^1, *and for every* $\Omega = h(\Omega_0)$ *in a* C^3 *neighborhood of* Ω_0 *there is a solution* (u,λ) *of*

$$\Delta u + f(x,u,\nabla u,\lambda) = 0 \text{ in } \Omega, \qquad \frac{\partial u}{\partial N} = g(x,u,\lambda) \text{ on } \partial\Omega,$$

with $h^*u \in W_0$, $\lambda \in \Lambda$, *such that*

$$\phi(x,u,\lambda) = 0 \text{ on } \partial\Omega \cap V.$$

Then there is such a solution (u,λ) *which also satisfies*

$$\phi'(x,u,\lambda)\nabla_{\partial\Omega}u = 0 \quad \text{and} \quad \phi_x(x,u,\lambda) + N_\Omega(x)\phi'(x,u,\lambda)g(x,u,\lambda) = 0$$

on $\partial\Omega \cap V$.

Proof. Condition 2(β) of Theorem 5.4 must fail so we conclude

$$\sigma \mapsto \sigma\frac{\partial}{\partial N}\phi(\cdot,u,\lambda) + \phi'(\cdot,u,\lambda)\, C_{L,M}\left(\nabla_{\partial\Omega}\sigma\cdot\nabla u - \sigma\frac{\partial}{\partial N}\left(\frac{\partial u}{\partial N} - g\right)\right)\bigg|_{\partial\Omega\cap V}$$

is of finite rank. The principal part is of order zero, namely (after integration by parts).

$$\begin{aligned}\sigma \mapsto\ & \sigma(x)(N(x)\cdot\phi_x + \phi' g)\\ & - 2\phi'(x,u(x),\lambda)\,\text{P.V.}\int_{\partial\Omega}\frac{(y-x)\cdot\nabla_{\partial\Omega}u(x)}{w_n|x-y|^n}\sigma(y)\bigg|_{x\in\partial\Omega\cap V},\end{aligned}$$

with the remainder an operator of order -1. Thus on $\partial\Omega\cap V$, $N\cdot\phi_x + \phi' g = 0$ and $\phi'\nabla_{\partial\Omega}u = 0$. Since $\phi(x,u(x),\lambda) \equiv 0$ on $\partial\Omega\cap V$, differentiation in a tangential direction τ gives $\tau\cdot\phi_x(\cdot,u(\cdot),\lambda) = 0$ which (with the equation $N\cdot\phi_x = -\phi' g$) gives the result claimed.

Now we will try to formulate plausible sufficient conditions on f, g to ensure generic simplicity of solutions.

Special Hypotheses on *f*, *g*

$f(x,u)$ is locally C^2, $g(x,u)$ is locally C^3, on $\mathbb{R}^n\times\mathbb{R}$.

(1) There is a discrete set $\{c_j\}\subset\mathbb{R}$, possible empty, such that $f(x,c_j)\equiv 0$, $g(x,c_j)\equiv 0$ for all x, but $|f'(x,c_j)| + |g'(x,c_j)| \neq 0$ on a dense set of $\mathbb{R}^n$.

(2) For any $c\in\mathbb{R}\backslash\{c_j\}$, the set $\{x\in\mathbb{R}^n \mid \text{at } (x,c) \text{ we have } g=0,\ g_x=0,\ g_{xx}=0,\ g_{xxx}=0,\ f=0,\ f_x=0, \text{ and } f_{xx}=0\}$ does not contain a C^2

hypersurface; e.g., has dimension $< n - 1$. [If f, g are independent of x, this says $|f(c)| + |g(c)| \neq 0$ when $c \in \{c_j\}$.]

(3) $\{(x, u) \in \mathbb{R}^n \times \mathbb{R} \mid u \notin \{c_j\}$ and at (x, u) we have $g = 0$, $g_x = 0$, $g_{xx} = 0$, $g_{xxx} = 0$, $g' = 0$, $g'_x = 0$, $g'_{xx} = 0$, $f = 0$, $f_x = 0$, $f_{xx} = 0$, $f' = 0$, *and* $f'_x = 0$, does not contain a subject which is a C^1-graph over a C^2 hypersurface in $\mathbb{R}^n$, e.g., has dimension $< n - 1$.

Remark. These conditions are far from optimal, but seem adequate for most applications. (In the genetics problem, they do not allow the extreme cases "dominant" or "recessive;" but these always have non-simple solutions. Hypothesis (1) says there may be trivial constant solutions $u(x) \equiv c_j$, but these will be simple for most choices of Ω by Example 6.6. If we omit hypothesis (1), the conclusion is that all solutions other than the constant solutions $\{u \equiv c_j\}$ will be simple, in most regions Ω.

Theorem. *If f, g satisfy the "special hypotheses" above, then for most bounded C^2 regions $\Omega \subset \mathbb{R}^n$, all solutions u of*

$$\Delta u + f(x, u) = 0 \text{ in } \Omega, \qquad \frac{\partial u}{\partial N} = g(x, u) \text{ on } \partial\Omega,$$

are simple.

Proof. It is enough to treat the case where Ω is connected and f, g are uniformly bounded; then instead of "most regions" we can say "open, dense set in $\mathrm{Diff}^2(\Omega)$." Let $F_\Omega(u) = (\Delta u + f(\cdot, u), \partial u/\partial N_\Omega - g(\cdot, u)|_{\partial\Omega})$, $F_\Omega : W^{2,p}(\Omega) \to L_p(\Omega) \times W^{1/p,p}(\partial\Omega)$. Then F_Ω *is* C^2, Fredholm with index zero, and $(h, u) \mapsto h^* F_{h(\Omega)} h^{*-1}(u)$ is C^1 on $\mathrm{Diff}^2(\Omega) \times W^{2,p}(\Omega)$. Suppose $h_\nu \to h$ *in* $\mathrm{Diff}^2(\Omega)$ and there exists $u_\nu \in W^{2,p}(\Omega)$ so $h_\nu^* F_{h_\nu(\Omega)} h_\nu^{*-1}(u_\nu) = (0, 0)$ and (u_ν, h_ν) is a critical point. Then the u_ν are bounded in $W^{2,p}(\Omega)$ so we may assume they converge in C^1 to some u, and it follows that $u \in W^{2,p}(\Omega)$, $h^* F_{h(\Omega)} h^{*-1}(u) = (0, 0)$ and (u, h) is a critical point. Thus the set h such that there is a non-simple solution in $h(\Omega)$, is closed and its complement (all solutions simple) is open in $\mathrm{Diff}^2(\Omega)$.

We prove denseness by contradiction. Suppose for every Ω in some C^2 neighborhood of a given Ω_0 that there is a non-simple solution. This implies the weaker claim with C^3 in place of C^2 (supposing, as we may, Ω_0 is C^3). Then we have the hypothesis of the previous calculation: there exists u so $F_\Omega(u) = (0, 0)$ and (u, i_Ω) is a critical point of $(u, h) \mapsto h^* F_{h(\Omega)} h^{*-1} u$ [$h \in \mathrm{Diff}^3(\Omega)$], and this holds for every region Ω C^3-close to Ω_0. Then there exist (u, ψ) with $F_\Omega(u) = (0, 0)$, $L_u \psi = 0$ in Ω, $M_u \psi = 0$ on $\partial\Omega$, $\psi \not\equiv 0$, and all the conditions listed above (before the "special hypothesis") must hold. And again, all this is still true for any regions near Ω; so we may assume, for a start, that all the constant solutions $\{c_j\}$ are simple. Suppose $u|\partial\Omega$ is constant on a neighborhood of some $x_0 \in \partial\Omega$, say $u(x) \equiv c$. If $c \notin \{c_j\}$, we contradict hypothesis (2); if $c = c_j$ for some j, then by uniqueness in the Cauchy problem, $u(x) \equiv c_j$ in Ω, which is false since u is not a simple solution. Thus, $|\nabla_{\partial\Omega} u| \neq 0$ on a dense open set of $\partial\Omega$, in particular on some $\partial\Omega \cap B$ with B an open ball. Then $\{(x, u(x)) \mid x \in \partial\Omega \cap B\}$ is a set of the kind prohibited by (3). In every case, we have a contradiction, so the proof is complete.

Example 6.8. Generic Simplicity of Complex Eigenvalues of a Dirichlet Problem. Let $b_j(x, \lambda), c(x, \lambda)$ $(j = 1, \dots, n)$ be polynomials in $\lambda \in \mathbb{C}$ with coefficients C^2 functions of $x \in \mathbb{R}^n$ possibly complex valued. [c need only be C^1.] Define

$$A(\lambda) = \Delta + \sum_{j=1}^{n} b_j(x, \lambda)\frac{\partial}{\partial x_j} + C(x, \lambda) = \Delta + b \cdot \nabla + c$$

and consider the eigenvalue problem

$$A(\lambda)u = 0 \text{ in } \Omega, \qquad u = 0 \text{ on } \partial\Omega, \qquad u \not\equiv 0$$

(with $u \in H^2 \cap H_0^1(\Omega, \mathbb{C})$). An *eigenvalue* is any $\lambda \in \mathbb{C}$ for which a solution $u \not\equiv 0$ exists; and the eigenvalue λ is *simple* if the kernel of the operator $A(\lambda) \mid H^2 \cap H_0^1(\Omega, \mathbb{C})$ is one-dimensional ($= \mathbb{C} \cdot u$) and if $(\frac{d}{d\lambda}A(\lambda))u$ is not contained in the range of the operator. This implies λ and u vary smoothly with perturbations of the coefficients or the domain Ω, λ remaining simple, and reduces to the usual definition in the usual case: $b = b(x)$, $c = c(x) + \lambda$. λ is a simple eigenvalue of $A(\lambda)$ if and only if it is a simple eigenvalue of $A^*(\lambda)$ or of $e^{-\phi}Ae^{\phi}$, for any smooth scalar $\phi(x, \lambda)$.

We will assume $n \geq 2$, but the one-dimensional case should be mentioned. Let $U(\cdot; \alpha, \lambda)$ be the solution of $U_{xx} + bU_x + cU = 0$ such that $U = 0$, $U_x = 1$ for $x = \alpha$. Then, for the interval (α, β), λ is an eigenvalue when $U(\beta; \alpha, \lambda) = 0$ and it is simple when, in addition, $\frac{\partial}{\partial\lambda}U(\beta; \alpha, \alpha) \neq 0$. This notion of "simplicity" is equivalent to that above when $n = 1$.

An important function associated with $L = \Delta + b \cdot \nabla + C$ is

$$D(L) = C - \frac{1}{2}\operatorname{div} b - \frac{1}{4}b^2 = C - \frac{1}{2}\sum_{1}^{n}\frac{\partial b_j}{\partial x_j} - \frac{1}{4}\sum_{1}^{n} b_j^2.$$

(I call this the "discriminant" of L, but perhaps another name has priority; surely it has been noticed before.) In case $n = 1$, if b, c are constants, the corresponding quadratic (in the form suggested by Fourier transformation) is $(iz)^2 + b(iz) + c$, with roots $z = ib/2 \pm \sqrt{D(L)}$. In general, $D(L^*) = D(L)$ where $L^* = \Delta - b \cdot \nabla + c - \operatorname{div} b$. (For the complex sesquilinear inner product, we should take conjugates: then $D(\overline{L^*}) = \overline{D(L)}$.) For any smooth scalar function $\phi(x,\lambda)$, $D(e^{\phi}Le^{-\phi}) = D(L)$

$$e^{\phi}Lc^{-\phi} = \Delta + (b - 2\nabla\phi)\cdot\nabla + c - b\cdot\nabla\phi + (\nabla\phi)^2 - \Delta\phi.$$

$D(L)$ also appears in the fundamental solution for L: in the expansion of the singularity (see Theorem 7.5.1) when 'c' first appears, it is in the combination $D(L)$. Similarly, $D(L)$ appear in $\frac{\partial}{\partial N}\mathcal{B}_L(g)$ – see Theorem 7.6.4 (In these cases, the b_j, c may even be matrix-valued!)

For the Dirichlet problem in Ω – assuming the b_j, c and Ω are smooth – L is self-adjoint in $L_2(\Omega,\mathbb{C})$ if and only if $\operatorname{Re} b \equiv 0$ and $D(L)$ is real-valued in Ω. Here we use the customary inner product of L_2, but if we include a weight (measure $e^{2\phi}dx$ instead of dx, for some smooth real-valued ϕ) the condition is: $\operatorname{Re} b \equiv 2\nabla\phi$ and $D(L)$ is real valued. We could change variables ($u \mapsto ue^{\phi}$) to get an operator $e^{\phi}Le^{-\phi}$ which is self-adjoint in the usual sense.

If $D(L)$ is real and $\operatorname{Im} b = 2\nabla\phi$ is a gradient, we could change variables ($u \mapsto ue^{i\phi}$) to get an operator $e^{i\phi}Le^{-i\phi}$ with real coefficients.

We treated the real case in Example 6.3. In Example 4.3, we considered a self-adjoint case with complex coefficients but with simple dependence on the (real) eigenvalue parameter. The general second-order self-adjoint case – b purely imaginary and $c - \frac{1}{2}\operatorname{div} b$ real, with b, c polynomials in λ – remains open. (I managed to treat the case: b independent of λ, $\partial c/\partial\lambda > 0$ on a dense set for each real λ.)

The non-real and non-self-adjoint case is: $D(L)$ is not real, or it is real but neither $\operatorname{Re}b$ nor $\operatorname{Im}b$ is a gradient. Here we only use the first condition. $\operatorname{Im} D(L) \neq 0$, but will give a more general result in 8.4. We will assume: for certain given open sets $\Lambda \subset \mathbb{C}$, $V \subset \mathbb{R}^n$, when $\lambda \in \Lambda$

(*) $\operatorname{Im} D(A(\lambda)) = \operatorname{Im}(c - \frac{1}{2}\operatorname{div} b - \frac{1}{4}b^2) \neq 0$ on a dense set of $x \in V$

We will conclude, for most bounded C^2 open $\Omega \subset \mathbb{R}^n$ such that $\partial\Omega$ meets V, all eigenvalues in Λ are simple and "freely moveable," i.e., they can be moved in any complex direction by perturbation of $\partial\Omega$. Thus, if the imaginary axis is in Λ, we may also say, generically, all eigenvalues have real part different from zero. (If the b_j, c are real-valued when λ is real, any real simple eigenvalue is moveable by perturbation of $\partial\Omega$ but it is not *freely* moveable, since it must remain on the real line. In this case, the real line cannot meet Λ.)

Here we assume $n \geq 2$. The case $n = 1$ always needs special treatment since the space of domains (intervals) is finite dimensional. For $n = 1$, the coefficient of du/dx is always a gradient, and may be removed; and the kernel is always one dimensional, for any eigenvalue; but whether eigenvalues are generically simple, for most intervals, is not known. (Except in the self-adjoint case, which coincides with the real case.)

Example. Suppose m, k_j, r, b_j, c are smooth real-valued functions of $x \in \mathbb{R}^n$ and we seek solutions of the form $u(x,t) = v(x)e^{\lambda t} \not\equiv 0$ for

$$mu_{tt} + k \cdot \nabla u_t + ru_t = \Delta u + b \cdot \nabla u + cu \text{ in } \partial\Omega$$

with $u = 0$ on $\partial\Omega$, i.e.,

$$A(\lambda)v \equiv \Delta v + (b - \lambda k) \cdot \nabla v + (c - m\lambda^2 - r\lambda)v = 0 \text{ in } \Omega,$$

with $v = 0$ on $\partial\Omega$, $v \not\equiv 0$. Then

$$\operatorname{Im} D(A(\lambda)) = -\operatorname{Im} \lambda\{M \operatorname{Re} \lambda + R\}$$

with $M = 2m + |k|^2$, $R = r - \frac{1}{2} \operatorname{div} k - \frac{1}{2} b \cdot k$. Consider three cases, for some open $V \subset \mathbb{R}^n$ (perhaps the whole space):

(i) $M \equiv 0$, $R \neq 0$ on a dense set of V, with $\Lambda = \mathbb{C} \backslash \mathbb{R}$;
(ii) $M \neq 0$ on a dense set and R/M is not cosntant on any open set in V, with $\Lambda = \mathbb{C} \backslash \mathbb{R}$;
(iii) $M \neq 0$ on a dense set and $R/M = \text{constant} = \mu_0$ in V, with $\Lambda = \mathbb{C} \backslash \{\mathbb{R} \cup (-\mu_0 + i\mathbb{R})\}$.

In each case, hypothesis $(*)$ holds so, when $\partial\Omega$ meets V, eigenvalues in Λ are (generically) simple and freely moveable. When λ is real, we can apply 6,3: generically in Ω, the real eigenvalues are simple and moveable, though not freely moveable. In case (iii), supposing $\bar{\Omega} \subset V$ and $k \equiv 0$, for eigenvalues of the form $\lambda = -\mu_0 + i\sigma$ (σ real), the equation becomes

$$\Delta v + b \cdot \nabla v + (c + m\mu_0^2 + m\sigma^2)v = 0$$

with real coefficients, so these eigenvalues are also generically simple and moveable. (If $k \not\equiv 0$, or k is not a gradient, these eigenvalues are still not understood.)

If zero is a regular value of

$$(u, \lambda) \to A_\Omega(\lambda)u : (H^2 \cap H_0^1(\Omega, \mathbb{C}) \backslash \{0\}) \times \Lambda \to L_2(\Omega, \mathbb{C}),$$

(for a given open $\Lambda \subset \mathbb{C}$) then all eigenvalues in Λ are simple. This holds for most choices of $\Omega = h(\Omega_0)$ provided zero is a regular value of

$$(u, \lambda, h) \to h^* A_{h(\Omega)}(\lambda) h^{*-1} u.$$

If Λ_0 is any compact subset of Λ, the set of Ω (near a certain Ω_0) where this last condition fails (with $\lambda \in \Lambda_0$) is closed, consisting of Ω such that

$$(**) \qquad \begin{cases} A(\lambda)u = 0 \text{ in } \Omega, & u = 0 \text{ on } \partial\Omega, \quad u \not\equiv 0 \\ A^*(\lambda)\psi = 0 \text{ in } \Omega, & \psi = 0 \text{ on } \partial\Omega, \quad \psi \not\equiv 0 \end{cases}$$

and $\mathrm{Re}(\partial\psi/\partial N \partial u/\partial N) = 0$ on $\partial\Omega$ has a solution (u, ψ, λ) with $\lambda \in \Lambda_0$. We show it has no interior.

Suppose, for contradiction, it has interior: we may suppose Ω_0 is in the interior and $\partial\Omega_0$ is C^4, and $(**)$ is solvable with $\lambda \in \Lambda$ for every Ω in some C^4-neighborhood of Ω_0. (Note that u, ψ are then in $C^{3,\alpha}(\Omega)$ for any $\alpha < 1$.) We will impose the final condition only in $\partial\Omega \cap V$, for some given open $V \subset \mathbb{R}^n$ – which may be only a subset of the V in hypothesis $(*)$ – and we suppose V meets $\partial\Omega_0$, so it also meets any $\partial\Omega$ for Ω near Ω_0. Then apply the transversality theorem to

$$\begin{aligned} F(u, \psi, \lambda, h) = (h^* A_{h(\Omega)}(\lambda) h^{*-1} u, h^* A^*_{h(\Omega)}(\lambda) h^{*-1} \psi, \\ h^* K_{h(\Omega)} h^{*-1}(u, \psi)) \end{aligned}$$

where $K_\Omega(u, \psi) = \mathrm{Re}(\nabla u \cdot \nabla\psi | \partial\Omega \cap V) \in L_1(\partial\Omega \cap V)$. (Since $u = \psi = 0$ on $\partial\Omega$, $K_\Omega(u, \psi) = 0$ on $\partial\Omega$.) For each h near i_Ω (and Ω near Ω_0) there is a solution u, ψ, λ of $F(u, \psi, \lambda, h) = (0, 0, 0) \in L_2(\Omega, \mathbb{C})^2 \times L_1(\partial\Omega \cap V)$ with $\lambda \in \Lambda$ so condition $(2.\beta)$ of Theorem 5.4 must fail. Thus we have solutions (u, ψ, λ) of $(**)$ which also satisfy: Γ is of finite rank, where

$$\Gamma(\sigma) = \mathrm{Re}\left\{\psi_N \frac{\partial}{\partial N} \mathcal{B}_{A(\lambda)}(\sigma u_N) + u_N \frac{\partial}{\partial N} \mathcal{B}_{A^*(\lambda)}(\sigma \psi_N)\right\}\Bigg|_{\partial\Omega\cap v},$$

$\sigma = \dot{h} N$ is the normal velocity of the perturbation, $u_N = \partial u/\partial N$, $\psi_N = \partial/\partial N$, and $\mathcal{B}_A, \mathcal{B}_{A^*}$ are the solution operators for the Dirichlet problem (see 7.6). We apply Theorems 2, 3 and 4 of Section 7.6. [An alternative argument is in Section 8.3.] Many terms vanish because $\mathrm{Re}(u_n \psi_n) \equiv 0$ on $\partial\Omega$, hence

$$\begin{aligned} &\mathrm{Re}\,(\psi_N(x) u_N(y) + \psi_N(y) u_N(x)) \\ &= \mathrm{Re} \begin{pmatrix} 2\psi_N(x) u_N(y) + (y - x) D(u_N \psi_N)(x) \\ + \tfrac{1}{2}(y - x)^2 : D^2(u_N \psi_N)(x) \\ -(y - x) \cdot D u_N(x)(y - x) \cdot D\psi_N(x) + O(|x - y|^{3-\epsilon}) \end{pmatrix} \\ &= -\,\mathrm{Re}((y - x) \cdot D u_N(x)(y - x) \cdot D\psi_N(x)) + O(|x - y|^{3-\epsilon}) \\ &= O(|x - y|^2), \end{aligned}$$

recalling u_N, ψ_N are $C^{3-\epsilon}$, for any $\epsilon > 0$. The part remaining is

$$\begin{aligned}\Gamma(\sigma)(x) &= \int_{\partial\Omega} \frac{2\sigma(y)dA_y}{\omega_n(x-y)^{n-2}} \operatorname{Re}(\tau\cdot\nabla_{\partial\Omega}u_N(x) + \tfrac{1}{2}b\cdot\tau u_N(x)) \\ &\quad\times(\tau\cdot\nabla_{\partial\Omega}\psi_N(x) - \tfrac{1}{2}b\cdot\tau\psi_N(x)) \\ &\quad+\int_{\partial\Omega} 2\sigma(y)dA_y E_n(y-x) \\ &\quad\times\operatorname{Re}(u_N(y)\psi_N(y)(c-\tfrac{1}{2}\operatorname{div} b - \tfrac{1}{4}b^2)(y,\lambda)) \\ &\quad+\int_{\partial\Omega}\{O(|x-y|^{3-\epsilon-n}) + \text{(smooth kernel)}\}\,\sigma(y)dA_y,\end{aligned}$$

for $x \in \partial\Omega \cap V$ and any smooth real σ. We have written τ for $\frac{y-x}{|y-x|}$ in the first intergral, which is nearly a unit tangent vector for y near x.

In case $n = 2$, the kernel in the first integral is smooth, but the coefficient of $E_\lambda(y-x) = \frac{1}{2\pi}\log(y-x)$ must vanish. Since $u_N\psi_N$ is purely imaginary and non-zero on a dense set, we conclude $\operatorname{Im} D(A(\lambda)) = 0$ on $\partial\Omega \cap V$.

In case $n \geq 3$, at each $x \in \partial\Omega\cap V$ and for each $\tau \perp N_x$ with $|\tau| = 1$,

$$\begin{aligned}\operatorname{Re}\left(\frac{\partial u_N}{\partial\tau} + \tfrac{1}{2}b\cdot\tau u_N\right)\left(\frac{\partial\psi_N}{\partial\tau} - \tfrac{1}{2}b\cdot\tau\psi_N\right)& \\ = \frac{1}{n-2}\operatorname{Re}(u_N\psi_N D(A))&\end{aligned}$$

(see 7.3.7). Consider a point of $\partial\Omega\cap V$ where $u_N \neq 0$ and $\psi_N \neq 0$ – almost any point. We may define (locally) real functions R, S, ϕ by

$$u_N = Re^{i\phi}, \qquad \psi_N = iSe^{-i\phi}$$

and then the equation becomes

$$(\tau\cdot\nabla_{\partial\Omega}\phi + \tfrac{1}{2}\operatorname{Im} b\cdot\tau)(\tau\cdot\nabla_{\partial\Omega}\log|S/R| - \operatorname{Re} b\cdot\tau) = \frac{1}{n-2}\operatorname{Im} D(A).$$

At any point of $\partial\Omega\cap V$, we may choose $\tau \perp N$ so also $\tau \perp (\nabla_{\partial\Omega}\phi + \tfrac{1}{2}\operatorname{Im} b_{\tan})$, since $n \geq 3$, to conclude:

$$\operatorname{Im} D(A(\lambda)) = 0 \text{ on } \partial\Omega\cap V, \qquad \text{when } n \geq 2.$$

(We also have, for $n \geq 3$, where $u_N\psi_N \neq 0$ in $\partial\Omega\cap V$,

$$0 = |\operatorname{Im} b_{\tan} + \nabla_{\partial\Omega}(2\arg u_N)|\cdot|\operatorname{Re} b_{\tan} - \nabla_{\partial\Omega}\log|\psi_N/u_N||.$$

We will use this condition in 8.3.)

This is not yet a contradiction to (*): for each Ω near Ω_0 there exists a solution (u, ψ, λ) of (**) with $\lambda \in \Lambda$ which satisfies also $\operatorname{Im} D(A(\lambda,\cdot)) = \operatorname{Im}(c - \tfrac{1}{2}\operatorname{div} b - \tfrac{1}{4}b^2) = 0$ on $\partial\Omega\cap V$. But the lemma below shows there is a nonempty open $V_0 \subset V-$ we may suppose it is near $\partial\Omega_0-$ and some $\lambda_0 \in \wedge$ such that Im $D(A(\lambda_{0,}\cdot)) \equiv 0$ in V_0. This contradiction proves (**) has not solutions for most Ω; in fact, restricting λ to a compact set of Λ, there are no solutions except perhaps in a closed set of infinite codimension. (In Theorem 5.4, m is arbitrary.)

The same condition – nonsolvability of (**) – implies the eigenvalue λ, assumed simple, is freely moveable. Indeed consider $h(x,t) = x + tV(x)$ for any smooth $V : \mathbb{R}^n \to \mathbb{R}^n$; the eigenvalue $\lambda(t)$ near λ, for the Dirichlet problem in $h(\Omega, t)$ is well-defined for small t and

$$\frac{d}{dt}\lambda(t)|_{t=0} = \int_{\partial\Omega} \sigma u_N \psi_N \Big/ \int_{\Omega} \psi \left(\frac{d}{d\lambda}A(\lambda)\right) u, \qquad \sigma = V \cdot N.$$

The denominator is non-zero, since λ is simple, and the set of possible "velocities," for various choices of V or σ, is a real-linear subpace of $\mathbb{C}$. It fails to be the whole complex plane only if $u_N\psi_N|\partial\Omega$ has constant argument, i.e., for some real constant θ, $\mathrm{Re}(e^{-i\theta}u_N\psi_N)|_{\partial\Omega} \equiv 0$. But this implies $(u, e^{-i\theta}\psi, \lambda)$ solves (**), which has no solutions. Thus λ is freely moveable.

It remains only to prove the lemma.

Lemma. *Let V be an open set in $\mathbb{R}^n$, $n \geq 2$, and M any set in $\mathbb{R}^m$, and suppose $F(x,\mu) = \sum_{|\alpha|\leq k} F_\alpha(x)\mu^\alpha$ is a polynomial in $\mu \in \mathbb{R}^m$ with coefficients $F_\alpha : V \to \mathbb{R}^p$ continuous. Let Ω_0 be a C^r domain ($r \geq 1$) in $\mathbb{R}^n$ whose boundary $\partial\Omega_0$ meets V and assume: for every Ω in some C^r-neighborhood of Ω_0 such that $\Omega\backslash V = \Omega_0\backslash V$ (i.e., perturbed only inside V) there exists $\mu = \mu(\Omega)$ in M such that*

$$F(x,\mu) = 0 \qquad \text{for all } x \in \partial\Omega \cap V.$$

Then there exists $\mu_0 \in M$ and a nonempty open set $V_0 \subset V$ such that $F(x,\mu_0) = 0$ for all $x \in V_0$.

Remark. The result is entirely reasonable, but the hypothesis is not. Our argument is algebraic – hence, the polynomial dependence – but it seems to miss the point: it should be a result of topological dimension theory. We satisfy an infinite-dimensional condition ($F(\cdot,\mu)|\partial\Omega \cap V \equiv 0$) by choice of a finite-dimensional parameter μ, and this is only possible if it is trivially satisfied for some μ_0. But I haven't managed to prove it by topological dimension theory. (Our argument probably also applies with only analytic dependence on μ, but this still seems excessive.)

If we apply the transversality theorem directly, assuming F and M are smooth, we conclude that the hypothesis still holds for $(x,\mu) \mapsto (F(x,\mu), \frac{\partial}{\partial x_1}F(x,\mu), \ldots, \frac{\partial}{\partial x_n}F(x,\mu))$ in place of F. If $x \mapsto F(x,\mu)$ were analytic for each μ with all $(\partial/\partial x)^\alpha F(x,\mu)$ continuous, we could continue to arbitrarily high-order derivatives and eventually find $\mu_0 \in M$ with $F(\cdot,\mu_0) \equiv 0$ in V, supposing V is connected. But analyticity in x is even less acceptable than analyticity in μ, for typical applications; and analyticity seems totally irrelevant.

Proof. Though not essential, it is convenient to reduce to the case of linear dependence. Let N be the number of multi-indices $\alpha = (\alpha_1, \ldots, \alpha_m)$ with $|\alpha| \leq k$, so $\mu \mapsto (\mu^\alpha)_{|\alpha|\leq k} : \mathbb{R}^m \to \mathbb{R}^N$ defines an imbedding of M onto $A \subset \mathbb{R}^N \backslash \{0\}$, and instead of $F(x,\mu)$ we use $\sum_{|\alpha|\leq k} a_\alpha F_\alpha(x) = a \cdot F(x)$ for $a \in A$.

We suppose, for contradiction, that $a \cdot F(x) \neq 0$ on a dense set of V for every choice of $a \in A \subset \mathbb{R}^N \backslash \{0\}$.

Choose $a_1 \in A$; then $a_1 \cdot F(x_1) \neq 0$ for some x_1 in V arbitrarily near $\partial\Omega_0$. Choose such x_1, so that there exists Ω_1 sufficiently near Ω_0 whose boundary contains x_1. ("Sufficiently near" means being inside the C^r-neighborhood of Ω_0 in our hypothesis.) Define $L_1 \subset \mathbb{R}^N$ by

$$L_1 = \{a \in \mathbb{R}^N : a \cdot F(x_1) = 0\}$$

so $a_1 \notin L_1$ but L_1 meets A: for some $a_1' \in A$, $a_1' \cdot F \equiv 0$ on $\partial\Omega_1 \cap V$, in particular at x_1.

Choose a_2 in $A \cap L_1$ and $\hat{x} \neq x_1$ on $\partial\Omega_1 \cap V$; $a_2 \cdot F \neq 0$ on a dense set so $a_2 \cdot F(x_2) \neq 0$ at points x_2 near $\hat{x}$. Choose such x_2 and Ω_2 near Ω_1 (hence, near Ω_0) such that $\partial\Omega_2$ contains both x_1 and x_2. Then define $L_2 = \{a \in \mathbb{R}^N : a \cdot F(x_1) = 0 \text{ and } a \cdot F(x_2) = 0\}$. We have $L_1 \supsetneqq L_2$ since $a_2 \in L_1 \backslash L_2$, and also L_2 meets A: for some $a_2' \in A$, $a_2' \cdot F = 0$ on $\partial\Omega_2 \cap V \supset \{x_1, x_2\}$.

We may continue thus indefinitely, in particular for N steps: $\mathbb{R}^N \supsetneqq L_1 \supsetneqq L_2 \supsetneqq \cdots \supsetneqq L_N$, with each L_k a linear space which meets A. But $N > \dim L_k > \dim L_{k+1}$ for each $k \geq 1$ so we find a contradiction: $L_N = \{0\}$ meets $A \subset \mathbb{R}^N \backslash \{0\}$.

Some Obvious Questions

In Example 6.5, are the solutions generically hyperbolic? What about a system of equations, not just one? How about the Navier-Stokes equation! If not a system, why not a fourth-order scalar problem? Introduce some parameters; as the parameters change, we can expect bifurcation of solutions, but is the bifurcation "generic" for most Ω?

In Example 6.7 can we allow dependence on ∇u in f? If so, are the resulting solutions hyperbolic? And what about systems? Higher-order equations? Generic bifurcation? Mixed boundary conditions?

In Example 6.2, how about other symmetries? Continuous symmetries? Symmetries in nonlinear problems? Bifurcation with symmetry? Rework the whole chapter with symmetry!

Why not add "genericity" of coefficients to generic domains? Surely this will give much stronger theorems!

What are generic properties of multi-parameter eigenvalue problems?

In Example 6.6, does the result hold for non-self-adjoint problems? Can one prove results like Examples 6.3 or 6.8 with other boundary conditions? Or for systems? Or higher-order equations? In the self-adjoint problem Example 4.4 can we allow polynomial dependence on λ, as in Examples 6.3 and 6.8?

Why not study some time-dependent problems? Say periodic solutions in time-dependent domains, or fixed generic domains. Are they generically simply. Hyperbolic? How do they bifurcate? Are the stable and unstable manifolds transversal, generically with respect to the domain?

How about control theory with the domain as parameter? Is this enough for local controllability? If we can move only a small part of the boundary, how much does this accomplish? Can we use perturbation of the boundary to supplement other control parameters, perhaps closing reachable sets? Or extending them in new directions? Can we optimize? Or will this lead to very irregular regions?

Author's Note: I only claimed to have questions, not answers.

Chapter 7

Boundary Operators for Second-Order Elliptic Equations

For the Laplace operator Δ in a smooth bounded region $\Omega \subset \mathbb{R}^n$, among the operators of interest are $\frac{\partial}{\partial N}\mathcal{B}_\Delta|_{\partial\Omega}$, $\mathcal{C}_\Delta|_{\partial\Omega}$, $\frac{\partial}{\partial N}\mathcal{A}_\Delta\mathcal{B}_\Delta|_{\partial\Omega}$ where

$$
\begin{aligned}
&u = \mathcal{A}_\Delta f \text{ is the solution of } \Delta u = f \text{ in } \Omega, && u = 0 \text{ on } \partial\Omega,\\
&v = \mathcal{B}_\Delta g \text{ is the solution of } \Delta v = 0 \text{ in } \Omega, && v = g \text{ on } \partial\Omega,\\
&w = \mathcal{C}_\Delta h \text{ is the solution of } \Delta w = \text{ constant in } \Omega, && \frac{\partial w}{\partial N} = h \text{ on } \partial\Omega,
\end{aligned}
$$

with $\int_\Omega w = 0$ (the constant is $\frac{1}{|\Omega|}\int_{\partial\Omega} h$).

Corresponding operators may be defined for other second-order elliptic equations (and for higher-order equations and systems). For applications to generic perturbation of the boundary, we need fairly explicit representations of these operators. One may write, for example,

$$
\mathcal{A}_\Delta f(x) = \int_\Omega G_\Omega(x, y) f(y) dy, \quad \mathcal{B}_\Delta g(x) = \int_{\partial\Omega} \frac{\partial G_\Omega}{\partial N_\Omega(y)}(x, y) g(y) dA_y
$$

where G_Ω is Green's function for the Dirichlet problem in Ω; but this is useless without fairly precise information about the singularity of G_Ω near the boundary. P. Levy [18] gives some information for the Laplacian in $\mathbb{R}^3$: he computes, in effect, the principal and subprincipal parts of $\frac{\partial}{\partial N}\mathcal{B}_\Delta$, by subtracting singularities of known harmonic functions. The method seems more difficult than ours when we attempt the next step, and we need more detail and more generality. In any case, this reference was found too late to be used here, though it gives a valuable verification for part of Theorem 7.6.7. Assuming everything is C^∞ one may use the "projection on the Cauchy data" of the theory of pseudo-differential operators [4, 37]. The principal symbols have been computed, but we need much more detail. This is presumably the best way to attack the problem – especially for more general elliptic systems – but I have not managed to complete the calculation with this method. (I did manage the calculation for a special type of rapidly-oscillating solution in 1985, after completing this manuscript. Chapter 8 presents the results. I hope Chapter 7 – with an important correction to Thm. 7.6.4 – is still of interest, presenting another method for a difficult problem.) We will use the more concrete (although more clumsy) method of integral operators. We can avoid singular integrals – until the final results – by integration by parts, so the detailed calculations use only weakly-singular (absolutely integrable) kernels.

In the first three sections, we study weakly-singular kernels with emphasis on differentiability properties (but without going to the C^∞ extreme). Then we treat limits of singular integrals and fundamental solutions of second-order equations having the Laplacian as principal part. We obtain integral equations for the operators of interest and "solve" these (to second or third order) to obtain fairly explicit representations of the operators. For example if Ω is a bounded C^4 region in $\mathbb{R}^n (n \geq 2)$ and $E(z)$ is the fundamental solution of the Laplacian

$$E(z) = \frac{|z|^{2-n}}{(2-n)\omega_n} \qquad \text{for } n \neq 2,$$

$$E(z) = \frac{1}{2\pi} \log |z| \qquad \text{for } n = 2,$$

where $\omega_n = |\mathcal{S}^{n-1}| = 2\pi^{n/2}/\Gamma(n/2)$, then for $x \in \partial\Omega$ and $g \in C^2(\partial\Omega)$,

$$\begin{aligned}\frac{\partial}{\partial N}\mathcal{B}_\Delta g(x) &= 2\,P.V.\int_{\partial\Omega} \frac{\partial^2 E(x-y)}{\partial N_x \partial N_y}(g(y)-g(x))dA_y - \frac{n-2}{n-1}\frac{H(x)}{2}g(x)\\ &\quad - \frac{n-1}{2\omega_{n-1}}\,\text{P.V.}\int_{\partial\Omega} \frac{(y-x).\hat{K}(x)(y-x)}{|y-x|^{n+1}} g(y)dA_y\\ &\quad + \int_{\partial\Omega} |x-y|^{2-n} Q_1\left(x, \frac{y-x}{|y-x|}\right) g(y) + \int_{\partial\Omega} R_1(x,y)g(y).\end{aligned}$$

Here $K(x) = DN_\Omega(x)$ is the curvature of $\partial\Omega$ at x (at least when restricted to the tangent space $T_x(\partial\Omega)$; see Th. 1.5), $H(x) =$ trace $K(x)$ is the mean curvature, $\hat{K}(x) = K(x) - \frac{H(x)}{n-1}P(x)$ has trace zero ($P(x) = I - N(x)\otimes N(x)$ is the projection on the tangent space), $\sigma \mapsto Q_1(x,\sigma)$ is a fourth-order polynomial computed explicitly below (Section 7.6), and the "remainder" R_1 is a kernel of class $C^1_*(2)$ in $\partial\Omega$, in the notation of Section 7.1 below (or $C^{r-3}_*(2)$, if $\partial\Omega$ is C^r). "P.V." indicates the Cauchy principal value when the integrand is singular at x,

$$\text{P.V.}\int_{\partial\Omega} \cdots = \lim_{\epsilon\to 0}\int_{\partial\Omega\setminus B_\epsilon(x)} \cdots.$$

If $n = 2$ then $\partial\Omega$ is a finite union of curves and (for definiteness) suppose it is a single curve of length L. Functions on $\partial\Omega$ may be written as functions of the arc length, extended as L-periodic functions on the real-line and then the operator has a simple form:

$$\begin{aligned}\frac{\partial}{\partial N}\mathcal{B}_\Delta g(t) &= \int_{\partial\Omega} \lambda(t-s)g''(s)ds + \int_{\partial\Omega} R_1(t,s)g(s)ds \\ &= \text{P.V.}\int_{\partial\Omega} \lambda'(t-s)g'(s) + \int_{\partial\Omega} R_1(t,s)g(s) \\ &= \text{P.V.}\int_{\partial\Omega} \lambda''(t-s)(g(s)-g(t)) + \int_{\partial\Omega} R_1(t,s)g(s)\end{aligned}$$

where $\lambda(t)$ is L-periodic, C^∞ except at $\{0, \pm L, \pm 2L_1, \ldots\}$, with

$$\lambda(t) = \frac{1}{\pi}\log|t|, \qquad \lambda'(t) = \frac{1}{\pi t}, \qquad \lambda''(t) = \frac{-1}{\pi + 2}$$

near $t = 0$, while $R_1(t,s)$ is L-periodic in each variable and of class $C^{1-\epsilon}$ (or $C^{r-3-\epsilon}$, if $\partial\Omega$ is C^r) for any $\epsilon > 0$. ($\frac{\partial}{\partial N}\mathcal{B}_\Delta g = Hg'$, where H is the Hilbert transform associated with the curve $\partial\Omega$.)

7.1 Weakly-Singular Integral Operators with $C^r_*(\alpha)$ Kernels

We study integral operators

$$u \mapsto \int_M K(x,y)u(y)dM_y, \qquad x \in M,$$

where M is an m-dimensional differentiable manifold without boundary ($m \geq 1$) having a positive measure dM locally equivalent to m-dimensional Lebesgue measure, and $K(x,\cdot)$ is absolutely integrable. (In our applications $M = \partial\Omega$ for some $\Omega \subset \mathbb{R}^n$, $m = n-1$, and dM is the surface-area measure dA in $\partial\Omega$ induced from n-dimensional Lebesgue measure in $\mathbb{R}^n$.) We study only a special class of kernels $K(x,y)$ (class $C^r_*(\alpha)$, $r =$ integer ≥ 0, $\alpha > 0$, defined below) r-times differentiable for $x \neq y$ with certain growth-restrictions on the derivatives as $x - y \to 0$. Standard tricks – partition of unity and local change of variables – permit us to work in Euclidean $\mathbb{R}^m$, and we do so until Assumptions 7.1.4. We will see later that fundamental solutions of second order elliptic equations are kernels of this class, as are the kernels of many integral operators associated with elliptic boundary value problems.

The space of value of the kernels $K(x,y)$ has little importance here – we will take the values to be real or complex matrices of given dimensions, and whenever products occur, there is the implicit assumption that the matrices are compatible. The transpose of a matrix B is written ${}^\top B$.

Definition 7.1.1. Let A be an open set in $\mathbb{R}^m$; a function $K(x, y)$ defined for $x \neq y$ in A is a *kernel of class* $C_*^r(\alpha)$ *in* A (r = integer $\geq 0, \alpha > 0$) if it is C^r for $x \neq y$ and for any $\delta > 0$ and $|i| + |j| + |k| \leq r$

$$\partial_x^i \partial_y^j (\partial_x + \partial_y)^k K(x, y) = O\left(1 + |x - y|^{\alpha - m - |i| - |j| - \delta}\right)$$

uniformly for $x \neq y$ in compact subsets of A. If $\alpha > m + |i| + |j|$, we require also that $\partial_x^i \partial_y^j (\partial_x + \partial_y)^k K(x, y)$ extend continuously to $\{x = y\}$; if $m > 1$ and $|i| + |j| < r$, this condition is superfluous, since $A \times A$ is not disconnected by the diagonal. We say the kernel K has *exponent* α, and we will sometimes use "operator of order $-\alpha$" to mean an integral operator with kernel of exponent α. (See Theorems 7.1.2 and 7.1.3 for partial justification of this usage.)

Remark. If $\phi : \mathbb{R}^m \backslash \{0\} \to \mathbb{R}$ is differentiable, $(\partial_x + \partial_y)\phi(x, y) = 0$. Thus if ϕ is C^r and homogeneous of degree $\alpha - m$ on $\mathbb{R}^m \backslash \{0\}$, $\phi(x - y)$ is a C_*^r kernel. $K(x, y)$ is of class $C_*^r(\alpha)$ if and only if $\partial_x K(x, y)$ and $\partial_y K(x, y)$ are of class $C_*^{r-1}(\alpha - 1)$ and $(\partial_x + \partial_y) K(x, y)$ is $C_*^{r-1}(\alpha)$. If $\alpha > r + m$, a "kernel of class $C_*^r(\alpha)$" simply means *a* C^r function. If $K(x, y)$ is a $C_*^r(\alpha)$ kernel in $A \subset \mathbb{R}^m$, $h : A_0 \subset \mathbb{R}^m \to A$ is a C^{r+1} diffeomorphism and $\psi : A_0 \times A_0 \to \mathbb{R}$ is of class C^r, then $(x, y) \mapsto \psi(x, y) K(h(x), h(y))$ is a $C_*^r(\alpha)$ kernel in A_0. This permits definition of $C_*^r(\alpha)$ kernels on C^{r+1} manifolds (Assumptions 7.1.4 below).

Example. Let $Q(x, y, \sigma)$ be C^r on $(\mathbb{R}^m)^3$, $\alpha > 0$, j = integer ≥ 0 and

$$K(x, y) = |x - y|^{\alpha - m} (\log |x - y|)^j Q\left(x, y, \frac{y - x}{|y - x|}\right) \qquad \text{for } x \neq y;$$

then K is a $C_*^r(\alpha)$ kernel in $\mathbb{R}^m$. We will see many kernels of this kind later. Here we have used the Euclidean norm $|\cdot|$ but we have the same conclusion with $|\cdot|_{x,y}$ in place of $|\cdot|$, where

$$|v|_{x,y} = (v \cdot G(x, y) v)^{1/2},$$

$G(x, y)$ is a C^r function whose values are symmetric positive matrices. For this case, in a neighborhood of a given $x_0 \in \mathbb{R}^m$, let $|v|_0 = |v|_{x_0, x_0}$ and write $K(x, y)$ as a finite sum of kernels $|x - y|_0^{\alpha + i - m} (\log |x - y|_0)^k Q_{ik}(x, y, \frac{y - x}{|y - x|_0})$ [where $i \geq 0, 0 \leq k \leq j$, $Q_{ik}(x, y, \sigma)$ is C^r], plus a C^r function of x, y.

Theorem 7.1.2. *Let K be a $C_*^r(\alpha)$ kernel in $\mathbb{R}^m$ with compact support and values in $\mathbb{C}^{a \times b}$ (= the $a \times b$ complex matrices), and let $\tilde{K}$ denote the corresponding integral operator*

$$\tilde{K} u(x) = \int_{\mathbb{R}^m} K(x, y) u(y) dy, \qquad x \in \mathbb{R}^m.$$

Then $\tilde{K}$ is a compact operator in any of the following pairs of spaces (with j, k integers in $0 \le j \le k \le r; 1 \le p, q \le \infty; 0 \le \sigma, \tau \le 1$),

(1) $W^{j,p}(\mathbb{R}^m, \mathbb{C}^b) \to W^{k,q}(\mathbb{R}^m, \mathbb{C}^a)$ *if* $j - m/p + \alpha > k - m/q$
(2) $C^{j,\sigma}(\mathbb{R}^m, \mathbb{C}^b) \to W^{k,q}(\mathbb{R}^m, \mathbb{C}^a)$ *if* $j + \sigma + \alpha > k - m/q$
(3) $W^{j,p}(\mathbb{R}^m, \mathbb{C}^b) \to C^{k,\tau}(\mathbb{R}^m, \mathbb{C}^a)$ *if* $j - m/p + \alpha > k + \tau, k < r$
(4) $C^{j,\sigma}(\mathbb{R}^m, \mathbb{C}^b) \to C^{k,\tau}(\mathbb{R}^m, \mathbb{C}^a)$ *if* $j + \sigma + \alpha > k + \tau, k < r, k < j + \alpha$.

Recalling that $j - m/p$ is the "net smoothness" of functions in $W^{j,p}(\mathbb{R}^m)$, an approximate summary is: for $0 < \alpha \le r$, $\tilde{K}$ is smoothing of order α, or an operator of order $-\alpha$.

Proof. The identities

$$\begin{aligned}\frac{\partial}{\partial x}\int K(x,y)u(y)dy &= \int \partial_x K(x,y)u(y) && \text{(when } \alpha > 1, r \ge 1)\\ &= \int (\partial_x + \partial_y)K(x,y)u(y)\\ &\quad + \int K(x,y)\partial_y u(y) && \text{(when } \alpha > 0, r \ge 1)\end{aligned}$$

allow us to reduce the case $(\alpha, r; j, k)$ to $(\alpha - 1, r - 1; j, k - 1)$ when $\alpha > 1$, $r \ge 1$, or to $(\alpha, r - 1; j, k - 1)$ and $(\alpha, r; j - 1, k - 1)$ when $\alpha > 0$, $r \ge 1$. Using this and the Sobolev embeddings (in the cases $\alpha > r; r \ge \alpha > k; r \ge k \ge \alpha$), we may reduce to one of the following cases:

(1) $\tilde{K} : L_p \to L_q$ is compact if $\alpha > 0, r \ge 0, \alpha - m/p > -m/q, p \le q$;
(2) $\tilde{K} : L_p \to C^\tau$ is compact if $0 < \alpha \le 1 = r, \alpha - m/p > \tau$;
(3) $\tilde{K} : C^\sigma \to C^\tau$ is compact if $0 < \alpha \le 1 = r, \alpha + \sigma > \tau$. Here $C^\sigma = C^{0,\sigma}$ or $C^{0,\sigma+}$ for $\sigma < 1$, and for $\sigma = 1$ may denote C^1 or $C^{0,1}$ (= Lipschitz), whichever gives the stronger conclusion.

In cases (2) and (3), it suffices to prove continuity; compactness follows from the compactness of the support of K and the compactness of the inclusion $C^{\sigma+\epsilon} \subset C^\sigma (\epsilon > 0)$.

(1) Continuity follows immediately from Young's inequality ($\|f * g\|_{L_q} \le \|f\|_{L_s}\|g\|_{L_p}$ when $1 \le p, q, s \le \infty$ and $1/q' = 1/p' + 1/s'$ for the conjugate exponents). If $\theta : \mathbb{R}^n \to \mathbb{R}$ is C^∞ with compact support, $\theta \equiv 1$ near 0, and $K_\epsilon(x, y) = (1 - \theta(\frac{x-y}{\epsilon}))K(x, y)$ for $\epsilon > 0$, then $\tilde{K}_\epsilon$ is certainly compact for any $\epsilon > 0$ and Young's inequality shows $\|\tilde{K} - \tilde{K}_\epsilon\|_{\mathcal{L}(L_p, L_q)} \to 0$ as $\epsilon \to 0$, so $\tilde{K}$ is compact.

(2) For $\lambda < \alpha - m$ (arbitrarily close) we have

$$\begin{aligned}|K(x', y) - K(x, y)| &\leq C \min\{|x - y|^\lambda + |x' - y|^\lambda, \\ &\quad |x' - x|(|x' - y|^{\lambda-1} + |x - y|^{\lambda-1}\} \\ &\leq C'|x - x'|^\theta(|x - y|^{\lambda-\theta} + |x' - y|^{\lambda-\theta}) \\ &\quad \text{for } 0 \leq \theta \leq 1 \text{ (constants } C, C').\end{aligned}$$

Take $\theta = \tau$ and (2) follows from Hölder's inequality.

(3) Say $K(x, y) = 0$ if $|x| > R$ or $|y| > R$. For $|x| \leq R$, $|x + h| \leq R$

$$\begin{aligned}\tilde{K}u(x + h) - \tilde{K}u(x) &= \int_{|y|<3R} (K(x + h, y) - K(x, y))(u(y) - u(x))dy \\ &\quad + \int_{|y|<3R} (K(x + h, y + h) - K(x, y))dy\, u(x)\end{aligned}$$

If $\tau \leq \min\{1, \alpha + \sigma - \delta\}$ for some $\delta > 0$, this implies $|\tilde{K}u(x + h) - \tilde{K}u(x)) \leq \text{const.}\ |h|^\tau\ \|u\|_{C^\sigma}$. In case $\alpha + \sigma > 1$ this gives continuity into $C^{0,1}$ (= Lipschitz); but it is easy to see $\tilde{K}(C^1) \subset C^1$ so in fact $\tilde{K}(C^\sigma) \subset C^1$.

Theorem 7.1.3. *Let K, L be kernels with compact support of class $C^r_*(\alpha)$, $C^r_*(\beta)$ respectively in $\mathbb{R}^m$; we suppose their values can be multiplied. Then $K \circ L$ defined by*

$$K \circ L(x, y) = \int_{\mathbb{R}^m} K(x, z)L(z, y)dz$$

is a kernel of class $C^r_*(\alpha + \beta)$ in $\mathbb{R}^m$ with compact support. If $\alpha + \beta > r + m$, $K \circ L$ is a C^r function on $\mathbb{R}^m \times \mathbb{R}^m$.

Proof. The case $r = 0$ is well-known (see, for example, W. Pogorzelski, *Integral Equations and Their Applications*, vol. I (Pergamon Press, 1966)).

Suppose $r \geq 1$, $\alpha > 1$; we have

$$\begin{aligned}(\partial_x + \partial_y)K \circ L(x, y) &= \int (\partial_x + \partial_y)K(x, z)L(z, y) \\ &\quad + \int K(x, z)(\partial_z + \partial_y)L(z, y) \\ \partial_x K \circ L(x, y) &= \int \partial_x K(x, z)L(z, y) \\ \partial_y K \circ L(x, y) &= \int \partial_z K(x, z)L(z, y) + \int K(x, z)(\partial_z + \partial_y)L(z, y)\end{aligned}$$

so the case (r, α, β) reduces to $(r-1, \alpha-1, \beta)$, when $r \geq 1, \alpha > 1$. Similarly (r, α, β) reduces to $(r-1, \alpha, \beta-1)$ when $r \geq 1, \beta > 1$. Thus we reduce eventually to either $r = 0$ – which is known – or to $r \geq 1$ with $0 < \alpha, \beta \leq 1$, which we now examine.

Let $\theta : \mathbb{R}^n \to \mathbb{R}$ be C^∞, $\theta(x) = 0$ for $|x| \leq 1, \theta(x) = 1$ for $|x| \geq 2$, and for any $\epsilon > 0$ define

$$K_\epsilon(x, y) = \theta\left(\frac{y-x}{\epsilon}\right)K(x, y), \quad \tilde{K}_\epsilon(x, y) = \left(1 - \theta\left(\frac{y-x}{\epsilon}\right)\right)K(x, y),$$

and similarly define $L_\epsilon, \tilde{L}_\epsilon$. We estimate the derivatives of $K \circ L$ near a given (x_0, y_0), $x_0 \neq y_0$. For any $\epsilon > 0$,

$$K \circ L = K_\epsilon \circ L_\epsilon + K_\epsilon \circ \tilde{L}_\epsilon + \tilde{K}_\epsilon \circ L_\epsilon + \tilde{K}_\epsilon \circ \tilde{L}_\epsilon$$

and we choose $\epsilon = \frac{1}{5}(x_0 - y_0)$. For (x, y) in a neighborhood of (x_0, y_0) we have $4\epsilon < |x-y| < 6\epsilon$ so $\tilde{K}_\epsilon \circ L_\epsilon(x, y) = 0$; and since K_ϵ, L_ϵ are C^r functions, we see $K \circ L$ is C^r near (x_0, y_0). Now for $|i| + |j| + |k| \leq r$, $\partial_x^i \partial_y^j (\partial_x + \partial_y)^k \tilde{K}_\epsilon \circ L_\epsilon(x, y)$ is a sum of terms

$$\text{const.} \int \left(1 - \theta\left(\frac{x-z}{\epsilon}\right)\right)\theta\left(\frac{y-z}{\epsilon}\right)(\partial_x + \partial_z)^{k'+i'} K(x, z) \\ \times \partial_z^{i''} \partial_y^j (\partial_y + \partial_z)^{k''} L(y, z)$$

with $k' + k'' = k,\ i' + i'' = i$. The terms containing some derivative of $\theta(\frac{y-z}{\epsilon})$ vanish, since this gives disjoint supports. Since $0 < \alpha, \beta \leq m$ and $|y - z| > \epsilon$, $|x - z| < 2\epsilon$ on the support of the integrand, we may estimate the integral by

$$\begin{aligned} \text{const.} \int_{|x-z|<2\epsilon} |x-z|^{\alpha-m-\delta} \epsilon^{\beta-m-i-j-\delta} dz &= O(\epsilon^{\alpha+\beta-m-i-j-2\delta}) \\ &= O(|x-y|^{\alpha+\beta-m-i-j-2\delta}), \end{aligned}$$

for any $\delta > 0$. Similarly for $K_\epsilon \circ \tilde{L}_\epsilon$. In the case $K_\epsilon \circ L_\epsilon$, we have a sum of terms

$$\text{const.} \int \theta\left(\frac{x-z}{\epsilon}\right)\theta\left(\frac{z-y}{\epsilon}\right)\partial_x^i(\partial_x + \partial_z)^{k'} K(x, z) \partial_y^j (\partial_z + \partial_y)^{k''} L(z, y)$$

$(k' + k'' = k)$, in addition to integrals where one or more of the derivatives $\partial_x^i \partial_y^j$ falls on $\theta(\frac{x-z}{\epsilon})$ or $\theta(\frac{z-y}{\epsilon})$. But in the latter, $|x - z|$ or $|z - y|$ is comparable to ϵ in the support of the integrand, and the estimates are like the previous case. In the remaining integrals we have $|x - z| \geq \epsilon$, $|y - z| \geq \epsilon$ and we write the integrals as $I + II = \int_{\substack{\epsilon<|x-z|<8\epsilon \\ \epsilon<|y-z|}} \cdots + \int_{|x-z|\geq 8\epsilon} \cdots$. Then

$$|I| \leq \int_{|x-z|<8\epsilon} \text{const. } \epsilon^{\alpha+\beta-2m-i-j-2\delta} = O(\epsilon^{\alpha+\beta-m-i-j-2\delta})$$

and

$$|II| \leq \int_{|x-z|\geq 8\epsilon} \text{const. } |x-z|^{\alpha+\beta-2m-i-j-2\delta} = O(\epsilon^{\alpha+\beta-m-i-j-2\delta}),$$

and ϵ is comparable to $|x-y|$, so the proof is complete.

Now we transfer our results to a differentiable manifold M.

Assumptions 7.1.4. *M is a compact m-dimensional C^{r+1} manifold without boundary, provided with a positive measure dM which is C^r-equivalent to m-dimensional Lebesgue measure; that is, for any system of coordinates x in M there is a positive C^r density $x \mapsto w(x)$ so $dM = w(x)dx$.*

In our applications, $M = \partial\Omega$ where Ω is a bounded C^{r+1} region in $\mathbb{R}^n$, $m = n-1$, and $dM = dA =$ surface area measure. Thus if $\partial\Omega$ is given locally as $\{x \mid x_n = \psi(x_1, \ldots, x_{n-1})\}$, then ψ is C^{r+1} and $dA = \sqrt{1 + \sum_1^{n-1}(\frac{\partial\psi}{\partial x_j})^2}\, dx_1 \cdots dx_{n-1}$.

We may define the spaces $W^{j,p}(M, \mathbb{C}^q)$, $C^{j,\sigma}(M, \mathbb{C}^q)$ in the obvious ways $(0 \leq j \leq r)$, and a function $K(x,y)$ defined for $x \neq y$ in M is a kernel of class $C^r_*(\alpha)$ on M provided it is C^r for $x \neq y$ and is a $C^r_*(\alpha)$ kernel when written in any local coordinate system. Thus if h : (open set) $\subset \mathbb{R}^m \to M$ is a C^{r+1} embedding, $(\xi, \eta) \mapsto K(h(\xi), h(\eta))$ is a $C^r_*(\alpha)$ kernel on the domain of h.

Choose an open cover $\{V_j\}_{j=1}^N$ for M such that, for each j, there is a C^{r+1} diffeomorphism $h_j : U_j \subset \mathbb{R}^m \to V_j$ and a C^r $w_j : U_j \to \mathbb{R}_+$ with $dM_{hj(\xi)} = w_j(\xi)d\xi$ $(\xi \in U_j)$. Choose a corresponding C^{r+1} partition of unity $\{\theta_j\}_{j=1}^N$, $\theta_j \geq 0$, supp $\theta_j \subset V_j$, $\sum_1^N \theta_j = 1$ on M. Then

$$\begin{aligned}\int_M K(x,y)u(y)dM_y &= \sum_{j=1}^n \int_{V_j} K(x,y)\theta_j(y)u(y)\,dM_y \\ &= \sum_{i,j=1}^N \int_{U_j} \theta_i(x)K(x, h_j(\eta))\theta_j(h_j(\eta))w_j(\eta)u(h_j(\eta))\,d\eta.\end{aligned}$$

Now for $x \notin V_j$, $(x,\eta) \mapsto \theta_i(x)K(x,h_j(\eta))\theta_j(h_j(\eta))w_j(\eta)$ is C^r_j and for $x = h_j(\xi) \in V_j$, $(\xi,\eta) \mapsto \theta_i(h_j(\xi))K(h_j(\xi), h_j(\eta))\theta_j(h_j(\eta))w_j(\eta)$ is a $C^r_*(\alpha)$ kernel on U_j (vanishing if η nears ∂U_j). Thus $K(x,y) = \sum_{i,j=1}^N K_{ij}(x,y)$ where $K_{ij}(x,y) = \theta_i(x)\theta_j(y)K(x,y)$ is a $C^r_*(\alpha)$ kernel supported in $V_i \times V_j$. Using this representation, we may easily transfer our results above to M and conclude, in particular, that when K, L are kernels of class $C^r_*(\alpha)$, $C^r_*(\beta)$ respectively on

M, then $K \circ L$ is kernel of class $C^r_*(\alpha+\beta)$ on M, where

$$K \circ L(x,y) = \int_M K(x,z)L(z,y)dM_z.$$

If $K(x,y)$ is a $C^r_*(\alpha)$ kernel on M and $\phi : M \to \mathbb{C}$ is C^{r+1}, then the commutator $[\tilde{K},\phi] = \tilde{K}\phi - \phi\tilde{K}$ is an integral operator with kernel $C^r_*(\alpha+1)$

$$[\tilde{K},\phi]u(x) = \tilde{K}(\phi u)(x) - \phi(x)\tilde{K}u(x) = \int_M K(x,y)(\phi(y)-\phi(x))u(y)dM_y.$$

If $K(x,y)$ is a $C^r_*(\alpha)$ kernel on M, $r \geq 1$, and V is a C^r tangent vector field on M with $L = V \cdot \nabla_M$ the corresponding first-order differential operator, the commutator $[\tilde{K},L]$ is an integral operator with kernel of class $C^{r-1}_*(\alpha)$

$$[\tilde{K},L]u(x) = -\int_M \{K(x,y)\mathrm{div}_M V(y) + (L_x + L_y)K(x,y)\}u(y)dM_y.$$

If also $\alpha > 1$ then $\tilde{K}L$ and $L\tilde{K}$ are integral operators with kernel of class $C^{r-1}_*(\alpha-1)$.

7.2 Integral Equations with $C^r_*(\alpha)$ Kernels

Suppose E, F are Banach spaces and $T \in \mathcal{L}(E,F)$ is a Fredholm operator, so the image $\mathcal{R}(T)$ is closed with finite codimension and also $\mathcal{N}(T)$ is finite-dimensional. (We will avoid the term "kernel" for the null-space $\mathcal{N}(T)$, at least in this chapter; "kernel" means the kernel of some integral operator.) Choose projections $P \in \mathcal{L}(E)$, $Q \in \mathcal{L}(F)$ so $\mathcal{R}(P) = \mathcal{N}(T)$, $\mathcal{N}(Q) = \mathcal{R}(T)$; P, Q are of finite rank, and are not unique except when they vanish. Given such P, Q, there is unique $S \in \mathcal{L}(F,E)$ such that $TS = I - Q$ and $PS = 0$ (see below). For any $f \in F$, $x = Sf$ is the unique solution of $Tx = f - Qf$ with $Px = 0$; and $Tx = f$ has a solution x if and only if $Qf = 0$.

We will treat in particular the case $E = F$, $T = I - K$ for some compact operator $K \in \mathcal{L}(E)$; then $S = I - Q + KS = I + R$ for some compact R. The main result of this section is that, if K is an integral operator with kernel of class $C^r_*(\alpha)$, then we may choose P, Q to be integral operators with C^r degenerate kernels and the corresponding R is an integral operator with $C^r_*(\alpha)$ kernel (the *resolvent kernel*). In fact $R - (K + K^2 + \cdots + K^j)$ has kernel of class $C^r_*((j+1)\alpha)$ for each $j \geq 1$.

This result is more or less well-known in the case $r = 0$, $\mathcal{N}(I-K) = \{0\}$. (W. Pogorzelski, *Integral Equations*, gives the result for $r = 0$ or 1 with $\alpha = 0$, $\mathcal{N}(I-K) = \{0\}$, and the case $r = 0$, $\alpha \geq 0$ is at least implicit.) The method of argument is also fairly standard, with some complications arising from the fact that $\mathcal{N}(I-K)$ may not be trivial. After some general remarks on the Fredholm case, we treat operators of finite rank (degenerate kernels), then continuous kernels and then C^r or $C_*^r(\alpha)$ kernels.

Returning to the Fredholm operator $T \in \mathcal{L}(E, F)$ and the associated projections P, Q, observe the spaces E, F split $E = E_1 \oplus E_2$, $F = F_1 \oplus F_2$, $E_1 = \mathcal{R}(P) = \mathcal{N}(T)$, $F_2 = \mathcal{N}(Q) = \mathcal{R}(T)$, with $E_2 = \mathcal{N}(P)$, $F_1 = \mathcal{R}(Q)$. In matrix form

$$T = \begin{pmatrix} 0 & 0 \\ 0 & L \end{pmatrix} : \begin{pmatrix} E_1 \\ E_2 \end{pmatrix} \to \begin{pmatrix} F_1 \\ F_2 \end{pmatrix}$$

where $L : E_2 \to F_2$ is an isomorphism,

$$P = \begin{pmatrix} I & 0 \\ 0 & 0 \end{pmatrix} \text{ in } E, \qquad Q = \begin{pmatrix} I & 0 \\ 0 & 0 \end{pmatrix} \text{ in } F.$$

It is then easy to see $S \in \mathcal{L}(F, E)$ is $S = \begin{pmatrix} 0 & 0 \\ 0 & L^{-1} \end{pmatrix}$ and

$$\begin{aligned} TS &= I - Q \qquad & ST &= I - P \\ SQ &= 0 & PS &= 0. \end{aligned} \tag{7.2.1}$$

The projections P, Q are not in general unique, so it is of interest to know the change in S resulting from other choices of P, Q. More generally, let $\tilde{Q}$ be an operator of finite rank in $\mathcal{L}(F)$ – not necessarily a projection – and suppose $\tilde{S} \in \mathcal{L}(F, E)$ satisfies $T\tilde{S} = I - \tilde{Q}$. Then $\tilde{S} - S$ is of finite rank, and is computable from $\tilde{S}, P, Q, \tilde{Q}$. In fact $T(\tilde{S} - S) = Q - \tilde{Q} = (1 - Q)(Q - \tilde{Q}) = TS(Q - \tilde{Q})$, and $PS = 0$ so $P\tilde{S} = \tilde{S} - S - S(Q - \tilde{Q})$, or $S(I - \tilde{Q}) = (I - P)\tilde{S}$. If $\tilde{Q}$ is a projection with $\mathcal{N}(\tilde{Q}) = \mathcal{R}(T)$, then $\tilde{Q}(I - Q) = 0$ so

$$S = (I - P)\tilde{S}(I - Q) \tag{7.2.2}$$

Theorem 7.2.1. *(Operators of finite rank) Let $K \in \mathcal{L}(E)$ be of finite rank, $\dim \mathcal{R}(K) = m < \infty$. Then the adjoint $K^* \in \mathcal{L}(E^*)$ also has rank m, and choosing any basis $\{e_1, \ldots, e_m\}$ for $\mathcal{R}(K)$ and any basis $\{\phi_1, \ldots, \phi_m\}$ for $\mathcal{R}(K^*)$, it follows that*

$$K = \sum_{i,j=1}^{m} K_{ij} e_i \oplus \phi_j, \qquad K^* = \sum_{i,j} K_{ji} \phi_i \otimes \tau e_j$$

for some $m \times m$ matrix $[K_{ij}]_{i,j=1}^m$, where τ is the canonical inclusion $E \subset E^{**}$. We use the standard notation: $e \otimes \phi \in \mathcal{L}(E)$ when $e \in E$, $\phi \in E^*$, is defined by $e \otimes \phi(x) = e\phi(x)$, $x \in E$.

Further we may choose the projections P, Q and the operator R (corresponding to $T = I - K, S = I + R$ in the notation (7.2.1) to be in

$$\text{span}\{e_i \otimes \phi_j \mid 1 \le i, j \le m\} \subset \mathcal{L}(E).$$

Proof. Kato [14, p. 161] gives an elementary proof that $\dim \mathcal{R}(K^*) = m$. The rest of the result (except the final remarks) is a problem of linear algebra, left to the reader.

Theorem 7.2.2. *(Continuous kernels) Let M be a compact metric space with a finite positive measure dM, and let $K : M \times M \to \mathbb{C}^{a\times a}$ be continuous. Let $\tilde{K}$ denote the corresponding integral operator in $L_2(M, \mathbb{C}^a)$*

$$\tilde{K}u(x) = \int_M K(x, \cdot)u(\cdot)dM, \qquad x \in M.$$

Using the bilinear duality $(\psi, \phi) \mapsto \int_M \psi \cdot \phi dM = \sum_{j=1}^a \int_M \psi_j \phi_j dM$ on $L_2(M, \mathbb{C}^a)$, the adjoint $\tilde{K}^*$ of $\tilde{K}$ is the integral operator with kernel $(x, y) \mapsto {}^\top K(y, x)$; $\mathcal{N}(I - \tilde{K})$ and $\mathcal{N}(I - \tilde{K}^*)$ are in $C(M, \mathbb{C}^a)$; and we may choose continuous $P, Q, R : M \to \mathbb{C}^{a\times a}$ such that the corresponding integral operators $\tilde{P}, \tilde{Q}, \tilde{R}$ satisfy the condition of (7.2.1).

$$\begin{gathered}\tilde{P}, \tilde{Q} \text{ are projections}, \mathcal{R}(\tilde{P}) = \mathcal{N}(I - \tilde{K}), \mathcal{N}(\tilde{Q}) = \mathcal{R}(I - \tilde{K}),\\ (I - \tilde{K})(I + \tilde{R}) = I - \tilde{Q}, \qquad (I + \tilde{R})(I - \tilde{K}) = I - \tilde{P},\\ (I + \tilde{R})\tilde{Q} = 0, \qquad \tilde{P}(I + \tilde{R}) = 0.\end{gathered}$$

More briefly: a continuous kernel has a continuous resolvent kernel.

Remark. For any choice of the projections $\tilde{P}, \tilde{Q}$ with continuous kernels, the corresponding resolvent kernel will be continuous, by (7.2.2) above.

Proof. If $K(x, y)$ is a continuous degenerate kernel, of the form $\sum_{j=1}^n f_j(x) \otimes g_j(y)$ for some continuous $f_j, g_j : M \to \mathbb{C}^a$, the result follows from Theorem 7.2.1, and we may choose P, Q, R in $\text{span}\{f_j(x) \otimes g_k(y) \mid 1 \le j, k \le N\}$.

If $K(x, y)$ is continuous and small,

$$\int_M |K(x, \cdot)| dM \le \theta < 1 \qquad \text{for all } x \in M,$$

then the Neumann series $K + K \circ K + K \circ K \circ K + \cdots = R$ converges in $C^0(M \times M)$. In fact if $K^j = K \circ K^{j-1} (j \ge 2)$ with $K^1 = K$, then $\|K^{j+1}\|_{C^0} = \sup_{x,y} | \int_M K(x, \cdot) K^j(\cdot, y) dM| \le \theta \; \|K^j\|_{C^0}$ so $\sum_1^\infty \|K^j\|_{C^0} \le \|K\|_{C^0} / (1 - \theta) < \infty$. Then $P = 0$, $Q = 0$ and R are the desired kernels.

In the general case let L be a continuous degenerate kernel such that $K - L$ is uniformly small on $M \times M$,

$$\sup_x \int_M |K(x,\cdot) - L(x,\cdot)| dM < 1.$$

Combining the previous cases, we find a continuous resolvent kernel for K. If $u = \tilde{K}u \in L_2(M)$ in fact u is continuous (after possible correction in a null set) so $\mathcal{N}(I - \tilde{K}) \subset C(M)$ and similarly $\mathcal{N}(I - \tilde{K}^*) \subset C(M)$. Since $C(M)$ is dense in $L_2(M)$ we may choose the projections $\tilde{P}, \tilde{Q}$ with continuous kernels. Since $L - K$ is small, there is a continuous kernel S so that $I + \tilde{S} = (I + \tilde{L} - \tilde{K})^{-1}$. If $u = \tilde{K}u + f - \tilde{Q}f$, $\tilde{P}u = 0$, then $u + (\tilde{L} - \tilde{K})u = \tilde{L}u + (1 - \tilde{Q})f$, $u = \tilde{L}_1 u + (1 - \tilde{Q}_1)(1 + \tilde{S})f$ where $L_1 = L + S \circ L$, $\tilde{Q}_1 = (1 + \tilde{S})\tilde{Q}(1 + \tilde{S})^{-1}$ is a projection and L_1, Q_1 are continuous degenerate kernels. Also $\mathcal{N}(\tilde{Q}_1) = (I + \tilde{S})\mathcal{N}(\tilde{Q}) = (1 + \tilde{S})\mathcal{R}(1 - \tilde{K}) = (I + \tilde{S})\mathcal{R}(I + \tilde{S})^{-1} - \tilde{L}) = \mathcal{R}(I - \tilde{L}_1)$ so there is a continuous degenerate resolvent kernel R_1 for L_1 with $(I - \tilde{L}_1)(I + \tilde{R}_1) = I - \tilde{Q}_1$. Thus $(1 - \tilde{L}_1)u = (1 - \tilde{L}_1)(1 + \tilde{R}_1)(1 + \tilde{S})f$, $\tilde{P}u = 0$. Define the continuous kernel R by

$$I + \tilde{R} = (I - \tilde{P})(I + \tilde{R}_1)(I + \tilde{S})$$

and it is easily verified that $\tilde{P}(I + \tilde{R}) = 0$ and $(I - \tilde{K})(I + \tilde{R}) = I - \tilde{Q}$, i.e., is a continuous resolvent kernel.

Theorem 7.2.3. *Suppose M is a compact C^{r+1} manifold without boundary provided with a positive measure dM, C^r-equivalent to Lebesgue measure, and let $K : M \times M \to \mathbb{C}^{a \times a}$ be a kernel of class $C^r_*(\alpha)$. Then we may choose the kernel P, Q of the projections to be C^r and it follows that the resolvent kernel R is of class $C^r_*(\alpha)$ and in fact $R - (K + K^2 + \cdots + K^N)[K^{j+1} = K \circ K^j]$ is a C^r kernel for N large, $N > (r + m)/\alpha$. If K is C^r then P, Q, R are C^r.*

Given any C^r kernels P_1, Q_1 for the projections, the corresponding resolvent kernel R_1 is $C^r_*(\alpha)$ and $R_1 - R$ is a C^r degenerate kernel, by (7.2.2).

Proof. Consider $\tilde{K}$ as an operator in $\mathcal{L}(L_2(M, \mathbb{C}^a))$. If $u = \tilde{K}u \in L_2(M)$, then $u = \tilde{K}^N u$ for every N, and K^N is a C^r kernel for large N so (after possible correction on a set of measure zero) u is C^r. Similarly for$\tilde{K}^*$, so we may choose the projections to have degenerate C^r kernels$P$, Q.

Now suppose K is a C^r kernel. We know the resolvent kernel R is at least continuous and $(I - \tilde{K})(I + \tilde{R}) = I - \tilde{Q}$ so

$$R(x, y) = \int_M K(x, \cdot)R(\cdot, y) + K(x, y) - Q(x, y) \qquad \text{for all } x, y \text{ in } M.$$

Thus for any i-order derivative D^i on $M (i \le r)$

$$\tilde{D}_y^j D_x^i R(x, y) \text{ is continuous on } M \times M$$

and R is a C^r kernel.

Now suppose K is a $C_*^r(\alpha)$ kernel, and choose N so $N\alpha > r + m$ and $N + 1$ is a sufficiently large prime so that $\exp(2\pi ij/(N + 1))$ is not an eigenvalue of $\tilde{K}$ for $1 \le j \le N$. (There are finitely many eigenvalues $\{\exp(2\pi ip/q)\}$ which are roots of unity – consider the prime factors of the q's.) This ensures $\mathcal{N}(I - \tilde{K}^{N+1}) = \mathcal{N}(I - \tilde{K})$ and $\mathcal{N}(I - \tilde{K}^{*N+1}) = \mathcal{N}(I - \tilde{K}^*)$, so we may use the same projections $\tilde{P}$, $\tilde{Q}$ for $\tilde{K}$ and for $\tilde{K}^{N+1}$. Let R_N be the C^r resolvent kernel for K^{N+1}. There exists a unique compact operator $\hat{R} \in (L_2(M))$ such that $(I - \tilde{K})(I + \hat{R}) = I - \tilde{Q}$, $\tilde{P}(I + \hat{R}) = 0$, and we must show $\hat{R}$ is an integral operator with $C_*^r(\alpha)$ kernel. Now $\hat{R} = \tilde{K}\hat{R} + \tilde{K} - \tilde{Q}$ so $\hat{R} = \tilde{K}^{N+1}\hat{R} + (1 + \tilde{T})$ $(\tilde{K} - \tilde{Q})$, $T = K + K^2 + \cdots + K^N$, so

$$\begin{aligned}(1 - \tilde{K}^{N+1})\hat{R} &= (1 + \tilde{T})(\tilde{K} - \tilde{Q}) = (1 - \tilde{Q})(1 + \tilde{T})(\tilde{K} - \tilde{Q}) \\ &= (1 - \tilde{K}^{N+1})(1 + R_N)(1 + \tilde{T})(\tilde{K} - \tilde{Q}).\end{aligned}$$

But $1 - \tilde{K}^{N+1} = (1 + \tilde{T})(1 - \tilde{K})$ and $1 + \tilde{T}$ is injective, by choice of N. It follows that $\hat{R} = (I + \tilde{R}_N)(I + \tilde{T})(\tilde{K} - \tilde{Q}) - \tilde{P}$ which is an integral operator whose kernel R is of class $C_*^r(\alpha)$ and

$$R - T = K^{N+1} - Q - P + R_N \circ (K - Q) - T \circ Q + R_N \circ T \circ (K - Q)$$

is C^r.

7.3 Fourier Transform and Composition of Weakly-Singular Kernels

The preceding theorems will be used to estimate remainders, but here we show how to calculate explicitly compositions of weakly-singular kernels of the form

$$|x - y|^{\alpha - m}(\log|x - y|)^j Q\left(x, y, \frac{y - x}{|y - x|}\right).$$

These compositions are (approximately) convolutions, so the Fourier transform is a convenient tool. We define the transform $f \mapsto \hat{f}$ by

$$\hat{f}(\xi) = \int_{\mathbb{R}^m} e^{-ix\cdot\xi} f(x)dx, \qquad f(x) = (2\pi)^{-m} \int_{\mathbb{R}^m} e^{ix\cdot\xi} \hat{f}(\xi)d\xi$$

for "nice" f, extended to tempered distributions $T \in \mathcal{S}'(\mathbb{R}^m)$ by: $\hat{T} \in \mathcal{S}'(\mathbb{R}^m)$ satisfies

$$\langle \hat{T}, \phi \rangle = \langle T, \hat{\phi} \rangle \qquad \text{for all } \phi \in \mathcal{S}(\mathbb{R}^m).$$

[$\mathcal{S}(\mathbb{R}^m)$ is the Schwartz space of C^∞ functions ϕ on $\mathbb{R}^m$ such that $x^j \partial_x^k \phi(x)$ is bounded on $\mathbb{R}^m$ for all j, k and $\mathcal{S}'(\mathbb{R}^m)$ is the dual space.]

Let $\mathcal{H}_k(\mathbb{R}^m)$ denote the space of harmonic polynomials on $\mathbb{R}^m (\Delta p(x) \equiv 0)$ which are homogeneous of degree k, the "solid spherical harmonics." The restrictions to S^{m-1} of the harmonic polynomials give all polynomials on S^{m-1}; given any homogeneous k-order polynomial $p(x)$ on $\mathbb{R}^m$, we have $p(x) = \sum_{0 \le j \le k/2} |x|^{2j} p_j(x)$ for some $p_j \in \mathcal{H}_{k-2j}(\mathbb{R}^m)$. (See Stein [38] for this and other properties of harmonic polynomials.) If $p \in \mathcal{H}_j(\mathbb{R}^m)$, $q \in \mathcal{H}_k(\mathbb{R}^m)$ and $j \neq k$, then $\int_{S^{m-1}} p(x)q(x)dA_x = 0$.

For $\alpha > 0$ and $p \in \mathcal{H}_k(\mathbb{R}^m)$, $|x|^{\alpha-m} p(x/|x|)$ or more precisely

$$\phi \to \int_{\mathbb{R}^m} |x|^{\alpha-m} p(x/|x|)\phi(x)dx, \qquad \phi \in \mathcal{S}(\mathbb{R}^m), \tag{7.3.1}$$

is a tempered distribution which is an analytic function of α and has analytic continuation for all complex α with $\alpha \notin \{-k, -k-2, -k-4, \cdots\}$. In fact, for any integer $N \ge 0$

$$\begin{aligned}
\int_{\mathbb{R}^m} & |x|^{\alpha-m} p(x/|x|)\phi(x) \\
&= \int_{|x|>1} |x|^{\alpha-m} p(x/|x|)\phi(x) + \int_{|x|<1} |x|^{\alpha-m} p(x/|x|) \\
&\quad \times \left\{ \phi(x) - \sum_{|j|\le N} \frac{1}{j!} x^j \partial^j \phi(0) \right\} \\
&\quad + \sum_{|j|\le N} \frac{1}{j!} \frac{\partial^j \phi(0)}{|j| + \alpha} \int_{S^{m-1}} p(\omega)\omega^j dA_\omega
\end{aligned} \tag{7.3.2}$$

and $\int_{S^{m-1}} p(\omega)\omega^j = 0$ unless $|j| \in \{k, k+2, k+4, \ldots\}$. This formula gives $\langle |x|^{\alpha-m} p(x/|x|), \phi(x) \rangle$ for any $\alpha \in \mathbb{C} \backslash \{-k, -k, -2, \ldots\}$, provided we choose N large enough. If $\alpha = -k-2i$ for some integer $i \ge 0$ the terms with $|j| = k+2i$ in the sum are not defined; we merely throw away these terms, and the rest is (by definition) the *finite part*, $\langle \text{f.p.}(|x|^{-m-k-2i} p(x/|x|), \phi(x) \rangle$. We see that, when $\alpha = -k-2i+\epsilon$

$$\langle |x|^{\alpha-m} p(x/|x|), \phi(x) \rangle - \frac{1}{\epsilon} \sum_{|j|=k+2i} \frac{1}{j!} \partial^j \phi(0) \int_{|\omega|=1} p(\omega)\omega^j \tag{7.3.3}$$

converges to $\langle \text{f.p.}(|x|^{-m-k-2i} p(x/|x|)), \phi(x) \rangle$ as $\epsilon \to 0$.

Now we find the Fourier transform (following Stein [38]).

Theorem 7.3.1. $\{|x|^{\alpha-m}p(x/|x|)\}^\wedge(\xi) = \gamma_{\alpha,\kappa}|\xi|^{-\alpha}p(\xi/|\xi|)$, *where*

$$\gamma_{\alpha,k} = i^{-k}\pi^{m/2}2^{\alpha}\Gamma\left(\frac{k+\alpha}{2}\right)\Big/\Gamma\left(\frac{m+k-\alpha}{2}\right)$$

$p \in \mathcal{H}_k(\mathbb{R}^m)$ *and* α *is complex with*

$$\alpha \notin \{-k, -k-2, -k-4, \ldots; m+k, m+k+2, \ldots\}$$

Remark. If $0 < \operatorname{Re}\alpha < m$, both distributions make "classical" sense; and if $0 < \operatorname{Re}\alpha < m/2$, then $|x|^{\alpha-m}p(x/|x|)$ is the sum of a function in $L_1(\mathbb{R}^m)$ – supported in $(\xi) \leq 1$ – and a function in $L_2(\mathbb{R}^m)$, so the Fourier transform also makes sense (in $L_\infty + L_2$). But in general, we must interpret these as distributions, obtained by analytic continuation outside the "classical strip," $0 < \operatorname{Re}\alpha < m$.

The left-side of the equation is also analytic for $\alpha = m+k, m+k+2, \ldots$; we study the behavior of the transform near these points, after this proof.

Proof. By the mean-value property of harmonic functions, for any $x \in \mathbb{R}^m$, the average value of $p(x+y)$ on any sphere $|y| = \text{const.}$ is $p(x)$, therefore $\int_{\mathbb{R}^m} e^{-|y|^2}p(x+h)dy = p(x)\int e^{-y^2}dy = \pi^{m/2}p(x)$. By analytic continuation, this holds for all $x \in \mathbb{C}^n$. Thus for any $\lambda > 0$, $\xi \in \mathbb{R}^m$,

$$\begin{aligned}
&\int_{\mathbb{R}^m} \exp(-\lambda|x|^2 - i\xi\cdot x)p(x)dx \\
&\quad = \int_{\mathbb{R}^m} dz\exp(-z^2 - \xi^2/4\lambda)\lambda^{-(m+k)/2}p(z - i\xi/2\sqrt{\lambda}) \\
&\quad = \pi^{m/2}\lambda^{-m/2-k}e^{-\xi^2/4\lambda}p(-i\xi/2).
\end{aligned}$$

Let $\phi \in C_c^\infty(\mathbb{R}^m\backslash\{0\})$; then for $0 < \delta < (m+k)/2$,

$$\begin{aligned}
&\int_{\mathbb{R}^m} \Gamma(\delta)|x|^{-2\delta}p(x)\hat{\phi}(x)dx \\
&\quad = \int_0^\infty \lambda^{\delta-1}d\lambda\int_{\mathbb{R}^m} e^{-\lambda x^2}p(x)\hat{\phi}(x)dx \\
&\quad = \pi^{m/2}\int_{\mathbb{R}^m}\left(\int_0^\infty \lambda^{\delta-\frac{m}{2}-k-1}e^{-\xi^2/4\lambda}d\lambda\right)p(-i\xi/2)\phi(\xi)d\xi,
\end{aligned}$$

and we have used the fact that $|\xi| \geq \text{const.} > 0$ on supp ϕ. If also $\delta > k/2$, the last integral converges and the equation holds for all $\phi \in \mathcal{S}(\mathbb{R}^m)$, by continuity. Now let $\delta = \frac{1}{2}(m+k-\alpha)$; $k/2 < \delta < (m+k)/2$ when $0 < \alpha < m$, and this gives the result claimed for $0 < \alpha < m$. Analytic continuation gives the general case.

Now we examine the transform for α near $\{m+k, m+k+2, \ldots\}$. If $\alpha = m+k+2j$ (j = integer ≥ 0), $|x|^{\alpha-m} p(x/|x|) = |x|^{2j} p(x)$ is a polynomial and the transform is well-known:

$$\langle |x|^{2j} p(x)\}^\wedge(\xi), \phi(\xi)\rangle = (2\pi)^m - (-\Delta_\xi)^j p(-i\partial_\xi)\phi(\xi)|_{\xi=0}$$

so the transform is a derivative of Dirac's δ, with support $\{0\}$. We compute the derivative with respect to α:

$$\begin{aligned}&\langle\{|x|^{2j} p(x)\log|x|\}^\wedge(\xi), \phi(\xi)\rangle\\ &= \lim_{\alpha\to m+k+2j} \big\langle(\gamma_{\alpha,k}\log|\xi|^{-1} + \gamma'_{\alpha,k})|\xi|^{-\alpha} p(\xi/|\xi|), \phi(\xi)\big\rangle.\end{aligned}$$

Using formula (7.3.2) – or its derivative with respect to α – we see that this is

$$\gamma'\langle \text{f.p.}|\xi|^{-m-k-2j} p(\xi/|\xi|), \phi(\xi)\rangle + \lim_{\alpha\to m+k+2j} \sum_{|S|=k+2j} \frac{1}{S!}\partial^S\phi(0)$$
$$\int_{|\omega|=1} P(\omega)\omega^2 \left\{\frac{\gamma'_{\alpha,k}}{m_k+2j-\alpha} + \frac{\gamma_{\alpha,k}}{(m+k+2j-\alpha)^2}\right\}$$

where $\gamma'_{\alpha,k} = \frac{\partial}{\partial\alpha}\gamma_{\alpha,k}$, and γ' is $\gamma'_{\alpha,k}$ for $\alpha = m+k+2j$. If $\alpha = m+k+2j+\delta$, the final coefficient above is

$$\frac{\gamma' + \delta\gamma'' + O(\delta^2)}{-\delta} + \frac{\delta\gamma' + \frac{1}{2}\delta^2\gamma'' + O(\delta^3)}{\delta^2} \to -\frac{1}{2}\gamma''.$$

Thus there is a polynomial $q(x)$, homogeneous of degree $k+2j$, such that

(7.3.4) $$\{|x|^{2j} p(x)\log|x| + q(x)\}^\wedge(\xi) = \gamma' \,\text{f.p.}\,(|\xi|^{-m-k-2j} p(\xi/|\xi|))$$

and

(7.3.5) $$\gamma' = \frac{\partial}{\partial\alpha}\gamma_{\alpha,k}|_{\alpha=m+k+2j} = j!(-1)^{j+1} i^{-k} 2^{\alpha-1}\pi^{m/2}\Gamma\left(\frac{k+\alpha}{2}\right) \neq 0.$$

In fact, $q(x) = (\gamma''/2\gamma')|x|^{2j} p(x)$ – but we do not need to know q explicitly.

Differentiating j times with respect to α gives

$$\left\{|x|^{\alpha-m}(\log|x|)^j p\left(\frac{x}{|x|}\right)\right\}^\wedge(\xi) = \sum_{S=0}^{j}\binom{j}{s}\gamma^{(s)}_{\alpha,k}(\log|\xi|^{-1})^{j-s}|\xi|^{-\alpha} p(\xi/|\xi|)$$

where $\gamma^{(s)}_{\alpha,k} = (\frac{\partial}{\partial\alpha})^s\gamma_{\alpha,k}$, so we may handle integral powers of $\log|x|$ with equal ease (or difficulty, depending on α).

We may also treat more general "direction functions" $\phi(x/|x|)$ by expansion in harmonic polynomials. Choose an orthonormal basis for $L_2(S^{m-1})$,

$\{p_j|S^{m-1}\}_{j=0}^{\infty}$, with each p_j a homogeneous harmonic polynomial ($p_0 =$ constant) and $j \mapsto d_j = \deg p_j$ non-decreasing. (We assume $m > 1; m = 1$ is trivial.) Then $d_j \approx (\frac{(m-1)!}{2} j)^{1/m-1}$ as $j \to \infty$, $\Delta_S p_j = -d_j(d_j + m - 2)p_j$ on S^{m-1}, and the Sobolev spaces on S^{m-1} may be written

$$H^\sigma(S^{m-1}) = \mathcal{D}(1-\Delta_S)^{\sigma/2} = \left\{C_j p_j \middle| \sum_0^\infty \sum_0^\infty |C_j|^2(1+d_j(d_j+m-2))^\sigma < \infty\right\}$$

for all $\sigma \geq 0$. If α is not an integer and $\phi = \sum_0^\infty \phi_j p_j \in H^\sigma(S^{m-1})$ then

$$\left\{|x|^{\alpha-m}\phi\left(\frac{x}{|x|}\right)\right\}^\wedge(\xi) = |\xi|^{-\alpha} T_\alpha \phi\left(\frac{\xi}{|\xi|}\right) \tag{7.3.6}$$

where $T_\alpha\phi(\omega) = \sum_0^\infty \phi_j \gamma_{\alpha,d_j} p_j(\omega)$, $\omega \in S^{m-1}$. If α is an integer, there may be finitely many j for which γ_{α,d_j} is zero or ∞ (the troublesome terms have $d_j \leq \max(\alpha - m, -\alpha)$); such terms are treated separately, and we define T_α only in the orthogonal complement. Since $|\gamma_{\alpha,d}| \approx (2\pi)^m d^{\alpha-m/2}$ as $d \to \infty$, it follows that $T_\alpha : H^\sigma(S^{m-1}) \to H^{\sigma-\alpha+m/2}(S^{m-1})$ [modulo, perhaps, a finite-dimensional subspace] is an isomorphism. For example, if α, β and $\alpha+\beta$ are not integers (to simplify) and $\sigma > -1/2$, $\phi \in H^{\sigma+\alpha}(S^{m-1})$, $\psi \in H^{\sigma+\beta}(S^{m-1})$, then

$$\left\{|x|^{\alpha-m}\phi\left(\frac{x}{|x|}\right)\right\} * \left\{|x|^{\beta-m}\psi\left(\frac{x}{|x|}\right)\right\} = |x|^{\alpha+\beta-m}\theta\left(\frac{x}{|x|}\right)$$

where $\theta = T_{\alpha+\beta}^{-1}(T_\alpha\phi \cdot T_\beta\psi) \in H^{\sigma+\alpha+\beta}(S^{m-1})$. Here we use the fact that $H^\sigma(S^{m-1})$ is an algebra under pointwise multiplication when $\sigma > (m-1)/2$.

Fortunately our explicit calculations involve only polynomials, hence finite sums. To facilitate these calculations, we include a table of values of $\gamma_{\alpha,k}$ and introduce $n = m+1$ (which occurs naturally in our calculations).

$$n = m+1, \quad \omega = \omega_n, \quad \bar{\omega} = \omega_{n-1} = \omega_m = 2\pi^{m/2}/\Gamma(m/2).$$

$\gamma_{\alpha,k}$	$k=0$	$k=1$	$k=2$	$k=3$	$k=4$
$\alpha=1$	$\frac{1}{2}(n-2)\omega$	$-i\bar{\omega}$	$-\frac{1}{2}\omega$	$2i\bar{\omega}(n-1)$	$3\omega/2n$
$\alpha=2$	$(n-3)\bar{\omega}$	$-\frac{1}{2}(n-2)i\omega$	$-2\bar{\omega}$	$\frac{3}{2}i\omega$	$8\bar{\omega}/(n-1)$
$\alpha=3$	$\frac{1}{2}(n-2)(n-4)\omega$	$-2(n-3)i\bar{\omega}$	$-\frac{3}{2}(n-2)\omega$	$8i\bar{\omega}$	$\frac{15}{2}i\omega$
$\alpha=4$	$2(n-3)(n-5)\bar{\omega}$	$-\frac{3}{2}(n-2)(n-4)i\omega$	$-8(n-3)\bar{\omega}$	$\frac{15}{2}(n-2)i\omega$	$48\bar{\omega}$

When $\gamma_{\alpha,k} = 0$, this table cannot be used to find the inverse transform; but in that case $\gamma'_{\alpha,k} = \frac{\partial}{\partial\alpha}\gamma_{\alpha,k} \neq 0$ and we can use (7.3.5). We complement this table by giving the derivative $\gamma'_{\alpha,k}$ for the points where $\gamma_{\alpha,k} = 0$.

n=2 (m=1)

$\gamma'_{\alpha,k}$	$k=0$	1	2	3
$\alpha=1$	$-\pi$			
$\alpha=2$	·	πi		
$\alpha=3$	12π	·		3π
$\alpha=4$	·	$-36\pi i$	·	$-15\pi i$

$n=3$

	$k=0$	1	2
$\alpha=1$			
2	-2π		
3	·	$4\pi i$	
4	8π	·	16π

n=4

	$k=0$	1
$\alpha=1$		
2		
3	$-2\pi^2$	
4	·	$6\pi^2 i$

$n=5$

	$k=0$
$\alpha=1$	
2	
3	
4	$-16\pi^2$

Another simpler method seems to always work, which we illustrate with an example. If $n \neq 3$, $(F.T.)^{-1}(|\xi|^{-2}) = \frac{1}{(n-3)\omega_{n-1}}|x|^{2-m} = -\frac{1}{\omega_{n-1}}\frac{|x|^{3-n}}{3-n}$; and the result for $n=3$ is obtained by replacing $|x|^{3-n}/(3-n)$ by $\log|x|$, i.e., $(F.T.)^{-1}(|\xi|^{-2}) = -\frac{1}{2\pi}\log|x|$. More precisely, $(F.T.)^{-1}(\text{f.p.}|\xi|^{-2})$ is $-\frac{1}{2\pi}\log|x|$ plus a constant, when $n=3$ $(m=2)$.

The Fourier transform shows the action of an integral operator on rapidly oscillating functions. Suppose, for example, Ω is a bounded C^{r+1} region in $\mathbb{R}^n$, $K(x,y)$ is a $C^r_*(\alpha)$ kernel in $\partial\Omega$, $0<\alpha\le r$, ϕ is C^r on $\partial\Omega$ and ψ is C^{r+1} and real-valued with $\nabla_{\partial\Omega}\psi(y)\neq 0$ on supp ϕ, and consider

$$(7.3.7)\qquad J(t,x) \equiv \int_{\partial\Omega} K(x,y)\phi(y)\exp(it\psi(y))dAy, \qquad x\in\partial\Omega$$

as $t\to+\infty$. If x is outside supp ϕ, it is easy to see $J(t,x) = O(t^{-r})$, and for every x (and $\epsilon>0$) we show $J(t,x) = O(t^{-\alpha+\epsilon})$ as $t\to+\infty$. It is enough to treat the case when supp ϕ is small, and then we may write $\partial\Omega$ (in supp ϕ) as $\{h(\xi)\,|\,\xi\in V^{\text{open}}\subset\mathbb{R}^{n-1}\}$, $h: V\to\partial\Omega\subset\mathbb{R}^n$ is a C^{r+1} embedding, and

$$J(t,h(\xi)) = \int_V K(h(\xi),h(\eta))\phi(h(\eta))\exp(it\psi(h(\eta)))\sqrt{\det{}^{\top}h_\eta h_\eta}\,d\eta.$$

We may choose new coordinates ζ in $\mathbb{R}^{n-1}$ so that $\zeta_1 = \psi(h(\eta))$ and the problem reduces to: if $\tilde{K}$ is a $C^r_*(\alpha)$ kernel with compact support in $\mathbb{R}^m$ $(m=n-1)$

$$\int_{\mathbb{R}^m}\tilde{K}(x,y)e^{ity_1}dy = O(t^{-\alpha+\epsilon}) \quad \text{as } t\to+\infty.$$

If $r \geq \alpha > 1$ then $\int_{\mathbb{R}^m} \tilde{K}(x,y)e^{ity_1}dy = -\frac{1}{it}\int_{\mathbb{R}^m} \partial_{y_1}\tilde{K}(x,y)e^{ity_1}dy$ and we eventually reduce to the case $0 < \alpha \leq 1 \leq r$. But in this case, if $e_1 = (1, 0, \ldots, 0)$,

$$\begin{aligned}\int_{\mathbb{R}^m} \tilde{K}(x,y)e^{ity_1}dy &= -\int_{\mathbb{R}^m} \tilde{K}(x, y + \pi e_1/t)e^{ity_1}dy \\ &= \tfrac{1}{2}\int_{\mathbb{R}^m} (\tilde{K}(x,y) - \tilde{K}(x, y+\pi e_1/t))e^{ity_1}dy\end{aligned}$$

so

$$\begin{aligned}&\left|\int_{\mathbb{R}^m} \tilde{K}(x,y)e^{ity_1}dy\right| \\ &\quad\leq \text{Const.}\int_{|y-x|<5/t} (|y-x|^\sigma + |y+\pi e_1/t - x|^\sigma) \\ &\quad\quad + \text{Const.}\int_{R\geq|y-x|\geq 5/t} t^{-1}(|y-x|^{\sigma-1} + |y+\pi e_1/t - x|^{\sigma-1}) \\ &\quad\leq O\big(t^{-(\sigma+m)}\big) = O\big(t^{-\alpha+\epsilon}\big),\end{aligned}$$

where $\sigma = \alpha - m - \epsilon$.

Suppose $K(x,y) - K_0(x, y-x)$ is a $C_*^S(\beta)$ kernel on $\partial\Omega$ with $r \geq s \geq \beta > \alpha$, $K_0(x,z)$ depends only on the component of z in the tangent space $T_x(\partial\Omega)$, and the Fourier transform $\hat{K}_0(x,\cdot)$ of $K_0(x,\cdot)|_{T_x(\partial\Omega)}$ is computable. Then $\int_{\partial\Omega} K(x,y)\phi(y)e^{it\psi(y)}dAy = \hat{K}_0(x, -t\nabla_{\partial\Omega}\psi(x))\phi(x)e^{it\psi(x)} + o(t^{-\alpha})$ as $t \to +\infty$. If $z \to K_0(x,z)$ is homogeneous of degree $\alpha - m$, and if $\tilde{K}$ is of finite rank, it follows that $K_0(x,z) = 0$ for all $z \in T_x(\partial\Omega)$.

Now consider a bounded open set $\Omega \subset \mathbb{R}^n$ with C^{r+1} boundary $\partial\Omega$ ($r \geq 2$) and a $C_*^r(\alpha)$ kernel $K(x,y)$ on $\partial\Omega$ of the form

$$K(x,y) = |x-y|^{\alpha-m}(\log(x-y))^j Q\left(x, y, \frac{y-x}{|y-x|}\right)$$

where $m = n-1 = \dim\partial\Omega$, $\alpha > 0$ and $Q(x,y,\omega)$ is a C^r function. For $y \in \partial\Omega$ near x, $y = x + \eta - \nu_x(\eta)N_x$, $\eta \in T_x(\partial\Omega)$; i.e., we may write $\partial\Omega$ as a graph over the tangent plane at x, $\eta = P_x(y-x)$ is the orthogonal projection of $y - x$ onto $T_x(\partial\Omega)$, and $\nu_x(\eta) = \tfrac{1}{2}K_x(\eta^2) + O(|\eta|^3)$ as $\eta \to 0$, where K_x is the curvature at x. Then $|x-y| = |\eta|(1 + \nu_x(\eta)^2/|\eta|^2)^{1/2} = |\eta|(1 + O(|\eta|^2))$ and

$$\begin{aligned}K(x, x+\eta-\nu_x(\eta)N_x) &= |\eta|^{\alpha-m}(\log|\eta|)^j Q(x,x,\eta/|\eta|) \\ &\quad + |\eta|^{\alpha+1-m}(\log|\eta|)^j(\omega Q_y(x,y,\omega) \\ &\quad - \tfrac{1}{2}K_x\omega^2 N_x Q_\omega(x,y,\omega))|_{y=x,\omega=\eta/|\eta|} + \cdots \\ &= \sum_{j=0}^{N-1}\tilde{K}_j(x,\eta) + \tilde{K}_N(x,\eta)\end{aligned}$$

or $K(x, y) = \sum_0^N K_j(x, y)$, $K_j(x, y) = \tilde{K}_j(x, P_x(y-x))$, where K_j is C_*^{r-j} $(\alpha + j)$ $(0 \le j \le N)$ and the Fourier transforms of $\eta \mapsto \tilde{K}_j(x, \eta)$ $(0 \le j < N)$ are explicitly computable when $\omega \mapsto Q(x, y, \omega)$ is a polynomial.

Suppose L is another such $C_*^r(\beta)$ kernel on $\partial\Omega$, and we compute the composition $K \circ L$,

$$K \circ L(x, z) = \int_{\partial\Omega} K(x, y) L(y, z) dA_y.$$

Let $\theta : \mathbb{R}^n \to \mathbb{R}$ be C^∞ with small support, but $\theta \equiv 1$ near 0 : a "cutoff" function. If K, L are $C_*^r(\alpha)$, $C_*^r(\beta)$ respectively, $K_\theta(x, y) = K(x, y)\theta(y - x)$ and $L_\theta(y, z) = L(y, z)\theta(z - y)$, then $K - K_\theta, L - L_\theta$ are C^r kernels and $K \circ L - K_\theta \circ L_\theta$ is also a C^r kernel. Without loss of generality, we shall assume K and L are supported in a small neighborhood of the diagonal of $\partial\Omega \times \partial\Omega$.

Define $h_x : T_x(\partial\Omega) \to \partial\Omega : \eta \mapsto x + \eta - \nu_x(\eta) N_x$ (η small) and then for z close to x

$$K \circ L(x, z) = \int_{T_x(\partial\Omega)} K(x, h_x(\eta)) \sqrt{1 + |\nu_x'(\eta)|^2} \cdot L(h_z(\tilde{\eta}), z)|_{h_z(\tilde{\eta}) = h_x(\eta)} d\eta.$$

We can solve $h_z(\tilde{\eta}) = h_x(\eta)$ for $\tilde{\eta} \in T_z(\partial\Omega)$ as a function of x, z and $\eta \in T_x(\partial\Omega)$. In fact, let $z = h_x(\zeta)$, $\zeta \in T_x(\partial\Omega)$ near 0, and define $J_{x,z} = I - N_x \otimes \nu_x'(\zeta) \in \mathbb{R}^{n \times n}$, where the gradient $\nu_x'(\zeta)$ is considered as a vector in $T_x(\partial\Omega) \perp N_x$. $J_{x,z}$ is an isomorphism of $\mathbb{R}^n$ and it carries $T_x(\partial\Omega)$ onto $T_z(\partial\Omega)$: given a curve $\zeta = \zeta(t)$ in $T_x(\partial\Omega)$ we have $h_x(\zeta(t)) \in \partial\Omega$ and

$$\frac{d}{dt} h_x(\zeta(t)) = J_{x,z} \dot{\zeta}(t) \in T_z(\partial\Omega)$$

if $h_x(\zeta(t)) = z$. Now $h_z(\tilde{\eta}) = h_x(\eta)$ with $\tilde{\eta} = J_{x,z}\sigma$, $z = h_x(\zeta)$, $(\sigma, \zeta \in T_x(\partial\Omega)$, becomes

$$\eta - \zeta - (\nu_x(\eta) - \nu_x(\zeta)) N_x = J_{x,z}\sigma - \nu_z(J_{x,z}\sigma) N_z.$$

Since $P_x J_{xz} \sigma = \sigma$ for $\sigma \in T_x(\partial\Omega)$, we have

$$\eta - \zeta = \sigma - \nu_z(J_{xz}\sigma) P_x N_z.$$

But $\tau = \sigma - \nu_z(J_{xz}\sigma) P_x N_z$ $(\sigma, \tau \in T_x(\partial\Omega))$ is solvable for $\sigma = F_{x,z}(\tau) = \tau + O(|\tau|^2 |\zeta|)$ so $h_z(\tilde{\eta}) = h_x(\eta)$ is solvable for $\tilde{\eta} = J_{xz} F_{xz}(\eta - \zeta)$ and the integral above becomes

$$K \circ L(x, z) = \int_{T_x(\partial\Omega)} K(x, h_x(\eta)) \sqrt{1 + |\nu_x'(\eta)|^2} L(h_z(J_{xz} F_{xz}(\eta - \zeta)), z) d\eta,$$

which is almost a convolution. Let

$$(7.3.8)\qquad \begin{gathered} K(x, h_x(\eta))\sqrt{1+|\nu_x'(\eta)|^2} = \sum_0^N \tilde{K}_j(x,\eta), \\ \tilde{K}_j(x,\eta) = O(|\eta|^{\alpha+j-m-\delta}) \end{gathered}$$

as $\eta \to 0$ in $T_x(\partial\Omega)$, for any $\delta > 0$, and

$$(7.3.9)\qquad \begin{gathered} L\big(h_z\big(J_{xz}F_{xz}J_{xz}^{-1}\tau\big), z\big) = \sum_{j=0}^N \tilde{L}_j(x,z,\tau), \\ \tilde{L}_j(x,z,\tau) = O(|\tau|^{\beta+j-m-\delta}) \end{gathered}$$

as $\tau \to 0$ in $T_z(\partial\Omega)$, for any $\delta > 0$. We suppose $\eta \mapsto \tilde{K}_j(x,\eta)$ $(j < N)$ on $T_x(\partial\Omega)$ and $\tau \mapsto \tilde{L}_k(x,z,\tau)$ $(k < N)$ on $T_z(\partial\Omega)$ have explicitly-computable Fourier transforms. Then we can compute the sum of convolutions

$$(7.3.10)\qquad \sum_{j+k<N} \int_{T_x(\partial\Omega)} \tilde{K}_j(x,\eta)\tilde{L}_k(x,z,J_{xz}(\eta-\zeta))d\eta$$

(where x, z are treated as parameters, and only at the end do we recall that $\zeta = P_x(z-x)$). The difference between this sum and $K \circ L$ is a kernel of class $C_*^{r-N}(\alpha+N)$.

If $\phi \in S(T_x(\partial\Omega))$we have

$$\left\langle \int_{T_x(\partial\Omega)} \tilde{K}_j(x,\eta)\tilde{L}_k(x,z,J_{xz}(\eta-\zeta))\,d\eta, (\hat{\phi}\zeta)\right\rangle$$

$$= \int_{T_x(\partial\Omega)} (F.T.\tilde{K}_j(x,\cdot))(\xi)\cdot F.T.(\tilde{L}_k(x,z,-J_{xz}\cdot))(\xi)\phi(\xi)\,d\xi.$$

Further

$$(7.3.11)\qquad \begin{aligned} F.T.(\tilde{L}_k(x,z,-J_{xz}\cdot))(\xi) &= \int_{T_x(\partial\Omega)} e^{-i\xi\cdot\tau}\tilde{L}_k(x,z,-J_{xz}\tau)\,d\tau \\ &= \int_{T_z(\partial\Omega)} e^{i\xi\cdot\sigma}\tilde{L}_k(x,z,\sigma)\,d\sigma \Big/ \sqrt{1+|\nu_x'(\zeta)|^2}, \end{aligned}$$

using the fact that $\xi \cdot J_{xz}\tau = \xi\cdot\tau$ for $\tau, \xi \in T_x(\partial\Omega)$. Thus we may compute the transform of $\tilde{L}_k(x,z,\cdot)|_{T_z(\partial\Omega)}$, using the earlier formulas, substitute $-P_z\xi$ for ξ and divide by $\sqrt{1+|\nu_x'(\zeta)|^2}$ to get the transform of $\tilde{L}_k(x,z,-J_{xz})|_{T_x(\partial\Omega)}$. If $L(y,z) = |y-z|^{\beta-m}(\log|y-z|)^j R(y,z,\frac{z-y}{|z-y|})$ and $R(y,z,\omega)$ is a polynomial in ω, the transform of $\tilde{L}_k(x,z,-J_{xz})|_{T_x(\partial\Omega)}$ will be a sum of terms $|P_z\xi|^{-\beta-k}(\log|P_z\xi|)^i S_{ik}(x,z,P_z\xi/|P_z\xi|)$ with $S_{ik}(x,z,\omega)$ a polynomial in ω. Writing ω for $\xi/|\xi|$, $|P_z\xi| = |\xi|\,(1+\frac{(\omega\cdot\nu_x'(\zeta))^2}{1+|\nu_x'(\zeta)|^2})^{1/2}$ and so

we have

$$|\xi|^{-\beta-k}\sum_{k=0}^{i}\binom{i}{h}(\log(|\xi|)^{i-h}\left[{}^{1}\!/_{2}\log\left(1+\frac{\omega\cdot\nu_x')^2}{1+|\nu_x'|^2}\right)\right]^h$$

$$\times\left(1+\frac{(\omega\cdot\nu_x'(\zeta))^2}{1+|\nu_x'(\zeta)|^2}\right)^{-\beta-k}\cdot S_{ik}\left(x,z,\frac{\omega-N_z\omega\cdot\nu_x'(\zeta)/\sqrt{1+|\nu_x'|^2}}{\left(1+\frac{(\omega\cdot\nu_x')^2}{1+|\nu_x'|^2}\right)^{1/2}}\right)$$

$$=\sum_{k=0}^{i}|\xi|^{-\beta-k}(\log|\xi|)^{i-h}\tilde{S}_{i,k,h}(x,z,\omega).$$

Since $\zeta = P_x(z-x)$ is small, $\nu_x'(\zeta)$ is also small and we may expand $\tilde{S}_{i,k,h}(x,z,\omega)$ in powers of $\omega\cdot\nu_x'(\zeta)/\sqrt{1+|\nu_x'(\zeta)|^2}$: there results a polynomial in $\omega=\xi/|\xi|$, and a remainder to be estimated using (7.3.6) above. [In fact in our calculations, we don't go beyond two terms: the remarks above serve to show the neglected terms are in fact negligible.]

7.4 Limits of Singular Integrals

Theorem 7.4.1. *Let $\Omega\subset\mathbb{R}^n(n\geq 2)$ be a bounded open set with C^2 regular boundary and let $Q(x,y,\omega)$ be Lipschitz continuous on a neighborhood of $\partial\Omega\times\partial\Omega\times S^{n-1}$. Also assume*

$$\int_{\left\{\substack{\omega\perp N_x\\|\omega|=1}\right\}} Q(x,x,\omega)\,dA_\omega = 0 \quad \textit{for } x\in\partial\Omega.$$

Let $J(x)=\int_{\partial\Omega}Q(x,y,\frac{y-x}{|y-x|})\frac{dA_y}{|x-y|^{n-1}}$ for $x\in\Omega$ near $\partial\Omega$. Then J extends uniquely to be continuous to $\partial\Omega$, and in this extension

$$J(x)=q(x)+P.V.\int_{\partial\Omega}Q\left(x,y,\frac{y-x}{|y-x|}\right)\frac{dA_y}{|x-y|^{n-1}},\qquad x\in\Omega,$$

where

$$q(x)=\int_{\left\{\substack{\omega\cdot N_x>0\\|\omega|=1}\right\}}Q(x,x,\omega)/\omega_0 N_x\,dA_\omega$$

$$=\lim_{\epsilon\to 0^+}\int_{\left\{\substack{\omega\cdot N_x>\epsilon\\|\omega|=1}\right\}}Q(x,x,\omega)/\omega\cdot N_x.$$

[When $\omega\mapsto Q(x,x,\omega)$ is a polynomial, we may compute $q(x)$ using (7.4.3) *below.]*

Example. $(n=2)$ $\partial\Omega = Y \subset \mathbb{R}^2 \approx \mathbb{C}$ and using complex notation,

$$J(z) = \int_\gamma \frac{g(\zeta)}{\zeta - z} d\zeta \quad \text{for } z \in \Omega,$$

where γ is given the usual orientation. Then $\tau_\zeta = d\zeta/ds$ is the "positive" unit tangent vector at $\zeta \in \gamma$ and $-i\tau_\zeta$ is the outward normal, and with $Q(z,\zeta,\omega) = \overline{\omega}\tau_\zeta g(\zeta)$, $J(z) = \int_\gamma Q(z,\zeta,\frac{\zeta-z}{|\zeta-z|})\frac{ds}{|\zeta-z|}$. The "unit sphere" of $T_\zeta\gamma$ has two points, $\pm\tau_\zeta$, and $Q(\zeta,\zeta,\tau_\zeta) + Q(\zeta,\zeta,-\tau_\zeta) = 0$. Also

$$\begin{aligned} q(\zeta) &= \lim_{\epsilon\to 0} \int_{\left\{\substack{\omega\cdot N > \epsilon \\ |\omega|=1}\right\}} \frac{Q(\zeta,\zeta,\omega)}{\omega\cdot N} \\ &= \lim_{\epsilon\to 0} \int_\epsilon^{\pi-\epsilon} \frac{d\theta}{\sin\theta} g(\zeta)(\cos\theta + i\sin\theta) = i\pi g(\zeta) \end{aligned}$$

so

$$\lim_{\substack{z\to z_0 \\ z\in\mathbb{R}}} \int_\gamma \frac{g(\zeta)}{\zeta - z} d\zeta = i\pi g(z_0) + P.V. \int_\gamma \frac{g(\zeta)}{\zeta - z_0} d\zeta \quad \text{for } z_0 \in \gamma,$$

one of the formulas of Plemelj. We first treat a simpler case.

Lemma 7.4.2. *Let $m \geq 1$ and suppose $Q(x,t;\xi,\tau)$ is Lipschitzian on a neighborhood of $(x,t) = (0,0) \in \mathbb{R}^m \times \mathbb{R}$ and $(\xi,\tau) \in S^m_+$,*

$$S^m_+ = \{(\xi,\tau) \in \mathbb{R}^m \times \mathbb{R} : |\xi|^2 + \tau^2 = 1, \tau \geq 0\}.$$

Also assume $Q(x,t;\xi,\tau) = 0$ when $|x| \geq r_0$ and

$$\int_{S^{m-1}} Q(0,0;\xi,0)\,dA_\xi = 0.$$

Define $J(t) = \int_{\mathbb{R}^m} Q(x,t;{}^x/_R,{}^t/_R)R^{-m}dx$, $R = \sqrt{|x|^2+t^2}$, for small $t > 0$; then

$$\lim_{t\to 0^+} J(t) = q + P.V. \int_{\mathbb{R}^m} Q(x,0;{}^x/_{|x|},0)|x|^{-m}dx$$

where

$$\begin{aligned} q &= \int_{S^m_+} \frac{1}{\tau}\{Q(0,0;\xi,\tau) - Q(0,0;\xi/|\xi|,0)\}dA_{\xi,\tau} \\ &= \int_{S^m_+} \frac{1}{\tau} Q(0,0;\xi,\tau)dA_{\xi,\tau} \quad (\textit{if we integrate first with respect to } \xi) \\ &= \int_0^{\pi/2} \frac{(\sin\theta)^{m-1}}{\cos\theta} d\theta \int_{S^{m-1}} Q(0,0;\sigma\sin\theta,\cos\theta)\,dA_\sigma \end{aligned}$$

Proof. Let $q(t) = \int_{|x|<r_0} Q(0, 0, {}^x/_R, {}^t/_R)R^{-m}dx$; then as $t \to 0$,

$$J(t) - q(t) = \int_{|x|<r_0} \{Q(x, t, {}^x/_R, {}^t/_R) - Q(0, 0, {}^x/_R, {}^t/_R)\}R^{-m}dx$$

$$\to \int_{|x|<r_0} \{Q(x, 0, {}^x/_{|x|}, 0) - Q(0, 0, {}^x/_{|x|}, 0)\}|x|^{-m}dx$$

$$= \lim_{\epsilon\to 0} \int_{\epsilon<|x|<r_0} \{\cdots\}|x|^{-m}dx$$

$$= P.V. \int_{\mathbb{R}^m} Q(x, 0; {}^x/_{|x|}, 0)|x|^{-m}dx,$$

by Lebesgue's dominated convergence theorem (dominated by const.$/|x|^{m-1}$). Further, for $t > 0$,

$$q(t) = \int_{|y|<r_0/t} \left\{Q\left(0, 0, \frac{y}{\sqrt{1+|y|^2}}, \frac{1}{\sqrt{1+|y|^2}}\right)\right.$$
$$\left. - Q\left(0, 0, \frac{y}{|y|}, 0\right)\right\}(1+|y|^2)^{-m/2}dy$$

which converges as $t \to 0^+$ to a limit q. If $y = \sigma \tan\theta$ $(0 \le \theta < \pi/2$, $\sigma \in S^{m-1})$ we have

$$q = \lim_{t\to 0^+} q(t)$$
$$= \int_0^{\pi/2} \frac{(\sin\theta)^{m-1}}{\cos\theta} d\theta \int_{S^{m-1}} [Q(0, 0, \sigma\sin\theta, \cos\theta) - Q(0, 0, \sigma, 0)]dA_\sigma$$

which is equivalent to the forms given above.

Remark. If $(\xi, \tau) \mapsto Q(0, 0; \xi, \tau)$ is a polynomial – which is true for our applications – then q is given by

$$q = -\,\omega_m \sum_{j\ge 1} q_j^0 \frac{1}{(\nu+1)\dots(\nu+j)}\left(\frac{1}{\nu+1} + \cdots + \frac{1}{\nu+j}\right)$$
$$+\,\omega_m \sum_{k\ge 1}\sum_{j\ge 0} q_j^{2k} \frac{(k-1)!\nu!}{(j+k+\nu)!}$$
$$+\,\omega_{m+1} \sum_{k\ge 1}\sum_{j\ge 0} q_j^{2k-1} \frac{(k-{}^3/_2)!}{(-{}^1/_2)!}\frac{(\nu+{}^1/_2)!}{(j+k+\nu-{}^1/_2)!} \tag{7.4.1}$$

where $\nu = \frac{m-2}{2}$ and $q_j^k = (\Delta_\xi^j \partial_\tau^k Q(0,0;\xi,\tau)|_{\xi=0,\tau=0})/(j!k!2^{2j+1})$. Thus

$$\begin{aligned} q = &- \omega_m \left[\frac{1}{2m^2}(\Delta_\xi Q)_0 + \frac{m+1}{2m^2(m+2)^2}(\Delta_\xi^2 Q)_0 + \cdots\right] \\ &+ \omega_{m+1}\left[\frac{1}{2}(\partial_\tau Q)_0 + \frac{1}{4(m+1)}(\Delta_\xi \partial_\tau Q)_0 + \cdots\right] \\ &+ \frac{\omega_m}{2!}\left[\frac{1}{m}(\partial_\tau^2 Q)_0 + \frac{1}{2m(m+2)}(\Delta_\xi \partial_\tau^2 Q)_0 + \cdots\right] \\ &+ \frac{\omega_{m+1}}{3!}\left[\frac{1}{2(m+1)}(\partial_\tau^3 Q)_0 + \cdots\right] + \cdots \end{aligned} \tag{7.4.2}$$

Example. Let $Q(x,t;\xi,\tau) = \frac{2\tau}{\omega_{m+1}} u(x,t)$, the Poisson kernel; then

$$\frac{2}{\omega_{m+1}} \int_{\mathbb{R}^m} \frac{tu(x,t)dx}{(|x|^2+t^2)^{(m+1)/2}} \to u(0,0) \qquad \text{as } t \to 0^+ \qquad [q = u(0,0)].$$

Proof of Thm. 7.4.1. Define $J|\partial\Omega$ by the given formula; we show for each $x_0 \in \partial\Omega$, $J(x) \to J(x_0)$ when $x \to x_0$ $(x \in \Omega)$, or equivalently $J(x - \delta N_x) \to J(x_0)$ as $(x,\delta) \to (x_0, 0)$ with $x \in \partial\Omega$, $\delta > 0$. Without loss of generality, $Q(x,y,\omega) = 0$ when $|y - x_0| \geq r_0$, for some small $r_0 > 0$; otherwise multiply by a cut-off-function. Then for x near x_0

$$\begin{aligned} J(x - \delta N_x) = &\int_{T_x(\partial\Omega)} d\eta \sqrt{1 + |\nu_x'(\eta)|^2} \\ &\times Q\left(x - \delta N_x, y, \frac{\eta + (\delta - \nu_x(\eta))N_x}{|\eta + (\delta - \nu)N_x|}\right) \cdot |\eta| + (\delta - \nu)N|^{-m} \end{aligned}$$

with $y = x + \eta - \nu_x(\eta)N_x$. If $\theta : \mathbb{R} \to \mathbb{R}$ is C^∞, $\theta(t) = 1$ for $t \leq 2r_0$, $\theta(t) = 0$ for $t \geq 3r_0$, then $\theta(|\eta|) = 1$ on the support of the integrand (if δ, $|x - x_0|$ and r_0 are small) so

$$\begin{aligned} J(x - \delta N_x) = &\int_{T_x(\partial\Omega)} \theta(|\eta|) Q\left(x, x, \frac{\eta + \delta N_x}{|\eta + \delta N_x|}\right)(|\eta|^2 + \delta^2)^{-m/2} \\ &+ \int_{T_x(\partial\Omega)} \theta(|\eta|) F_0(x, \delta, \eta) \end{aligned}$$

where $|F_0(x,\delta,\eta)| \leq \text{const.}/|\eta|^{m-1}$ and $F_0(x,\delta,\eta)$ is continuous as $(x,\delta) \to (x_0, 0)$ when $\eta \neq 0$. The second integral is handled using dominated convergence, while the first is

$$\int_{T_x(\partial\Omega)} \theta(\delta|\eta|)Q\left(x,x,\frac{\eta+N_x}{\sqrt{|\eta|^2+1}}\right)(|\eta|^2+1)^{-m/2}d\eta$$
$$=\int_{T_x(\partial\Omega)} \theta(\delta|\eta|)\left(Q\left(x,x,\frac{\eta+N_x}{\sqrt{1+|\eta|^2}}\right)-Q\left(x,x,\frac{\eta}{|\eta|}\right)\right)(|\eta|^2+1)^{-m/2}d\eta$$
$$=\int_{T_x(\partial\Omega)} \theta(\delta|\eta|)F_1(x,\eta)d\eta.$$

Then using $J_{x_0,x}=I-N_{x_0}\otimes\nu'_{x_0}(\xi)$ as before ($x=x_0+\xi-\nu_{x_0}(\xi)N_x$) when $(x,\delta)\to(x_0,0)$ with $x\in\partial\Omega$, $\delta>0$

$$\begin{aligned}\sqrt{1+|\nu'_{x_0}(\xi)|^2}J(x-\delta N_x) &= \int_{T_{x_0}(\partial\Omega)} \theta(\delta|J_{x_0x}\eta|)F_1(x,J_{x_0x}\eta)d\eta \\ &\quad+\int_{T_{x_0}(\partial\Omega)} \theta(|J_{x_0x}\eta)|)F_0(x,\delta,J_{x_0x}\eta)d\eta \\ &\to \int_{T_{x_0}(\partial\Omega)} F_1(x_0,\eta)d\eta \\ &\quad+\int_{T_{x_0}(\partial\Omega)} \theta(|\eta|)F_0(x_0,0,\eta)d\eta = J(x_0)\end{aligned}$$

7.5 Fundamental Solutions

We study a differential operator of the form

$$L_x=\Delta+\sum_{j=1}^{n} b_j(x)\frac{\partial}{\partial x_j}+c(x),\qquad \Delta=\sum_{j=1}^{n}\frac{\partial^2}{\partial x_j^2}, \tag{7.5.1}$$

where the $b_j(x)$, $c(x)$ are smooth functions on $\mathbb{R}^n$ with values in $\mathbb{C}^{p\times p}$, the $p\times p$ complex matrices. F. Morgan [*Arch. Rat. Mech. Anal.* **78** (1982)] proves uniqueness in the Cauchy problem for such systems (diagonal in the principal part) when the b_j, c are bounded and measurable. The (formal) adjoint or transposed operator is

$$L_x^*=\Delta_x-\sum_{j=1}^{n} {}^{\top}b_j(x)\frac{\partial}{\partial x_j}+{}^{\top}c(x)-\sum_{1}^{n}\frac{\partial^T b_j(x)}{\partial x_j} \tag{7.5.2}$$

and for smooth $u, v : \Omega \to \mathbb{C}^p$ we have

$$\int_\Omega \{v \cdot Lu - (L^* v) \cdot u\}$$

$$= \int_{\partial\Omega} \left\{ v \cdot \frac{\partial u}{\partial N} - \frac{\partial v}{\partial N} \cdot u + v \left(\sum_1^n b_j N_j \right) u \right\}, \tag{7.5.3}$$

where $u \cdot v = \sum_{k=1}^p u_k v_k$.

Let Q_0 be some fixed bounded open set in $\mathbb{R}^n$ – we will always choose our regions Ω with closure $\overline{\Omega} \subset Q_0$. A *fundamental solution* for L in Q_0 is a C^2 function $(x, y) \mapsto S(x, y) \in \mathbb{C}^{p \times p}(x \neq y$ in $Q_0)$ such that $L_y^{*\top} S(x, y) = 0$ for $x \neq y$ in Q_0, $S(x, y) - E_n(x - y)I = O(|x - y|^{3-n} + 1)$, $\frac{\partial}{\partial y}(S(x, y) - E_n(x - y)I) = O(|x - y|^{2-n})$, where E_n is the fundamental solution for the Laplacian in $\mathbb{R}^n$. $E_n(z) = |z|^{2-n}/(2 - n)|S^{n-1}|$ for $n > 2$, $E_2(z) = \frac{1}{2\pi} \log |z|$ for $n = 2$. In this case, if Ω is a C^2 region with $\overline{\Omega} \subset Q_0$ and $u \in C^2(\Omega, \mathbb{C}^p)$ we have, for any $x \in Q_0$,

$$u(x)\chi_\Omega(x) = \int_\Omega S(x, y) Lu(y) dy + \int_{\partial\Omega} \left\{ \frac{\partial S}{\partial N_y}(x, y) u(y) - S(x, y) \frac{\partial u}{\partial N}(y) - S(x, y) b \cdot N(y) u(y) \right\} \tag{7.5.4}$$

where

$$\chi_\Omega(x) = \begin{cases} 1 & \text{if } x \in \Omega \\ 1/2 & \text{if } x \in \partial\Omega \\ 0 & \text{if } x \notin \overline{\Omega}. \end{cases}$$

A slightly weaker notion – but adequate for our purposes – we call an *approximate fundamental solution*, when ${}^\top(L_y^{*} {}^\top S(x, y)) = \sum_1^N f_j(x) \otimes g_j(y)$ $(x \neq y$ in $Q_0)$, for some given continuous $f_j, g_j : Q_0 \to \mathbb{C}^p$, with other conditions as above. In this case, the only change in (7.5.4) is to add $-\sum_{j=1}^N f_j(x) \int_\Omega g_j(y) \cdot u(y) dy$ to the right-hand side of the equation.

Theorem 7.5.1. *Assume the $b_j(x)$ are locally C^{r+2} on $\mathbb{R}^n$ and $c(x)$ is locally $C^{r+1}(r \geq 0)$ and let Q_0 be any bounded open set in $\mathbb{R}^n$. Then there exists an approximate fundamental solution $S(x, y)$ for L in Q_0 which is a kernel of class $C_*^{r+2}(2)$ in Q_0 – in fact S is the sum of a kernel $(E_n(x - y)I)$ of class $C_*^\infty(2)$ with kernels of class $C_*^{r+2}(3)$, $C_*^{r+1}(4)$ and $C_*^r(5)$, ${}^\top L_y^{*\top} S(x, y) = \sum_{j=1}^N f_j(x) \otimes \overline{g}_j(y)$ for some $f_j, g_j \in C^{r+2}(Q_0, \mathbb{C}^p)$ (any $\epsilon > 0$).*

Let $S_1(x, y) = E_n(z)\{I + \beta(x) \cdot z + \frac{1}{2} z \cdot B(x)z + K_n\gamma(x)|z|^2(\log|z|)^\sigma\}$, $z = y - x$, where $\beta(x) \cdot z = \sum_1^n \beta_j(x)z_j$, $z \cdot Bz = \sum_{j,k=1}^n B_{jk}z_jz_k$, $\beta_j(x) = \frac{1}{2}b_j(x)$, $B_{jk}(x) = B_{kj}(x) = \frac{1}{2}(\beta_j\beta_k + \beta_k\beta_j + \partial\beta_j/\partial x_k + \partial\beta_k/\partial x_j)$, $\gamma(x) = \frac{1}{2}(C(x) - \sum_1^n \partial\beta_j/\partial x_j - \sum_1^n \beta_j^2)$, $k_n = \frac{1}{n-4}$ and $\sigma = 0$ when $n \neq 4$; $k_n = -1$, $\sigma = 1$ when $n = 4$. Then $S(x, y) - E_n(x - y)I$ is $C_*^{r+2}(3)$, $S(x, y) - E_n(x - y)\{I + \beta(x) \cdot (y - x)\}$ is $C_*^{r+1}(4)$ and $S(x, y) - S_1(x, y)$ is $C_*^r(5)$.

Proof. Let $R_1(x, y)$ be defined by $L_y^{*\top} S_1(x, y) = {}^\top R_1(x, y)$, $x \neq y$ in $\mathbb{R}^n$; then R_1 is a kernel of class $C_*^r(3)$ on $\mathbb{R}^n$. Let B be a large ball in $\mathbb{R}^n$ such that $\overline{Q}_0 \subset B$ and let $\{g_1, \ldots, g_N\}$ be an $L^2(B)$-orthonormal basis for $\{g \mid L_g = 0 \text{ in } B, g = 0 \text{ on } \partial B\}$; we may suppose the $g_j \in C^{r+3-\epsilon}(B)$, for any $\epsilon > 0$, since the coefficients are C^{r+1}. We seek a solution $y \mapsto S_2(x, y)$ of $L_y^{*\top} S_2(x, y) = -{}^\top R_1(x, y) + \sum_1^N \overline{g}_j(y \otimes f_j(x)(y \in B)$ with ${}^\top S_2(x, y) = 0$ for $y \in \partial B$ and some appropriate f_j. Since $R_1(x, \cdot) \in L_p(B)$ when $1 < p < \frac{n}{n-3}$ (for $n > 3$), or any $1 < p < \infty$ (for $n = 2, 3$), there is a solution $S_2(x, \cdot) \in W^{2,p} \cap W_0^{1,p}(B)$ provided $f_j(x) = \int_B R_1(x, y)g_j(y)dy$ $(x \in B, 1 \leq j \leq N)$, which is at least C^{r+2} on B. If $G(x, y)$ is the Green function for the Laplacian in B, we may write an integral equation for S_2

$$S_2(x, y) = \int_B G(y, z)(F(x, z) + \sigma(x, z))$$

$$\sigma(x, y) = \int_B G_1(y, z)(F(x, z) + \sigma(x, z))$$

where

$$F(x, y) = -R_1(x, y) + \sum_1^N f_j(x) \otimes \overline{g}_j(y)$$

$$G_1(y, z) = \left(\sum_1^n b_j(y)\frac{\partial}{\partial y_j} - c(y) + \sum_1^n \partial_j b_j(y)\right) G(y, z).$$

Now G is $C_*^\infty(2)$ kernel while G_1 is $C_*^{r+1}(1)$, F is $C_*^r(3)$. Let Γ be a resolvent for kernel for G_1, class $C_*^{r+1}(1)$, and then $\sigma(x, y) = \int_B G_2(y, z)F(x, z)$ (plus, perhaps, something from the kernel), $G_2(y, z) = G_1(y, z) + \int_B \Gamma(y, \cdot)G_1(\cdot, z)$ is $C_*^{r+1}(1)$, so σ is $C_*^r(4)$. Substitution shows S_2 is $C_*^r(5)$. The other claims are proved similarly.

7.6 Calculation of Some Boundary Operators

Let $b_j : \mathbb{R}^n \to \mathbb{C}^{p\times p}$ be locally C^{r+2} $(1 \leq j \leq n)$, $C : \mathbb{R}^n \to \mathbb{C}^{p\times p}$ locally C^{r+1} and let $S(x, y)$ be an approximate fundamental solution (Th. 7.5.1) for

$$L_x = \Delta + \sum_1^n b_j(x)\frac{\partial}{\partial x_j} + c(x), \quad \Delta = \sum_1^n \frac{\partial^2}{\partial x_j^2}.$$

We will obtain representations of various solution operators associated with L (in a given C^{r+4} smooth region Ω) such as:

1) $\mathcal{A}_L f + \mathcal{B}_L g = u$ is a solution of

$$Lu - f|_\Omega \in \{\text{finite}\}, \quad u|_{\partial\Omega} = g$$

where $\{\text{finite}\}$ is some unspecified element of a specific finite-dimensional space – in this case, complementary to the image of $L \mid H^2 \cap H_0^1(\Omega, \mathbb{C}^p)$. We also require that u belongs to a given complement to the kernel of $L \mid H^2 \cap H_0^1$.

2) $\mathcal{A}_{L,M} f + \mathcal{C}_{L,M} g = v$ is a solution of

$$Lv - f|_\Omega \in \{\text{finite}\}, \qquad Mv|_{\partial\Omega} = g$$

where $Mv = \frac{\partial v}{\partial N_\Omega} + \mathcal{A}(x)v$, given $\mathcal{A} : \mathbb{R}^n \to \mathbb{C}^{p\times p}$ of class C^{r+2}.

For our applications, operators of finite rank are negligible and will be simply omitted, or indicated by a term "$\{\text{finite}\}$." The first step is to obtain integral equations for the operators.

1) Let $u = \mathcal{A}_L f + \mathcal{B}_{Lg}$ as above; by (7.5.4)

$$u(x) = \int_\Omega S(x,y)f(y)dy + \int_{\partial\Omega}\left(\frac{\partial S}{\partial N_y}(x,y)g(y) - S(x,y)b, N(y)g(y)\right)$$
$$-\int_{\partial\Omega} S(x,y)\frac{\partial u}{\partial N}(y) + \{\text{finite}\}, \qquad \text{for } x \in \Omega$$

so $\frac{\partial u}{\partial N}|_{\partial\Omega} \equiv \dot{u}$ satisfies

$$\begin{aligned} {}^1\!/\!_2\dot{u}(x) + \int_{\partial\Omega}\frac{\partial S}{\partial N_x}(x,y)\dot{u}(y)dA_y \\ = \int_\Omega \frac{\partial S}{\partial N_x}(x,y)f(y)dy + \frac{1}{2}b\cdot N(x)g(x) \\ - \int_{\partial\Omega}\frac{\partial S}{\partial N_x}(x,y)\, b\cdot N(y)g(y)dA_y \\ + \lim_{x\to\partial\Omega}\int_{\partial\Omega}\frac{\partial^2 S(x,y)}{\partial N_x \partial N_y}g(y) + \{\text{finite}\} \end{aligned} \tag{7.6.1}$$

for $x \in \partial\Omega$. The term $\int_{\partial\Omega}\frac{\partial^2 S(x,y)}{\partial N_x\partial N_y}g(y)\,dA_y$ has a limit as $x \to \partial\Omega$ (from the interior) provided g is C^2, since all the other limits exist, we examine this in more detail below.

2) Let $v = \mathcal{A}_{L,M} f + \mathcal{C}_{L,M} g$; by (7.5.4)

$$v(x) = \int_\Omega S(x, y) f(y) dy - \int_{\partial\Omega} S(x, y) g(y) dA_y$$
$$+ \int_{\partial\Omega} \left\{ \frac{\partial S}{\partial N_y}(x, y) + S(x, y)(\mathcal{A}(y) - b \cdot N(y)) \right\} v(y) dA_y + \{\text{finite}\}$$

for $x \in \Omega$, and $v|\partial\Omega \equiv v_0$ satisfies

$$\tfrac{1}{2} v_0(x) - \int_{\partial\Omega} K(x, y) v_0(y) dA_y$$
$$= \int_\Omega S(x, y) f(y) dy - \int_{\partial\Omega} S(x, y) g(y) + \{\text{finite}\} \tag{7.6.2}$$

for $x \in \partial\Omega$, where $K(x, y) = \frac{\partial S}{\partial N_y}(x, y) + S(x, y)(\mathcal{A}(y) b \cdot N(y))|_{\partial\Omega \times \partial\Omega}$.

Suppose $R(x, y)$ is the resolvent kernel for $2K(x, y)$ provided by Th. 7.2.3 – $(I - 2\tilde{K})^{-1} = I + \tilde{R}$, modulo operators of finite rank. Since K is a kernal of class $C_*^{r+1}(1)$ on $\partial\Omega (K(x, y) = O(|x - y|^{2-n}) = O(|x - y|^{1-m})$ on $\partial\Omega$, $m = n - 1 = \dim \partial\Omega)$, and $\partial\Omega$ is C^{r+2} (in fact, C^{r+4}), it follows that R is $C_*^{r+1}(1)$ and the solution of (7.6.2) is

$$v_0(x) = 2 \int_\Omega S(x, y) f(y) + 2 \int_{\partial\Omega} R(x, z) \left(\int_\Omega S(z, y) f(y) \right) dA_z$$
$$- 2 \int_{\partial\Omega} S(x, y) g(y) - 2 \int_{\partial\Omega} \left(\int_{\partial\Omega} R(x, z) S(z, y) dA_z \right) g(y) + \{\text{finite}\}.$$

In particular

$$\mathcal{C}_{L,M} g|_{\partial\Omega}(x) = -2 \int_{\partial\Omega} C(x, y) g(y) dA_y + \{\text{finite}\} \tag{7.6.3}$$

where

$$C(x, y) = S(x, y) + \int_{\partial\Omega} R(x, z) S(z, y) dA_z$$
$$= E_n(x - y) I + (\text{kernel of class } C_*^{r+1}(2) \text{ on } \partial\Omega)$$

To obtain more detail, note

$$C(x, y) = E_n(x - y) I + E_n(x - y) \beta(x) \cdot (y - x)$$
$$+ \int_{\partial\Omega} R_0(x, z) E_n(z - y) dA_z + (\text{class } C_*^r(3))$$

where $R_0(x, y) = 2x\{\frac{\partial E_n(x-y)}{\partial N_y} + E_n(x - y)(a(x) - \tfrac{3}{2} b \cdot N(x))\}$ so we want to compute $\int_{\partial\Omega} E_n(x - z) E_n(z - y)$ and $\int_{\partial\Omega} \frac{\partial E_n(x-z)}{\partial N_z} E_n(z - y)$ – or at least, their leading terms. Computation of the leading terms is easy, since we merely find the convolution in the tangent space. Thus $E_n(x-y) = \frac{|x-y|^{1-m}}{(1-m)\omega_n}$ so, from the table in 7.3,

$$\int_{\partial\Omega} E_n(x-z)E_n(z-y)dA_z \approx \frac{|x-y|^{2-m}}{(2-n)^2\omega_n^2}\frac{\gamma_{1,0}^2}{\gamma_{2,0}} = \frac{|x-y|^{3-n}}{4\omega_{n-1}(n-3)},$$

at least for $n>3$; if $n=3$, we have $\frac{-1}{8\pi_0}\log|x-y|$; if $n=2$, $-{}^1\!/_8|x-y|$ (which is the result obtained with $n=2$ in the general expression above). With the convention that "$\frac{|z|^p}{p}$" means $\log|z|$ when $p=0$, we have

$$\int_{\partial\Omega} E_n(x-z)E_n(z-y)dA_z = -\frac{|x-y|^{3-n}}{4\omega_{n-1}(3-n)}$$

(7.6.4)

$$+ \text{ (kernel of class } C_*^{r+3}(3)).$$

Now

$$\frac{\partial E_n(x-z)}{\partial N_z} = \frac{N_z\cdot(z-x)}{\omega_n|z-x|^n} = \frac{\zeta\cdot K(x)\zeta}{2\omega_n|\zeta|^n} + O(|\zeta|^{3-n})$$

where $\zeta = z-x$, and $K(x) = DN(x)$ is the (degenerate) curvature matrix which we write $K(x) = \hat{K}(x) + \frac{H(x)}{n-1}P(x)$, $P(x) = I - N(x)\otimes N(x)$, so trace $\hat{K}(x) = 0$. In the tangent space at x, this is

$$\frac{\zeta\cdot K(x)\zeta}{2\omega_n|\zeta|^n} = \frac{|\zeta|^{1-m}}{2\omega_n}\left(\frac{\zeta}{|\zeta|}\cdot\hat{K}(x)\frac{\zeta}{|\zeta|} + \frac{H(x)}{n-1}\right)$$

with Fourier transform

$$\frac{|\xi|^{-1}}{2\omega_n}\left(\gamma_{1,2}\frac{\xi\cdot\hat{K}\xi}{|\xi|^2} + \gamma_{1,0}\frac{H(x)}{n-1}\right),$$

so the leading term of the convolution is

$$\int_{\partial\Omega}\frac{\partial E_n}{\partial N_x}E_n(z-y)dA_z$$

$$= -\frac{|x-y|^{3-n}}{16\omega_{n-1}}\left\{\frac{(y-x)\cdot\hat{K}(x)(y-x)}{|y-x|^2} + \frac{2(n-2)}{(n-1)(n-3)}H(x)\right\}$$

(7.6.5) $$+ \text{ (kernel of class } C_*^{r+1}(3))$$

Thus the kernel $C(x, y)$ of (7.6.3) is, more exactly,

$$\begin{aligned} C(x,y) &= E_n(x-y)I + E_n(x-y)\beta(x)\cdot(y-x) \\ &\quad + \frac{|x-y|^{3-n}}{16\omega_{n-1}}\left(\frac{(y-x)\cdot\hat{K}(x)(y-x)}{|y-x|^2} + \frac{2(n-2)}{(n-1)(n-3)}H(x)\right)\cdot I \\ &\quad - \frac{1}{2\omega_{n-1}}(a(x) - {}^3\!/\!_2 b\cdot N(x))\frac{|x-y|^{3-n}}{3-n} + (\text{kernel of class } C^r_*(3)) \end{aligned}$$

(7.6.6)

The calculation for the "Dirichlet case," in the integral equation (7.6.1), are similar but more complicated. We first study the case $L = \Delta$.

Theorem 7.6.1. *If Ω is a bounded C^{r+4}-regular region in $\mathbb{R}^n (n \geq 2, r \geq 0)$ and $g \in C^2(\partial\Omega)$ then*

$$\begin{aligned} &\lim_{\substack{x\to\partial\Omega \\ x\in\Omega}} \int_{\partial\Omega} \frac{\partial^2 E_n(x-y)}{\partial N_x \partial N_y} g(y)dA_y \\ &= P.V.\int_{\partial\Omega} \frac{\partial^2 E(x-y)}{\partial N_x\partial N_y}(g(y)-g(x)) \\ &= P.V.\int_{\partial\Omega} \frac{(x-y)\cdot\nabla_{\partial\Omega}g(y)}{\omega_n|x-y|^n}dA_y + \int_{\partial\Omega} Q_0(x,y)g(y) \\ &= \int_{\partial\Omega} E_n(x-y)\Delta_{\partial\Omega}g(y)dA_y + \int_{\partial\Omega} Q_0(x,y)g(y) \end{aligned}$$

where

$$\begin{aligned} Q_0(x,y) &= \left.\left\{\frac{\partial^2 E}{\partial N_x\partial N_y} + \frac{\partial^2 E}{\partial N_y^2} + H(y)\frac{\partial E}{\partial N_y}\right\}\right|_{\partial\Omega\times\partial\Omega} \\ &= \frac{1}{2\omega_n|z|^n}(|K(x)z|^2 + H(x)z\cdot K(x)z - n(z\cdot k(x)z)^2/|z|^2) \\ &\quad + O(|z|^{3-n}), \quad z = y-x \end{aligned}$$

$K(x) = DN(x)$ is the (degenerate) curvature matrix (see Th. 1.5*). H = trace K is the mean curvature, and Q_0 is a kernel of class $C^{r+2}_*(1)$ on $\partial\Omega$. If $n = 2$, Q_0 is a C^{r+2} function.*

Proof. $\int_{\partial\Omega} \frac{\partial E_n(x-y)}{\partial N_y}dA_y = \chi_\Omega(x)$ so for $x \in \Omega$

$$\int_{\partial\Omega} \frac{\partial^2 E(x-y)}{\partial N_x\partial N_y}g(y) = \int_{\partial\Omega} \frac{\partial^2 E(x-y)}{\partial N_x\partial N_y}(g(y)-g(x)),$$

to which we apply Th. 7.4.1. We have

$$\frac{\partial^2 E(x-y)}{\partial N_x\partial N_y}(g(y)-g(x)) = \frac{Dg(x)\cdot z}{\omega_n|z|^n}\left(-1 + n\frac{(N(x)\cdot z)^2}{|z|^2}\right) + O(|z|^{2-n})$$

so, if $\frac{z}{|z|} = \xi + tN(x)(\xi \perp N(x))$ the first term becomes

$$|z|^{1-n}\frac{Dg(x)\cdot\xi + t\frac{\partial g}{\partial N}(x)}{\omega_n}(-1+nt^2),$$

and (7.4.3) gives $q(x) = 0$ and the first equality claimed above.

Now for $x \in \Omega$, $y \in \partial\Omega$,

$$\Delta_y E_n(x-y) = 0 = \Delta_{\partial\Omega_y} E(x-y) + H(y)\frac{\partial E(x-y)}{\partial N_y} + \frac{\partial^2 E(x-y)}{\partial N_y^2}$$

so

$$\int_{\partial\Omega}\left\{\frac{\partial^2 E}{\partial N_x \partial N_y} + \frac{\partial^2 E}{\partial N_y^2} + H(y)\frac{\partial E}{\partial N_y^2}\right\} g(y)$$
$$= -\int_{\partial\Omega} E(x-y)\Delta_{\partial\Omega} g(y) + \int_{\partial\Omega}\frac{\partial^2 E}{\partial N_x \partial N_y} g(y)$$

for $x \in \Omega$. Another application of Th. 7.4.1 (with $q = 0$ again) gives the third equality of the theorem, which implies the second on integration by parts. To obtain the form claimed for Q_0, write $\partial\Omega$ near x as a graph over the tangent plane: $y = x + \eta - \nu_x(\eta)N_x$ with $\nu_x(\eta) = \,^1\!/_2\eta\cdot K(x)\eta + O(|\eta|^3)$ as $\eta \to 0$ ($\eta \in T_x(\partial\Omega) \perp N(x)$). Then $N(y) = (\nu_x'(\eta) + N_x)/\sqrt{1+|\nu_x'|^2}$, considering $\nu_x'(\eta) = \frac{\partial}{\partial\eta}\nu_x(\eta) \in T_x(\partial\Omega)$, $N_{\dot x}(y-x) = -\,^1\!/_2\eta\cdot K(x)\eta + O(|\eta|^3)$, $N_{\dot y}(y-x) = \,^1\!/_2\eta\cdot K(x)\eta + O(|\eta|^3)$, $N_{\dot x}N_y = 1/\sqrt{1+|\nu_x'|^2} = 1 - \,^1\!/_2|K(x)\eta|^2 + O(|\eta|^3)$. Note $\nu_x(\cdot)$ is C^{r+4} for fixed x and $(x,n) \mapsto \frac{\partial}{\partial\eta}\nu_x(x)$ is C^{r+3}.

The two-dimensional case is sufficiently simple to deserve sparate treatment.

Theorem 7.6.2. *For $n = 2$, $\partial\Omega$ is a finite-union of closed curves and for definiteness we suppose it is a single curve of length L. We write functions on $\partial\Omega$ as functions of the arc-length, extended as L-periodic functions on $\mathbb{R}$. Let $\lambda : \mathbb{R} \to \mathbb{R}$ be L-periodic, C^∞ on $\mathbb{R}\backslash[0, \pm L, \pm 2L, \dots]$ with*

$$\lambda(t) = \frac{1}{\pi}\log|t|, \qquad \lambda'(t) = \frac{1}{\pi t}, \qquad \lambda''(t) = -\frac{1}{\pi t^2} \qquad \textit{near } t = 0.$$

Then there exists $R : \mathbb{R}\times\mathbb{R} \to \mathbb{R}$ of class $C^{r+1-\epsilon}$(any $\epsilon > 0$), L-periodic in each argument, so that

$$\frac{\partial}{\partial N}\mathcal{B}_\Delta g(t) = \int_{\partial\Omega}\lambda(t-s)g''(s)ds + \int_{\partial\Omega}R(t,s)g(s)ds$$
$$+ P.V.\int_{\partial\Omega}\lambda'(t-s)g'(s)ds + \int_{\partial\Omega}R(t,s)g(s)ds$$
$$= P.V.\int_{\partial\Omega}\lambda''(t-s)(g(s)-g(t))ds + \int_{\partial\Omega}R(t,s)g(s)ds$$

for any $C^2 g$. The middle-form is closest to the usual representation of the Hilbert transform H, $\frac{\partial}{\partial N}\mathcal{B}\Delta g = Hg'$, where $\phi + iH\phi$ is the restriction to $\partial\Omega$ of an analytic function of $x + iy$ for $(x, y) \in \Omega$, with $\int_{\partial\Omega} H\phi = 0$ to fix the constant.

Proof. The kernel Q_0 of the previous theorem is C^{r+2} on $\partial\Omega \times \partial\Omega$, as is $2\frac{\partial}{\partial N_x}E(x-y) = \frac{1}{\pi}\frac{N_x(x-y)}{|x-y|^2} = \frac{\nu_x(\eta)}{\pi(\eta^2+\nu_x(\eta)^2)}$ where $y = x + \eta - \nu_x(\eta)N_x$, $\eta \in T_x(\partial\Omega) \approx \mathbb{R}$.

Let $R_1(x, y)$ be the resolvent kernel for $-2\frac{\partial E(x-y)}{\partial N_x}$; then from Th. 7.6.1 and (7.6.1) – with $S = E_n$ – we find

$$\frac{\partial}{\partial N}\mathcal{B}_\Delta g(t) = \frac{1}{\pi}\int_{\partial\Omega} \log|t-s| g''(s)ds + \int_{\partial\Omega} R_2(t,s)g(s)ds$$

where

$$R_2(t,s) = 2Q_0(t,s) + 2\int_{\partial\Omega} R_1(t,\cdot)Q_0(\cdot,s) + \frac{1}{\pi}\frac{\partial^2}{\partial s^2}\int_{\partial\Omega} R_1(t,s)\log|\sigma - s|d\sigma.$$

Since Q_0, R_1 are C^{r+2} while $\log|\sigma - s|$ is a kernel of class $C_*^{r+3}(1)$ on $\partial\Omega$, $R_2(t, s)$ is at least C^r, and it is easily shown to be $C^{r+1-\epsilon}(\epsilon > 0)$.

Theorem 7.6.3. *Let $\Omega \subset \mathbb{R}^n$ be bounded and C^{r+4} $(n \geq 3,\ r \geq 0)$. For $g \in C^2(\partial\Omega)$,*

$$\begin{aligned}\frac{\partial}{\partial N}\mathcal{B}_\Delta g(x) = {} & 2\lim_{x\to\partial\Omega}\int_{\partial\Omega}\frac{\partial^2 E_n(x-y)}{\partial N_x \partial N_y}g(y) - \frac{n-2}{n-1}\frac{H(x)}{2}g(x)\\ & - \frac{n-1}{2\omega_{n-1}}P.V.\int_{\partial\Omega}\frac{(y-x)\cdot\hat{K}(x)(y-x)}{|y-x|^{n+1}}g(y)\\ & + \int_{\partial\Omega}|x-y|^{2-n}Q_1\left(x, \frac{y-x}{|y-x|}\right)g(y)\\ & + \int_{\partial\Omega}R_1(x,y)g(y), \qquad x \in \partial\Omega,\end{aligned}$$

where $R_1(x, y)$ is a kernel of class $C_*^r(2)$ on $\partial\Omega$ and

$$\begin{aligned}Q_1(x,\zeta) = {} & \frac{1}{\omega_{n-1}}\left\{-\frac{n-1}{4}\zeta\cdot D^2N(x)\zeta^2 + \frac{1}{4}\zeta\cdot\nabla_{\partial\Omega}H(x)\right\}\\ & + \frac{1}{\omega_n}\left\{\frac{n}{6}(\zeta\cdot\hat{K}(x)\zeta)^2 - \frac{2}{3}|\hat{K}(x)\zeta|^2 + \frac{n-2}{(n-1)}\zeta\cdot\hat{K}(x)\zeta\right.\\ & \left. + \frac{n-2}{2(n-1)^2}H(x)^2 + \frac{1}{3(n-2)}\operatorname{trace}(\hat{K}(x)^2)\right\}\end{aligned}$$

$K(x) = DN(x), H(x) = trace\ K(x)$, $\hat{K}(x) = K(x) - \frac{H(x)}{n-1}P(x)$, $P(x) = I - N_x \otimes N_x$. *Note* $K(x) = D^2t(x), D^2N(x) = D^3t(x)$ *where* $t(x)$ *is the "normal" coordinate of* Th. 1.5.

Remark. If $n = 2$ then $\hat{K} = 0$, $D^2N(x)\zeta^3 = \zeta \cdot \nabla_{\partial\Omega}H(x)$ for $\zeta \in T_x(\partial\Omega)$, $|\zeta| = 1$, and the above reduces to the previous theorem; but here we shall assume $n > 2$.

The first term above is treated in Th. 7.6.1. The singular integral (*P.V.*) above exists since $\hat{K}(x) \mid T_x(\partial\Omega)$ has trace zero.

Proof. Let $Pg(x) = \lim_{x\to\partial\Omega} \int_{\partial\Omega}(\partial^2 E/\partial N_x \partial N_y)g(y)dA_y$, the operator treated in (7.6.7) above, and $M\gamma(x) = \int_{\partial\Omega} \frac{\partial E(x-y)}{\partial N_x}\gamma(y)dA_y$ so the integral equation (7.6.1) for $\gamma = \frac{\partial}{\partial N}\mathcal{B}_\Delta g$ becomes $\frac{1}{2}\gamma + M\gamma = Pg$. Let $\gamma_0 = 2Pg - 4MPg + 8M^2Pg$ and $\gamma_1 = \gamma - \gamma_0$; then $\frac{1}{2}\gamma_1 + M\gamma_1 = -8M^3Pg$. We calculate MP, M^2P with sufficient accuracy, and see in particular that M^2P is an integral operator with kernel of class $C^r_*(1)$, so M^3P has kernel of class $C^r_*(2)$ and may be absorbed in the remainder.

Now $P = P_0 + \tilde{Q}_0$, $P_0g(x) = \int_{\partial\Omega} E_n(x-y)\Delta_{\partial\Omega}g(y)$ and MP_0 has $C^{r+2}_*(2)$ kernel – which we calculate to within a remainder $C^{r+2}_*(4)$, so that integration by parts still leaves a remainder $C^r_*(2)$. The kernel of MP_0 is

$$K_1(x, y) = \int_{\partial\Omega} \frac{\partial E(x-z)}{\partial N_x} E(z-y)dA_z$$

which is clearly C^{r+3} for $x \neq y$ and we are concerned only with the behavior near $x = y$. If $z = x + \zeta - \nu_x(\zeta)N_x$ then $(D^2N)_x\zeta^2 = \nu_x'''(0)\zeta^2 - N_x|K(x)\zeta|^2 + O(|\zeta|^3)$ so

$$\begin{aligned}\frac{\partial E(x-z)}{\partial N_x} &= \frac{N_x(x-z)}{\omega_n|x-z|^n} = \frac{\nu_x(\zeta)}{\omega_n(|\zeta|^2 + \nu_x(\zeta)^2)^{n/2}} \\ &= \frac{1}{\omega_n|\zeta|^n}\,\tfrac{1}{2}\zeta \cdot K(x)\zeta + \frac{1}{\omega_n(\zeta)^n}\left(\frac{1}{6}\zeta \cdot (D^2N_x\zeta^2)) + O(|\zeta|^{4-n}\right)\end{aligned}$$

We calculate the Fourier transform (of the first terms) as before:

$$\begin{aligned}&-\frac{1}{4}|\xi|^{-1}\frac{\xi \cdot \hat{K}\xi}{|\xi|^2} + \frac{n-2}{4(n-1)}H|\xi|^{-1} \\ &\qquad + \frac{3i}{2}|\xi|^{-2}\left(\frac{1}{6}\xi \cdot (D^2N\xi^2) - \frac{1}{2(n+1)}\frac{\xi \cdot \nabla_{\partial\Omega}H}{|\xi|}\right) \\ &\qquad - \frac{n-2}{4(n+1)}i|\xi|^{-2}\frac{\xi \cdot \nabla_{\partial\Omega}H}{|\xi|}.\end{aligned}$$

Similarly for E_n; multiply, and take the inverse transform to conclude

$$K_1(x,y) = \frac{1}{\omega_{n-1}}\Bigg(-\frac{|\eta|^{3-n}}{16}\frac{\eta\cdot\hat{K}\eta}{|\eta|^2} + \frac{n-2}{8(n-1)}H\frac{|\eta|^{3-n}}{3-n} - \frac{3}{32}|\eta|^{4-n}\left(\frac{1}{6}\frac{\eta\cdot D^2N\eta^2}{|\eta|^3} - \frac{1}{2(n+1)}\frac{\eta\cdot\nabla H}{|\eta|}\right) + \frac{n-2}{16(n+1)}\frac{|\eta|^{3-n}}{3-n}\eta\cdot\nabla_{\partial\Omega}H\Bigg) + (\text{kernel of class } C_*^{r+1}(4))$$

Similarly (details omitted)

$$\begin{aligned} K_2(x,y) &= \int_{\partial\Omega}\frac{\partial E(x-z)}{\partial N_x}K_1(z,y)dA_z \\ &= \frac{|\eta|^{4-n}}{\omega_n}\left(\frac{n}{48}\frac{(\eta\cdot\hat{K}\eta)^2}{|\eta|^4} - \frac{1}{12}\frac{|\hat{K}\eta|^2}{|\eta|^2} + \frac{n-2}{8(n-1)}\frac{\eta\cdot\hat{K}\eta}{|\eta|^2} + \frac{n-2}{16(n-1)^2}H^2 + \frac{1}{24(n-2)}\operatorname{tr}(\hat{K}^2)\right) \\ &\quad + (\text{kernel of class } C_*^{r+1}(4)) \end{aligned}$$

Now

$$\begin{aligned} MP_0g(x) &= \int_{\partial\Omega}K_1(x,y)\Delta_{\partial\Omega}g(y) = \lim_{\epsilon\to 0}\int_{\partial\Omega\setminus B_\epsilon(x)}K_1(x,y)\Delta_{\partial\Omega}g(y) \\ &= P.V.\int_{\partial\Omega}\Delta_{\partial\Omega_y}K_1(x,y)g(y) + \frac{n-2}{n-1}\frac{H(x)}{8}g(x) \\ M^2P_0g(x) &= \int_{\partial\Omega}\Delta_{\partial\Omega_y}K_2(x,y)g(y). \end{aligned}$$

To compute $\Delta_{\partial\Omega_y}$, note that this is $\Delta_\eta + O(|\eta|^2\frac{\partial^2}{\partial\eta^2}) + O(|\eta|\frac{\partial}{\partial\eta})$ where $y = x + \eta - \nu_x(\eta)N_x$, $\eta \in T_x(\partial\Omega)$ near 0. Straightforward calculation completes the proof.

Now we consider the general case: $L = \Delta + \sum_1^n b_j(x)\frac{\partial}{\partial x_j} + c(x)$. By Thm. 7.4.1

$$\begin{aligned} \lim_{x\to\partial\Omega}\int_{\partial\Omega}\frac{\partial^2(S-E)}{\partial N_x\partial N_y}g(y)dA_y &= -\beta\cdot N(x)g(x) + P.V.\int_{\partial\Omega}\frac{\partial^2(S-E)}{\partial N_x\partial N_y}g(y) \\ &= -\beta\cdot N(x)g(x) - P.V.\int_{\partial\Omega}\frac{z\cdot\beta(x)}{\omega_n|z|^n}g(y) \\ &\quad + \int_{\partial\Omega}Q_2\left(x,\frac{z}{|z|}\right)\frac{g(y)}{|z|^{n-2}} \\ &\quad + \int_{\partial\Omega}(C_*^r(2)\text{ kernel})g(y), \qquad z = y - x, \end{aligned}$$

with $Q_2(x,\zeta) = -\frac{1}{2\omega_n}\zeta\cdot B(x)\zeta + |\zeta|^{n-2}E_n(\zeta)\{\frac{\partial\beta}{\partial N}\cdot N - N\cdot BN + \gamma\}$. The singular integral above may be written

$$-P.V.\int_{\partial\Omega}\frac{z\cdot\beta(x)}{\omega_n|z|^n}g(y) = \int_{\partial\Omega}E(y-x)\mathrm{div}_{\partial\Omega}(\beta(y)g(y))$$
$$+\int_{\partial\Omega}\left\{Q_3\left(x,\frac{z}{|z|}\right)|z|^{2-n} + (C_*^r(2)\text{ kernel})\right\}g(y)$$

where $Q_3(x,\zeta) = \frac{1}{\omega_n}(\zeta\cdot(D\beta(x)\zeta) - {}^1\!/_2\beta\cdot N\zeta\cdot K(x)\zeta)$ and we recall $\beta\cdot N = \sum_{j=1}^n\beta_jN_j$ is $p\times p$ matrix. Thus

$$\lim_{x\to\partial\Omega}\int_{\partial\Omega}\frac{\partial^2S(x,y)}{\partial N_x\partial N_y}g(y) = \int_{\partial\Omega}E_n(x-y)(\Delta_{\partial\Omega}g(y) + \mathrm{div}_{\partial\Omega}(\beta(y)g(y))$$
$$+\int_{\partial\Omega}\left\{Q_4\left(x,\frac{z}{|z|}\right)|z|^{2-n} + C_*^r(2)\text{ kernel}\right\}g(y)$$

for some (computable) Q_4, and the calculation is now similar to that in Th. 7.6.3. Without more detail, we record the conclusion.

Theorem 7.6.4. *If $b_j:\mathbb{R}^n\to\mathbb{C}^{p\times P}$ are C^{r+2} $(1\le j\le n)$, $C:\mathbb{R}^n\to\mathbb{C}^{p\times p}$ is C^{r+1}, Ω is a bounded C^{r+4} region in $\mathbb{R}^n$ and $g\in C^2(\partial\Omega,\mathbb{C}^p)$ then for $x\in\partial\Omega$, $L=\Delta+\sum_1^n b_j(x)\frac{\partial}{\partial x_j}+c(x)$*

$$\frac{\partial}{\partial N}\mathcal{B}_Lg(x) - \frac{\partial}{\partial N}\mathcal{B}_\Delta g(x)$$
$$= -\frac{1}{2}b\cdot N(x)g(x) - P.V.\int_{\partial\Omega}\frac{b(x)\cdot(y-x)}{\omega_n|y-x|^n}g(y)$$
$$+\int_{\partial\Omega}\left\{\frac{1}{\omega_nr^{n-2}}Q_2\left(x,\frac{y-x}{r}\right) + \frac{1}{\omega_{n-1}r^{n-2}}Q_3\left(x,\frac{y-x}{r}\right)\right.$$
$$\left.+E_n(y-x)Q_4(x)\right\}g(y) + \int_{\partial\Omega}\tilde{R}(x,y)g(y)$$

where $\tilde{R}(x,y)$ is a $C_*^r(2)$ kernel in $\partial\Omega$,

$$Q_2(x,\zeta) = -\frac{1}{4}\sum_{j,k=1}^n\zeta_j\zeta_kb_jb_k - \frac{1}{2}\sum\zeta_j\zeta_k\frac{\partial b_j}{\partial x_k} + \frac{1}{2}\zeta\cdot\hat{K}(x)\zeta b\cdot N(x)$$
$$Q_3(x,\zeta) = \frac{1}{8}(b\cdot Nb\cdot\zeta - b\cdot\zeta b\cdot N) + \frac{1}{4}\sum_{j,k}\frac{\partial b_j}{\partial x_k}(\zeta_jN_k - \zeta_kN_j)$$
$$-\frac{n-1}{4}\zeta\cdot\hat{K}(x)\zeta b(x)\cdot\zeta$$
$$Q_4(x,\zeta) = c(x) - \frac{1}{2}\sum_j\frac{\partial b_j}{\partial x_j} - \frac{1}{4}\sum_j b_j^2.$$

In case $n = 2$, this simplifies slightly: taking $\partial\Omega$ as a curve of length L (as in Th. 7.6.2), and writing "$\log|t|$," "sgn t" and "$1/t$" for L-periodic functions having this form near $t = 0$, we have

$$\begin{aligned}
\frac{\partial}{\partial N}\mathcal{B}_L g(t) &= \frac{1}{\pi} P.V. \int_{\partial\Omega} \frac{ds}{t-s}\left(g'(s) + \frac{1}{2} b_{\tan}(t) g(s)\right) - \frac{1}{2} b \cdot N(t) g(t) \\
&+ \left\{\frac{1}{16}(b_2 b_1 - b_1 b_2) + \frac{1}{8}\left(\frac{\partial b_1}{\partial x_2} - \frac{\partial b_2}{\partial x_1}\right)\right\} \int_{\partial\Omega} \operatorname{sgn}(t-s) g(s) \\
&+ \left\{c(t) - \frac{1}{2}\operatorname{div} b(t) - \frac{1}{4}(b_1^2 + b_2^2)\right\} \frac{1}{2\pi} \int_{\partial\Omega} \log(t-s) g(s) \\
&+ \int_{\partial\Omega} \tilde{R}(t,s) g(s)
\end{aligned}$$

where $\tilde{R}(t,s)$ is a $C_*^r(2)$ kernel (in particular, C^1) $b \cdot N = b_1 N_1 + b_2 N_2$, $b_{\tan} = -N_2 b_1 + N_1 b_2$.

Chapter 8

The Method of Rapidly-Oscillating Solutions

8.1 Introduction

The theory of pseudo-differential operators (abbreviated ψDO_p) shows the essential behavior of many operators appears most clearly when applied to rapidly oscillating functions, studying the behavior as the frequency tends to infinity. This provides a more flexible and transparent approach to our "finite-rank" problems than the method of integral operators of Chapter 7. It is also often simpler, for simple problems – but for this reason, we are encouraged to attack harder problems. (If you are not familiar with ψDO_p, don't worry; our argument is direct. Here, the intention is only to show why we *don't* cite results from this theory.)

We have a ψDO_p which, by hypothesis – a hypothesis we hope to contradict – has finite rank. A ψDO_p of finite rank is trivial: its symbol is identically zero. So all we have to do is compute the symbol. But it must be computed in some detail. Often the (apparent) principal and subprincipal symbols both vanish identically and uselessly – we must go to the third stage, sometimes further. This is not surprising since we start with second order operators, so the coefficient of the zero-order part only begins to play a role at the third stage. We also deal with quite degenerate problems; after all, we expect to find a contradiction. The theory of ψDO_p serves only as inspiration, since it ordinarily computes explicitly only the principal symbol. One shows the other terms are, in principle, computable, but there is rarely any need to compute them – except in our problems.

Consider an example: the operator

$$\sigma \mapsto \Gamma(\sigma) = \mathrm{Re}\left\{\psi_N \frac{\partial}{\partial N}\mathcal{B}_A(\sigma u_N) + u_N \frac{\partial}{\partial N}\mathcal{B}_{A^*}\cdot(\sigma\psi_N)\right\}$$

was studied is 6.8 above, and will be treated by our new method in 8.4 below. (This calculation in the 1985 manuscript was incorrect, revealing an error in the earlier version of Theorem 7.6.10 – an error which only became clear on comparing with the results of the method of rapidly oscillating solutions.) Apparently, Γ is first-order; it *would* be first-order for most choices of u_N, ψ_N. But we treat the case where $\mathrm{Re}(u_N\psi_N) \equiv 0$ on the boundary, so the first-order and zero-order parts vanish and we are left with an operator of order -1:

$$\Gamma(\gamma \cos \omega\theta) = \omega^{-1}\gamma C(\theta) \cos \omega\theta + O(\omega^{-2}) \quad \text{as } \omega \to +\infty,$$

where the coefficient $C(\theta)$ is computed explicitly in 8.4 (We assume γ, θ are smooth real functions on $\partial\Omega$ and $|\nabla_{\partial\Omega}\theta| \equiv 1$ on the support of γ.) Γ is assumed to have finite rank and in consequence $\gamma C(\theta) = 0$ on $\partial\Omega$, for all permitted choices of γ, θ, which gives us new information about u_N, ψ_N and the coefficients of the elliptic operator A appearing in Γ. Here we use the simple

Lemma. *Suppose S is a C^1 manifold; $A, B \in L_2(S)$ with compact support; θ is real-valued and C^1 on S with $\nabla_s\theta \neq 0$ in supp $A \cup$ supp B; E is a finite dimensional subspace of $L_2(S)$; and $u(\omega) \in E$ for all large real ω, satisfying*

$$u(\omega) = A \cos \omega\theta + B \sin \omega\theta + o(1) \ \textit{in} \ \ L_2(S)$$

as $\omega \to +\infty$. Then $A = 0$, $B = 0$.

Proof. Let $\{e_1, \ldots, e_N\}$ be an orthonormal basis for E; then

$$u(\omega) = \sum_{k=1}^{N} C_k(\omega) e_k, \qquad C_k(\omega) = \int_S \bar{e}_k(A \cos \omega\theta + B \sin \omega\theta + o(1))$$

so each $C_k(\omega) \to 0$ as $\omega \to +\infty_{\mathrm{i}}$ by the Riemann-Lebesgue lemma, and $\|u(\omega)\|_{L_2(S)} \to 0$. But

$$\|u(\omega)\|^2_{L_2(S)} \to \int_S \frac{1}{2}(|A|^2 + |B|^2) \quad \text{as } \omega \to +\infty.$$

In Section 2, we find formal asymptotic expansions for, among others, $\mathcal{B}_L(ge^{i\omega\theta})$ as $\omega \to +\infty$, i.e., the solution u of

$$Lu = \Delta u + b \cdot \nabla u + cu = 0 \text{ in } \Omega, \qquad u = ge^{i\omega\theta} \text{ on } \partial\Omega,$$

where $|\nabla_{\partial\Omega}\theta| = 1$ in supp G. We give a summary of the formulas at the end of this chapter, for reference.

In Section 3, we show the formal solutions are close to exact solutions.

In Section 4, we consider the operator Γ mentioned above (already treated in 6.8), to illustrate our method. We extend the argument of 6.8 to allow a much weaker hypothesis when $n \geq 3$.

In Section 5, we treat a new problem: proving generic simplicity of solutions of a system

$$\Delta u_j + f_i(x, u_1, u_2, \ldots, u_p) = 0 \text{ in } \Omega, \quad u_j = 0 \text{ on } \partial\Omega \quad (1 \le j \le p).$$

Most of these results were stated, without proof, in "Generic properties of equilibrium solutions by perturbation of the boundary," D. B. Henry, in *Dynamics of Infinite Dimensional Systems,* ed. S. N. Chow and J. K. Hale, NATO ASI Series (Springer-Verlag, 1987).

8.2 Formal Asymptotic Solutions

Results will be summarized at the end of the chapter.

We study the differential operator

$$L = \Delta I + \sum_{j=1}^{n} b_j(x)\frac{\partial}{\partial x_j} + c(x), \quad x \in \mathbb{R}^n,$$

where I is the $p \times p$ identity matrix and $b_j(x)$, $c(x)$ are smooth functions with values in $\mathbb{C}^{p\times p}$ (the $p \times p$ complex matrices); we will be more precise about smoothness conditions in 8.3. We work in an open set $\Omega \subset \mathbb{R}^n$ with smooth boundary – typically, we will work in a small neighborhood of a given point of the boundary. We seek a formal solution u of

$$Lu = f = 2\omega e^{\omega S} F \text{ in } \Omega, \qquad u = g = e^{i\omega\theta} G \text{ on } \partial\Omega$$

$$F = \sum_{k\ge0} \frac{F_k(x)}{(2\omega)^k}, \qquad G = \sum_{k\ge0} \frac{G_k(x)}{(2\omega)^k}$$

$$S|\partial\Omega = i\theta, \quad \mathrm{Re}\frac{\partial S}{\partial N}|\partial\Omega > 0 \quad (\theta|\partial\Omega \text{ given})$$

and u has the form $u = e^{\omega S} \cdot \sum_{k\ge0} U_k(x)/(2\omega)^k$. Note Re $S < 0$ inside Ω so f and u both tend rapidly to zero as $\omega \to +\infty$, except on or very near $\partial\Omega$. Setting $U_{-1} \equiv 0$, we find on substitution

$$0 = \omega^2(\nabla S)^2 \sum_0^\infty U_k/(2\omega)^k$$
$$+ \sum_0^\infty (2\omega)^{-k}(\Lambda U_k + LU_{k-1} - F_k) \text{ in } \Omega, \quad \text{as } \omega \to \infty$$

with $U_k \mid \partial\Omega = G_k, k \ge 0$, where

$$\Lambda = \nabla S \cdot \nabla + \tfrac{1}{2}(\Delta S + b \cdot \nabla S).$$

(Scalar operators, like $\nabla S \cdot \nabla$, are implictly multiplied by an identity matrix, and $b \cdot \nabla S(x) = \sum_{j=1}^n b_j(x)\partial S/\partial x_j \in \mathbb{C}^{p\times p}$.)

We choose the complex-valued S so

$$(\nabla S)^2 = \sum_1^n (\partial S/\partial x_j)^2 = 0$$

and choose the U_k inductively, solving

$$\begin{cases} \Lambda U_k + LU_{k-1} = F_k & (k \geq 0) \\ U_k | \, \partial\Omega = G_k \end{cases}$$

with $U_{-1} \equiv 0$

Well . . . not quite. There ordinarily are no exact solutions, but we only need that $(\nabla S)^2$ and the $\nabla U_k + LU_{k-1} - F_k$ tend to zero rapidly as $x \to \partial\Omega$. (See 8.3 for exact conditions.) We find the approximate solutions using the "normal coordinates" near $\partial\Omega$ treated in Theorem 1.5, and we need more details about these.

Normal Coordinates

(1) $(y, t) \mapsto x = y + tN_\Omega(y)$ is a diffeomorphism of $(y, t) \in \partial\Omega \times (-r, r)$ onto a neighborhood $B_r(\partial\Omega)$ of $\partial\Omega$ in $\mathbb{R}^n$, for small $r > 0$, when $\partial\Omega$ is at least C^2, with inverse $x \mapsto (y, t)$:
- $y =$ the point of $\partial\Omega$ closest to x ($= \pi(x)$, in notation of Thm. 1.5)
- $t = \pm$ distance from x to $\partial\Omega$, sign $\begin{cases} +, & \text{outside } \Omega \\ -, & \text{inside } \Omega. \end{cases}$
- We extend the unit normal N by: $N(y + tN(y)) = N(y)$, $-r < t < r$, $y \in \partial\Omega$.

(2) $dx = (I + tK_\Omega(y))dy + N_\Omega(y)dy = (I + tK_\Omega(y))(Ndt + dy)$ where $K = DN$ is the (degenerate) curvature matrix, $K(x) = D^2 t(x)$, with $KN = 0$, and "dy" is tangent to $\partial\Omega$ at y.

(3) $\text{vol}(dx) = \det(I + tK_\Omega(y))dt\, dA_y = e^\lambda dt\, dA_y$ where $\lambda(t, y) = \log\det(I + tK(y)) = \sum_{m=1}^\infty \frac{(-1)^{m-1}}{m} t^m H_m(y)$ for small t, $H_m = \text{trace } K^m$, trace K is the mean curvature.

(4) Let $u(x)$ be C^1 near Ω and define $\tilde{u}(y, t) = u(y + tN(y))$ for small t; then $du = \nabla u \cdot dx = \tilde{u}_t dt + \tilde{u}_y \cdot dy$ so $\nabla u(y + tN(y)) = (I + tK(y))^{-1}\tilde{u}_y + N(y)\tilde{u}_t$. (We don't always distinguish $\tilde{u}$ from u, and sometimes write $\partial u/\partial N$ for $\tilde{u}_t$ and $\nabla_{\partial\Omega} u$ for $\tilde{u}_y$.)

(5) If u, v are C^2 near $\partial\Omega$ and v has small support, we change variables in $0 = \int (v\Delta u + \nabla v \cdot \nabla u)\text{vol}(dx)$ by (3), (4) and integrate by parts to obtain:

$$\Delta u(y + tN(y)) = \tilde{u}_{tt} + \lambda_t \tilde{u}_t + (I + tK)^{-2}\lambda_y \cdot \tilde{u}_y + \text{div}_{\partial\Omega}((I + tK)^{-2}\tilde{u}_y).$$

(On $\partial\Omega$, at $t = 0$, this reduces to $\Delta u = u_{NN} + Hu_N + \Delta_{\partial\Omega} u$.)

Calculation of S

$$S(y + tN(y)) = \sum_{k\geq 0} t^k S_k(y)/k!,$$

at least asymptotically, as $t \to 0$ with $y \in \partial\Omega$ (in the relevant portion of $\partial\Omega$, which is usually small). On $\partial\Omega$ we have

$$S = S_0 = i\theta(y) \qquad (\theta(\,\cdot\,) \text{ given, smooth, real})$$

and $\partial S/\partial N|_{\partial\Omega} = S_1$ has Re $S_1 > 0$, and we want $(\nabla S)^2(y + tN(y)) = 0$ (asymptotically, as $t \to 0$).

Now $\nabla S(y + tN(y)) = N(S_1 + tS_2 + t^2/_2 S_3 + \ldots) + (I - tK + t^2K^2 - \ldots)(S_0' + tS_1' + t^2/_2 S_2' + \ldots)$, writing ϕ' for $\phi_y = \nabla_{\partial\Omega}\phi$, so

$$(\nabla S)^2 = 0 = (S_0')^2 + S_1^2 + t(2S_1S_2 + 2S_0' \cdot S_1' - 2S_0' \cdot K S_0') + \ldots.$$

The zero-order term gives

$$S_1 = \sqrt{-(S_0')^2} = |\nabla_{\partial\Omega}\theta|,$$

since Re $S_1 > 0$; we assume $|\nabla_{\partial\Omega}\theta| \equiv 1$ in the region of interest, which makes things much simpler. In this case

$$S_0 = i\theta, \qquad S_1 = 1, \qquad S_2 = -\nabla_{\partial\Omega}\theta \cdot K\nabla_{\partial\Omega}\theta,$$

and we may compute as many terms as desired – assuming $\partial\Omega$ and θ are sufficiently smooth. (S_k, $k \leq 4$, are given explicitly in the summary at the end of this chapter.)

We note, for later use, that each S_k is the sum of a real part, even in θ, and a purely-imaginary part, odd in θ.

The hypothesis $|\nabla_{\partial\Omega}\theta| \equiv 1$ is very helpful in our problems (since the top-order operator is the Laplacian), so we pause to discuss this equation. For $n = 2$, $\partial\Omega$ is a curve (or several) and θ is arc-length along the curve. For $n \geq 3$, there will be many choices of θ. Let $x_0 \in \partial\Omega$ and let Σ be a C^2 hypersurface containing x_0 and transverse to $\partial\Omega$. Define $\theta_\Sigma = \pm$ {distance in $\partial\Omega$ from x to $\partial\Omega \cap \Sigma$}, changing sign as we cross $\partial\Omega \cap \Sigma$. Then for various choices of Σ and various choices of the real constant C, $\theta = C \pm \theta_\Sigma(x)$ yields all the C^2 solutions of $|\nabla_{\partial\Omega}\theta| \equiv 1$ near x_0. If Σ and $\partial\Omega$ are C^m ($m \geq 2$) then θ_Σ is C^m. In particular,.given $\tau \perp N(x_0)$, $|\tau| = 1$, we may choose a C^m solution of $|\nabla_{\partial\Omega}\theta| = 1$ near x_0 such that $\nabla_{\partial\Omega}\theta(x_0) = \tau$. (This may be proved directly, or we may appeal to the theory of first-order scalar PDEs [5, 6].) For $n = 2$, θ is essentially unique, so this case often differs from $n \geq 3$.

Calculation of Λ

$$\Lambda = \nabla S \cdot \nabla + \tfrac{1}{2}\sigma, \qquad \sigma = \Delta S + b \cdot \nabla S$$

and we will write these in normal coordinates.

$$\begin{aligned}\nabla S \cdot \nabla &= (I + tK)^{-2} S_y \cdot \partial_y + S_t \partial_t \\ &= i\theta' \cdot \partial_y + \partial_t + t(-2iK\theta' \cdot \partial_y + S_2 \partial_t) \\ &\quad + t^2 \left(3iK^2\theta' \cdot \partial_y + \tfrac{1}{2}S_2' \cdot \partial_y + \tfrac{1}{2}S_3 \partial_t\right) + \cdots\end{aligned}$$

$$\begin{aligned}\Delta S &= S_2 + tS_3 + t^2/_2 S_3 + \cdots + (H - tH_2 + \cdots)\left(1 + tS_2 + t^2/_2 S_3 + \cdots\right) \\ &\quad + (I - 2tK + \cdots)\left(tH' - t^2/_2 H_2' + \cdots\right)\left(i\theta' + t^2/_2 S_2' + \cdots\right) \\ &\quad + \mathrm{div}_{\partial\Omega}((I - 2tK + 3t^2K^2 - \cdots)\left(i\theta' + t^2/_2 S_2' + \cdots\right) \\ &= S_2 + H + i\Delta_{\partial\Omega}\theta \\ &\quad + t(S_3 + HS_2 - H_2 + i\theta' \cdot H' - 2i\,\mathrm{div}_{\partial\Omega}(K\theta')) \\ &\quad + t^2/_2 \left(S_4 + HS_3 - 2H_2S_2 + 2H_3 - 4iK\theta' \cdot H' - i\theta' \cdot H_2' \right.\\ &\quad \left. + \ 3i\,\mathrm{div}_{\partial\Omega}(K^2\theta') + \tfrac{1}{2}\Delta_{\partial\Omega} S_2\right) + O(t^3)\end{aligned}$$

Writing $b(y + tN(y)) = b(y) + t\dot{b}(y) + t^2/_2\ddot{b}(y) + \cdots$

$$b \cdot \nabla S(y + tN(y)) = b \cdot (i\theta' + N) + i(\dot{b}(i\theta' + N) + b \cdot (-iK\theta' + NS_2)) + \cdots$$

$$\sigma = \Delta S + b \cdot \nabla S = \sigma_0(y) + t\sigma_1(y) + t^2/_2\sigma_2(y)) + \cdots.$$

Calculation of U_0

$$\Lambda U_0 = F_0(y + tN(y)) = F_0(y) + t\dot{F}_o(y) + t^2/_2\ddot{F}_0(y)) + \cdots$$

with

$$U_0 \mid \partial\Omega = G_0, \qquad U_0(y + tN(y)) = G_0(y) + t\dot{U}_0(y) + t^2/_2\ddot{U}_0(y) + \cdots$$

and $\Delta U_0 = F_0$ becomes

$$\begin{aligned}F_0 + t\dot{F}_0 + t^2/_2\ddot{F}_0 + \cdots &= \dot{U}_0 + i\theta' \cdot \partial_y G_0 + \tfrac{1}{2}\sigma_0 G_0 \\ &\quad + t\left(\ddot{U}_0 + i\theta' \cdot \partial_y \dot{U}_0 + \left(\tfrac{1}{2}\sigma_0 + S_2\right)\dot{U}_0 \right.\\ &\quad \left. + \left(-2iK\theta' \cdot \partial_y G_0 + \tfrac{1}{2}\sigma_1 G_0\right)\right) \\ &\quad + t^2/_2\left(\dddot{U}_0 + i\theta' \cdot \partial_y \ddot{U}_0 + \left(\tfrac{1}{2}\sigma_0 + 2S_2\right)\ddot{U}_0 \right.\\ &\quad + (-4iK\theta' \cdot \partial_y \dot{U}_0 + \sigma_1 \dot{U}_0) \\ &\quad \left. + \left(6iK^2\theta' \cdot \partial_y G_0 + S_2' \cdot \partial_y G_0 + \tfrac{1}{2}\sigma_2 G_0\right)\right) + O(t^3)\end{aligned}$$

which determines $U_0 = G_0$, $\dot{U}_0$, $\ddot{U}_0$, $\dddot{U}_0$ on $\partial\Omega$. (These all vanish off the support of G_0, where F_0, $\dot{F}_0$, $\ddot{F}_0$ vanish.)

Calculation of U_1, U_2

$\Lambda U_1 = F_1 - LU_0, U_1 \mid \partial\Omega = G_1$ so we must find $LU_0 = \Delta U_0 + b \cdot \nabla U_0 + cU$

$$\begin{aligned} LU_o(y+tN(y)) = {} & \ddot{U}_0 + (H + b \cdot N)\dot{U}_0 + \Delta_{\partial\Omega} G_0 + b \cdot \nabla_{\partial\Omega} G_0 + cG_0 \\ & + t\{(\dddot{U}_0 + (H + b \cdot N)\ddot{U}_0 + \Delta_{\partial\Omega}\dot{U}_0 + b \cdot \nabla_{\partial\Omega}\dot{U}_0 \\ & + c\dot{U}_0 + (\dot{b} \cdot N\dot{U}_0 - H_2) - 2\,\mathrm{div}_{\partial\Omega}\,(KG_0') \\ & + (\dot{b} - bK + H') \cdot G_0' + \dot{c}G_0)\} + O(t^2), \end{aligned}$$

and substitution gives $\dot{U}_1$, $\ddot{F}_1$ on $\partial\Omega$ ($U_1 \mid \partial\Omega = G_1$). From these we compute $LU_1 \mid \partial\Omega$, hence $\dot{U}_2 \mid \partial\Omega$.

Thus we formally solve the Dirichlet problem to any desired order, assuming sufficient smoothness (and patience). If instead, we are given, for example,

$$\frac{\partial U}{\partial N} + \beta U|_{\partial\Omega} = e^{i\omega\theta} 2\omega \left(H_0 + \frac{1}{2\omega} H_1 + \frac{1}{(2\omega)^2} H_2 + \dots \right)$$

then (supposing β is independent of ω) we may determine the corresponding G's from

$$H_0 = \tfrac{1}{2} G_0, \quad H_k = \dot{U}_{k-1} + \tfrac{1}{2} G_k + \beta G_{k-1} \qquad (k \le 1),$$

or

$$G_0 = 2H_0, \qquad G_1 = 2H_1 - 4\beta H_0 - 2F_0 + 4i\theta' \cdot \partial_y H_0 + 2\sigma_0 H_0,$$

and so on.

We summarize our results at the end of the chapter for ease of reference.

8.3 Exact Solutions

Suppose the integer $N \ge 0$ and for $k = 0, 1, \dots, N$, $F(x)$ is C^{N+1-k} near $\partial\Omega$ and G_k is C^{N+2-k} on $\partial\Omega$, both with compact support. Also suppose $b_j(x) \in \mathbb{C}^{p\times p}$ is C^{N+1} and $c(x) \in \mathbb{C}^{p\times p}$ is C^N. For our "generic" problems, we may always suppose $\partial\Omega$ is as smooth as desired, but for this calculation, class C^{N+3} suffices, and we suppose θ has the same smoothness (as always, suppose $|\nabla_{\partial\Omega}\theta| \equiv 1$). We may then choose $S(y + tN(y))$ of class C^{N+3} such that

$$(\nabla S(x))^2 = O\big(d(x)^{N+2}\big) \qquad \text{as } d(x) = \mathrm{dist.}\,(x, \partial\Omega) \to 0.$$

We choose $U_k(x)$ of class C^{N+2-k} so that, for $0 \le k \le N$,

$$\Lambda U_k + LU_{k-1} - F_k(x) = O(d(x)^{N+1-k})$$

$$U_k|\partial\Omega = G_k \quad (U_{-1} \equiv 0)$$

Choose a C^∞ "cutoff" $\chi : \chi \equiv 1$ near $\partial\Omega \cap \cup_k \operatorname{supp} F_k \cup \operatorname{supp} G_k$, but χ is supported near this set.

We compare the formal solution

$$u_{\text{form.}} = \chi \cdot e^{\omega S} \sum_{k=0}^{N} U_k/(2\omega)^k$$

with the exact solution

$$u = \mathcal{B}_L(g) + A_L(f)$$

as $\omega \to +\infty$, where

$$\|g - e^{i\omega\theta} \sum_{k=0}^{N} G_k/(2\omega)^k \|_{C^2(\partial\Omega)} = o(\omega^{-N})$$

$$\|f - \chi \cdot 2\omega e^{\omega S} \sum_{k=0}^{N} F_k/(2\omega)^k \|L_{p(\Omega)} = o(\omega^{-N})$$

and we show $\|u - u_{\text{form}}\|_{W_p^2(\Omega)} = o(\omega^{-N})$ as $\omega \to +\infty$. (We might use the norm of $W_p^{2-1/p}(\partial\Omega)$ instead of $C^2(\partial\Omega)$ for the boundary values.) Here $1 < P < \infty$ is fixed and will ordinarily be chosen with $p > n$ so $W_p^2 \subset C^1$ and, in particular,

$$\left.\frac{\partial u}{\partial N}\right|_{\partial\Omega} - \chi \cdot e^{i\omega\theta} \left\{ \omega \sum_{0}^{N} G_k/(2\omega)^k + \sum_{0}^{N} \dot{U}_k/(2\omega)^k \right\} = o(\omega^{-N})$$

uniformly on $\partial\Omega$ as $\omega \to +\infty$, or even in $C^\alpha(\partial\Omega)$ provided $0 < \alpha < 1 - n/p$.

Sometimes we need a stronger norm, such as $W_p^2(\Omega)$; the calculations are similar, assuming a bit more smoothness for the b_j, c: one more derivative, for W_p^3.

There are various ways to define $\mathcal{B}_L, \mathcal{A}_L$, differing by operators of finite rank, but the above results holds in any case.

Suppose $L \mid H^2 \cap H_0^1(\Omega)$ has m-dimensional kernel, so its image is closed, with codimension m. Let $\{\psi_1, \ldots, \psi_m\}$ be a basis for any complement to the image; then if $\{\sigma_1, \ldots, \sigma_m\}$ is a basis for the kernel of $L^* \mid H^2 \cap H_0^1(\Omega)$, $L^* = \Delta - b \cdot \nabla + c - \operatorname{div} b$, we have $\det [\int_\Omega \sigma_j \cdot \psi_k]_{j,k=1}^m \neq 0$. Also choose a complement to the kernel of L, in the form $\{u : \int_\Omega u \cdot \phi_j = 0 \, (1 \le j \le m)\}$. We could suppose the $\{\bar{\phi}_j\}$ are a basis for the kernel of L, but anything nearby would work as well. The solutions σ_k of the adjoint equation are $C^{N+2-\epsilon}$, any $\epsilon > 0$, and we suppose the ψ_k, ϕ_j have similar smoothness. Then $u = \mathcal{B}_L(g) + \mathcal{A}_L(f)$

is the solution of

$$Lu = \Delta u + b \cdot \nabla u + c = f + \sum_{j=1}^{m} \alpha_j \psi_j \text{ in } \Omega$$

for some choice of the $\alpha_j \in \mathbb{C}$, such that

$$u = g \text{ on } \partial\Omega \quad \text{and} \quad \int_\Omega u \cdot \phi_j = 0, \quad 1 \le j \le m.$$

The α_j are uniquely determined, since for each $1 \le k \le m$

$$\begin{aligned} \sum_{j=1}^{m} \alpha_j \int_\Omega \sigma_k \cdot \psi_j &= \int_\Omega \sigma_k \cdot (Lu - f) \\ &= -\int_\Omega \sigma_k \cdot f - \int_{\partial\Omega} \frac{\partial \sigma_k}{\partial N} \cdot g, \quad 1 \le k \le m, \end{aligned}$$

with an invertible matrix.

Now

$$\begin{aligned} L &\left(e^{\omega S} \sum_0^N U_k/(2\omega)^k \right) - 2\omega e^{\omega S} \sum_0^N \frac{F_k}{(2\omega^k)} \\ &= \frac{e^{\omega S}}{(2\omega)^N} \left\{ (2d\omega)^{2+N} \frac{\nabla S \cdot \nabla S}{4d^{N+2}} \sum_0^N \frac{U_k}{(2\omega)^k} + LU_N \right. \\ &\quad \left. + \sum_{k=0}^{N} (2\omega)^{1-k+N} \frac{(\Lambda U_k + LU_{k-1} - F_k)}{d^{N+1-k}} \right\} \end{aligned}$$

and Re $S(x) < -½d(x)$ in Ω near $\partial\Omega$. Since $\chi \equiv 1$ near $\partial\Omega$ – at least, the part where the G_k and F_k, hence U_k, are nonzero – we see, for some constant C,

$$|Lu_{\text{form.}} - \chi \cdot 2\omega e^{\omega S} \sum_0^N F_k/(2\omega)^k| \le C\omega^{-N} e^{-\omega d/4}$$

in Ω, uniformly as $\omega \to +\infty$. Let $z \in W_p^2(\Omega)$ be defined by

$$z = u_{\text{form.}} - u - \sum_{k=1}^{m} \beta_k \tau_k$$

where $\{\tau_1, \ldots, \tau_m\}$ is a basis for the kernel of $L|H^2 \cap H_0^1(\Omega)$, with $\beta_k \in \mathbb{C}$ chosen to ensure $\int_\Omega z \cdot \phi_j = 0$ for all $j = 1, \ldots, m$.

Note det $[\int_\Omega \tau_k \cdot \phi_j]_{j,k=1}^m \neq 0$; this certainly holds when the $\{\bar{\phi}_j\}$ form a basis for the kernel, and it still is true when they are nearby. Thus the β_k are well-defined.

In Ω

$$|L_z| = |Lu_{\text{form}} - Lu|$$
$$\le |f - \chi \cdot 2\omega e^{\omega S} \sum_0^N F_k/(2\omega)^k| + C\omega^{-N}e^{-\omega d/4} + \sum_{j=1}^m |\alpha_j| \sup|\psi_j|$$

and on $\partial\Omega, \omega^N z = \omega^N(u_{\text{form.}} - u) = \omega^N(\epsilon^{i\omega\theta}\Sigma_0^N G_k/(2\omega)^k - g)|_{\partial\Omega}$ which tends to zero in $C^2(\partial\Omega)$ as $\omega \to +\infty$. Also

$$\int_\Omega \sigma_k \cdot f + \int_{\partial\Omega} \frac{\partial\sigma_k}{\partial N} \cdot g = o(\omega^{-N}) + \sum_{j=0}^N (2\omega)^{1-j} \int_\Omega \sigma_k \cdot F_j e^{\omega S}$$
$$+ \sum_{j=0}^N (2\omega)^{-j} \int_{\partial\Omega} \sigma_k \cdot G_j e^{i\omega\theta} = o(\omega^{-N}) \qquad as\ \omega \to +\infty,$$

by the Riemann-Lebesgue lemma, so the $|\alpha_j| = o(\omega^{-N})$. Similarly

$$\sum_{k=1}^m \beta_k \int_\Omega \tau_k \cdot \phi_j = \int_\Omega u_{\text{form}} \cdot \phi_j = \int_\Omega \sum_{i=0}^N (2\omega)^{-i} \chi U_i \cdot \phi_j e^{\omega S} = o(\omega^{-N})$$

so the $|\beta_k| = o(\omega^{-N})$ as $\omega \to \infty$. Thus

$$||Lz||_{L_p(\Omega)} = o(\omega^{-N}), \qquad ||z \mid \partial\Omega||_{C^2(\partial\Omega)} = o(\omega^{-N}),$$

so $||z||_{W_p^2(\Omega)} = o(\omega^{-N})$. Then

$$||u_{\text{form}} - u||_{W_p^2(\Omega)} \le ||z||_{W_p^2(\Omega)} + ||\sum_1^m \beta_k \tau_k||_{W_p^2(\Omega)} = o(\omega^{-N})$$

and the claim is proved.

8.4 Example 6.8 Revisited: Generic Simplicity of Complex Eigenvalues

In the example, $b_j(x, \lambda)$ and $c(x, \lambda)$ are polynomials in λ with smooth complex-valued coefficients. (It suffices that the coefficients of the b_j are C^2 and those of c are C^1, but we won't worry about smoothness in the calculation.) We often suppress the λ-dependence and treat

$$A = \Delta + b \cdot \nabla + c, \qquad A^* = A - b \cdot \nabla u + c - \text{div}\, b.$$

As in 6.8, our hypothesis for contradiction is that, for each Ω near Ω_0, there exists $\lambda \in \Lambda$ and solutions u, ψ of

$$Au = 0 \text{ in } \Omega, \qquad A^*\psi = 0 \text{ in } \Omega, \qquad u = 0 = \psi \text{ on } \partial\Omega$$
$$u \not\equiv 0, \qquad \psi \not\equiv 0$$

with $\operatorname{Re}(u_N\psi_N) \equiv 0$ on $\partial\Omega$. ($u_N = \partial u/\partial N, \psi_N = \partial\psi/\partial N$.) We have $u_N\psi_N \neq 0$, purely imaginary, on a dense set of $\partial\Omega$.

Solvability of this problem for each Ω near Ω_0 implies we have a solution (λ, u, ψ) which also satisfies:

$$\sigma \mapsto \Gamma(\sigma) = \operatorname{Re}\left\{\psi_N \frac{\partial}{\partial N}\mathcal{B}_A(\sigma u_N) + u_N \frac{\partial}{\partial N}\mathcal{B}_{A^*}(\sigma\psi_N)\right\}$$

has finite rank.

Instead of appealing to Chapter 7, we will show: if γ, θ are smooth, real functions on $\partial\Omega$ with $|\nabla_{\partial\Omega}\theta| = 1$ in supp γ, then uniformly on $\partial\Omega$,

$$\Gamma(\gamma\cos\omega\theta) = \frac{\gamma\cos\omega\theta}{\omega}C(\theta) + o(\omega^{-1}) \qquad \text{as } \omega \to \infty$$

where

$$C(\theta) = \operatorname{Re}\{(D^\theta u_N + \tfrac{1}{2}b^\theta u_N)\cdot(D^\theta\psi_N - \tfrac{1}{2}b^\theta\psi_N) - qu_N\psi_N\}$$

in supp γ, $D^\theta = \nabla_{\partial\Omega} - \nabla_{\partial\Omega}\theta\frac{\partial}{\partial\theta}$, $b^\theta = b - Nb\cdot N - \nabla_{\partial\Omega}\theta b\cdot\nabla_{\partial\Omega}\theta$ and $q = c - \tfrac{1}{2}\operatorname{div} b - \tfrac{1}{4}b^2\, (= D(A(\lambda)))$.

The lemma of the introduction (8.1) then shows $C(\theta) = 0$ for all such θ; we examine this condition later. Note when $n = 2$, $D^\theta = 0$ and $b^\theta = 0$ so $\operatorname{Im} q = 0$ on $\partial\Omega$. We construct formal asymptotic solutions

$$\mathcal{B}_A(\gamma u_N e^{i\omega\theta}) = e^{\omega S}\left(U_0 + \frac{1}{2\omega}U_1 + \cdots\right)$$
$$\mathcal{B}_{A^*}(\gamma\psi_N e^{i\omega\theta}) = e^{\omega S}\left(\Psi_0 + \frac{1}{2\omega}\Psi_1 + \cdots\right)$$

with the same "S" since this depends only on the highest derivative, but we write $\sigma = \Delta S + b\cdot\nabla S$ for A, $\sigma^* = \Delta S - b\cdot\nabla S$ for A^*.

It is convenient to use the complex exponential:

$$\Gamma(\gamma\cos\omega\theta) = \operatorname{Re}\tfrac{1}{2}(\Gamma_0(\gamma e^{i\omega\theta}) + \Gamma_0(\gamma e^{-i\omega\theta})) = \operatorname{Re}\cdot\operatorname{even}\Gamma_0(\gamma e^{i\omega\theta})$$

where

$$\Gamma_0(\gamma^{i\omega\theta}) = \psi_N \frac{\partial}{\partial N}\mathcal{B}_A(\gamma u_N e^{i\omega\theta}) + u_N \frac{\partial}{\partial N}\mathcal{B}_{A^*}\cdot(\gamma\psi_N e^{i\omega\theta})$$
$$= e^{i\omega\theta}\left\{2\omega\gamma u_N\psi_N + \psi_N\dot{U}_0 + u_N\dot{\Psi}_0 + \frac{1}{2\omega}(\psi_N\dot{U}_1 + u_N\dot{\Psi}_1) + o(\omega^{-1})\right\}$$

Thus the coefficient of "ω" in Γ $(\gamma \cos\omega\theta)$ is

$$\text{Re}\cdot\text{even}\, e^{i\omega\theta}2\gamma u_N\psi_N = \gamma\cos\omega\theta\,\text{Re}\,(u_N\psi_N) \equiv 0.$$

The coefficient of "ω^0" is

$$\begin{aligned}\text{Re}\cdot\text{even}\ & e^{i\omega\theta}(\psi_N\dot{U}_0 + u_N\dot{\Psi}_0)\\ &= -\cos\omega\theta_\gamma(q-H)\text{Re}\,u_N\psi_N\\ &\quad+\sin\omega\theta(\partial_\theta\gamma + \gamma\Delta_{\partial\Omega}\theta)\,\text{Re}\,u_N\psi_N + \gamma\partial_\theta(\text{Re}\,u_N\psi_N) \equiv 0\end{aligned}$$

The coefficient of ω^{-1} does *not* vanish trivially, but there is a lot of cancellation; we pause to organize the calculations.

We distinguish terms of type R or type I:

- type R = (real-valued, even in θ) + (pure imaginary, odd in θ)
- type I = (real, odd) + (imaginary, even).

Note the products: $R\times R = R$, $R\times I = I$, $I\times I = R$. Type I is also called "negligible" since

$$\text{Re}\cdot\text{even}\cdot e^{i\omega\theta}(R+I) = \text{Re}\cdot\text{even}\, e^{i\omega\theta}R.$$

(Indeed $e^{i\omega\theta} = \cos\omega\theta + i\sin\omega\theta$ is type R so $e^{i\omega\theta}I$ is type I with real-even-part zero.)

For example $u_N\psi_N$ is type I, as is $i\gamma\partial_\theta(u_N\psi_N)$ or $i\gamma\Delta_{\partial\Omega}\theta u_N\psi_N$, or $\psi_N\dot{U}_0 + u_N\dot{\Psi}_0$, where S and every S_k is type R. Also note $\nabla_{\partial\Omega}$, $i\frac{\partial}{\partial\theta}$, $\nabla S\cdot\nabla + ½\Delta S$ preserve the type.

Some terms, containing b or c, may have parts of both types; we will simply carry those along in the calculation but terms which are clearly negligible will be collected in "+ (negl.)". The coefficient of $(2\omega)^{-1}$ is

$$\begin{aligned}\psi_N\dot{U}_1 + u_N\dot{\Psi}_1 = &-(\psi_N\ddot{U}_0 + (H + b\cdot N)\psi_N\dot{U}_0 + \psi_N\Delta_{\partial\Omega}\gamma u_N\\ &+\psi_N b\cdot\nabla_{\partial\Omega}\gamma u_N + c\gamma u_N\psi_N + u_N\ddot{\Psi}_0 + (H - b\cdot N)u_N\dot{\Psi}_0\\ &+u_N\Delta_{\partial\Omega}\gamma\psi_N - u_N b\cdot\nabla_{\partial\Omega}\gamma\psi_N + (c - \text{div}\,b)\gamma u_N\psi_N).\end{aligned}$$

We have

$$\begin{aligned}\psi_N \Delta_{\partial\Omega}\gamma u_N + u_N \Delta_{\partial\Omega}\gamma \psi_N &= 2u_N\psi_N\Delta_{\partial\Omega}\gamma + \nabla_{\partial\Omega}\gamma \cdot \nabla_{\partial\Omega}(u_N\psi_N)\\ &\quad + \gamma(\psi_N\Delta_{\partial\Omega}u_N + u_N\Delta_{\partial\Omega}\psi_N)\\ &= -2\gamma\nabla_{\partial\Omega}\psi_N \cdot \nabla_{\partial\Omega}u_N + (\text{negl.}),\end{aligned}$$

so

$$\begin{aligned}\psi_N\dot{U}_1 + u_N\dot{\Psi}_1 = &-(\psi_N\ddot{U}_0 + u_N\ddot{\Psi}_0) - b\cdot N(\psi_N\dot{U}_0 - u_N\dot{\Psi}_0)\\ &+ 2\gamma\nabla_{\partial\Omega}\psi_N\nabla_{\partial\Omega}u_N - \gamma b\cdot(\psi_N\nabla_{\partial\Omega}u_N - u_N\nabla_{\partial\Omega}\psi_N)\\ &- (2c - \operatorname{div} b)\gamma u_N\psi_N + (\text{negl.}).\end{aligned}$$

Also

$$\begin{aligned}-(\psi_N\ddot{U}_0 + u_N\ddot{\Psi}_0) &= \psi_N\big(i\partial_\theta\dot{U}_0 + (\tfrac{1}{2}\sigma_0 - q)\dot{U}_0\\ &\quad + \big(-2ik\theta'\cdot\nabla_{\partial\Omega} + \tfrac{1}{2}\sigma_1\big)\gamma u_N\big)\\ &\quad + u_N\big(i\partial_\theta\dot{\Psi}_0 + (\tfrac{1}{2}\sigma_0^* - q)\dot{\Psi}_0\\ &\quad + \big(-2ik\theta'\cdot\nabla_{\partial\Omega} + \tfrac{1}{2}\sigma_1^*\big)\gamma\psi_N\big)\\ &= \underline{i\partial_\theta(\psi_N\dot{U}_0 + u_N\dot{\Psi}_0)} - i\partial_\theta\psi_N\dot{U}_0 - i\partial\theta u_N\dot{\Psi}_0\\ &\quad + \underline{\tfrac{1}{4}(\sigma_0 + \sigma_0^* - 2q)(\psi_N\dot{U}_0 + u_N\dot{\Psi}_0)}\\ &\quad + \tfrac{1}{2}b\cdot(N + i\theta')(\psi_N\dot{U}_0 - u_N\dot{\Psi}_0)\\ &\quad + \underline{\tfrac{1}{2}(\sigma_1 + \sigma_1^*)\gamma u_N\psi_N} + (\text{negl.}),\end{aligned}$$

with some negligible terms underlined. Further

$$\begin{aligned}&-i\partial_\theta\psi_N\dot{U}_0 - i\partial_\theta u_N\dot{\Psi}_0\\ &\quad = -2\gamma\partial_\theta\psi_N\partial_\theta u_N + \frac{i}{2}\gamma b\cdot(N + i\theta')(u_N\partial_\theta\psi_N - \psi_N\partial_\theta u_N) + (\text{negl.}),\end{aligned}$$

and more such calculations yield finally:

$$\begin{aligned}\psi_N\dot{U}_1 + u_N\dot{\Psi}_1 = &\,2\gamma\nabla_{\partial\Omega}\psi_N\cdot\nabla_{\partial\Omega}u_N - \gamma b\cdot(\psi_N\nabla_{\partial\Omega}u_N - u_N\nabla_{\partial\Omega}\psi_N)\\ &- 2\gamma\partial_\theta\psi_N\partial_\theta u_N + \gamma b\cdot\theta'(\psi_N\partial_\theta u_N - u_N\partial_\theta\psi_N)\\ &- 2\gamma\{c - \tfrac{1}{2}\operatorname{div} b - \tfrac{1}{4}(b\cdot N)^2 - \tfrac{1}{4}(b\cdot\theta')^2\}u_N\psi_N\\ &+ (\text{negl.})\end{aligned}$$

Writing D_θ for $\nabla_{\partial\Omega} - (\nabla_{\partial\Omega}\theta)\frac{\partial}{\partial\theta}$ and $b_\theta = b - Nb\cdot N - \nabla_{\partial\Omega}\theta b\cdot\theta'$. This becomes

$$\begin{aligned}&2\gamma\big(D_\theta\psi_N - \tfrac{1}{2}b_\theta\psi_N\big)\cdot\big(D_\theta u_N + \tfrac{1}{2}b_\theta u_N\big)\\ &\qquad - 2\gamma\big\{c - \tfrac{1}{2}\operatorname{div} b - \tfrac{1}{4}b^2\big\}u_N\psi_N + (\text{negl.}).\end{aligned}$$

The non-negligible part is even in θ, so we obtain

$$\Gamma(\gamma\cos\omega\theta) = \frac{\gamma\cos\omega\theta}{\omega}\operatorname{Re}\{(D_\theta\psi_N - \tfrac{1}{2}b_\theta\psi_N)\cdot(D_\theta u_N + \tfrac{1}{2}b_\theta u_N) - (c - \tfrac{1}{2}\operatorname{div} b - \tfrac{1}{4}b^2)u_N\psi_N\} + o(\omega^{-1}) \quad \text{as } \omega \to \infty,$$

and the coefficient of ω^{-1} must vanish for all allowed choices of γ, θ. If $n = 2$, $D_\theta = 0$ and $b_\theta = 0$ so we obtain $\operatorname{Im} D(A(\lambda)) = 0$ on $\partial\Omega \cap V$.

Let $A = \nabla_{\partial\Omega}\psi_N - \tfrac{1}{2}b_{\tan}\psi_N$, $B = \nabla_{\partial\Omega}u_N + \tfrac{1}{2}b_{\tan}u_N$, and $C = c - \tfrac{1}{2}\operatorname{div} b - \tfrac{1}{4}b^2 (= D(A(\lambda)))$. At each $x \in V \cap \partial\Omega$, for each $\tau \perp N_\Omega(x)$ with $|\tau| = 1$, we can choose θ with $|\nabla_{\partial\Omega}\theta| = 1$ near x, $\nabla_{\partial\Omega}\theta(x) = \tau$, γ supported near x with $\gamma(x) \neq 0$, and conclude

$$\operatorname{Re}\{(A - \tau A\cdot\tau)\cdot(B - \tau B\cdot\tau) - C\} = 0$$

or

$$\operatorname{Re}(A\cdot\tau B\cdot\tau) = \operatorname{Re}(A\cdot B - C) \text{ for all } \tau \perp N, |\tau| = 1.$$

The average value of the left-side, over all such τ, is $\frac{1}{n-1}\operatorname{Re} A\cdot B$, which implies $(n-2)$
$\operatorname{Re} A\cdot\tau B\cdot\tau = \operatorname{Re} C$ when $|\tau| = 1$, $\tau \perp N$, or

$$(n-2)\operatorname{Re}\left(\frac{\partial\psi_N}{\partial\tau} - \tfrac{1}{2}b\cdot\tau\psi_N\right)\left(\frac{\partial u_N}{\partial\tau} + \tfrac{1}{2}b\cdot\tau u_N\right) = \operatorname{Re} D(A(\lambda))u_N\psi_N,$$

the same equation obtained in 6.8 above. As in 6.8, we conclude:

$$\text{for } n \geq 2, \quad \operatorname{Im} D(A(\lambda)) = 0 \text{ on } \partial\Omega \cap V;$$

and for $n \geq 3$, where $u_N\psi_N \neq 0$ in $\partial\Omega \cap V$,

$$|\operatorname{Im} b_{\tan} - \nabla_{\partial\Omega}(2\arg u_N)|\cdot|\operatorname{Re} b_{\tan} - \nabla_{\partial\Omega}\log|\psi_N/u_N|| = 0.$$

We show this contradicts the hypothesis (for $n \geq 3$):

> for each $\lambda \in \Lambda$, if $c - \tfrac{1}{2}\operatorname{div} b - \tfrac{1}{4}b^2$ is real everywhere in a neighborhood of some $x \in V$, then neither $\operatorname{Re} b$ not $\operatorname{Im} b$ is a gradient on any neighborhood of x.

Differently stated: for $\lambda \in \Lambda$

$$|\operatorname{Im} D(A(\lambda))| + \prod_{\beta=\operatorname{Re} b, \operatorname{Im} b}\ \sum_{i,j,=1}^{n} |\partial\beta_i/\partial x_j - \partial\beta_j/\partial x_i| > 0$$

on a dense set of V.

The lemma of 6.8 shows Im $D(A(\lambda)) = 0$ in some nonempty open $V_0 \subset V$, so $\Gamma(\mathrm{Re}\, b)$, $\Gamma(\mathrm{Im}\, b)$ are both non-zero on a dense set of V_0, where $\Gamma(\beta) = [\partial\beta_i/\partial x_j - \partial\beta_j/\partial x_i]_{i,j=1}^n$ is an anti-symmetric matrix at each x. (Recall b generally also depends on λ.) But we also know (for $n \geq 3$) on a dense open set in $V \cap \partial\Omega$, each point has a neighborhood where Re $b_{\tan}$ or Im $b_{\tan}$ is a gradient on $\partial\Omega$; this will give a contradiction. (Here we use some simple topology: if two continuous real-valued functions f_1, f_2 have product $f_1 \cdot f_2 \equiv 0$, then interior $f_1^{-1}(0) \cup$ interior $f_2^{-1}(0)$ is dense in the domain.)

(1) Suppose $\beta(x) = (\beta_1(x), \ldots, \beta_n(x)) \in \mathbb{R}^n$ is C^1 and $\partial\Omega$ is C^2, and in some neighborhood of $x_0 \in \partial\Omega$ there is C^1 function ϕ such that $\beta_{\tan} = \nabla_{\partial\Omega}\phi$ on $\partial\Omega$ in this neighborhood. Then $\Gamma = \Gamma(\beta) = [\partial\beta_i/\partial x_j - \partial\beta_j/\partial x_i]_{i,j=1}^n$ satisfies $\Gamma = \Gamma N_\Omega \otimes N_\Omega - N_\Omega \otimes \Gamma N_\Omega$ on $\partial\Omega$, in this neighborhood.

In fact $\nabla_{\partial\Omega}\phi = \beta_{\tan}$ is C^1 so we may modify ϕ, if necessary, so that ϕ is locally C^2. Further, $\partial\Omega = \{x : \psi(x) = 0\}$ for some $\psi \in C^2$ with $|\nabla\psi(x)| \geq 1$ where $\psi(x) = 0$, and $\phi + (\text{constant}) \cdot \psi$ could be used in place of ϕ, so we may assume $\partial\phi/\partial N \neq 0$ in the neighborhood of interest. Now consider $\beta(x) - \nabla\phi(x)$, a C^1 function on a full ($\mathbb{R}^n$)-neighborhood of x_0, such that $\beta - \nabla\phi = \mu N$ on $\partial\Omega$ near x_0, for some non-zero C^1 function μ. We, may extend N, μ with $|N| \equiv 1$, $\mu \neq 0$, so this equation still holds off $\partial\Omega$, and then

$$\Gamma_{ij} = \partial_i \mu N_j - \partial_j \mu N_i + \mu(\partial_i N_j - \partial_j N_i)$$

on this neighborhood; so, on $\partial\Omega$ near x_0, $\Gamma = \tau \otimes N - N \otimes \tau$ with $\tau = \nabla_{\partial\Omega}\mu - \mu\partial N/\partial N \perp N$. It follows that $\Gamma N = \tau$ so $\Gamma = (\Gamma N) \otimes N - N \otimes \Gamma$. (This condition would be trivial for $n = 2$, but we assume $n \geq 3$.)

Remark. We will use the notation "$a \otimes b$" more generally below. It may be defined for any linear spaces E, F : if $e \in E$, $f \in F$, $e \otimes f$ is the bilinear form on the product of the duals $E' \times F'$ given by

$$(e \otimes f)(\phi, \psi) = \phi(e) \cdot \psi(f) \qquad \text{for } \phi \in E', \quad \psi \in F'.$$

Note $e \otimes f = 0$ if and only if $e = 0$ or $f = 0$.

In each $E = E' = \mathbb{R}^n$, $F = F' = \mathbb{R}^m$, we may write $e \otimes f$ as a matrix: $e = (e_1, \ldots, e_n)$, $f = (f_1, \ldots, f_m)$, $(e \otimes f)_{ij} = e_i f_j$ so $(e \otimes f)(x, y) = x \cdot ((e \otimes f)y) = x \cdot ef \cdot y$ for $x \in \mathbb{R}^n$, $y \in \mathbb{R}^m$.

(2) Let $\Gamma_1 = \Gamma(\mathrm{Re}\, b)$, $\Gamma_2 = \Gamma(\mathrm{Im}\, b)$, $Q_j(N) = \Gamma_j + N \otimes \Gamma_j N - (\Gamma_j N) \otimes N (j = 1, 2)$. On a dense set of $\partial\Omega \cup V$, either Re $b_{\tan}$ or Im $b_{\tan}$ is a gradient on $\partial\Omega$ near given any point, so $Q_1(N)$ or $Q_2(N)$ vanishes, or $Q_1(N) \otimes Q_2(N) = 0$ on $\partial\Omega \cap V$.

We note, for later use, if $n \geq 3$ and at some point of $\partial\Omega \cap V$, $Q_j(N) = 0$ and $Q'_j(N)\tau = 0$ for all $\tau \perp N$, then $\Gamma_j = 0$ at this point. In fact, $Q'_j(N)\tau = 0$ says $\tau \otimes \Gamma_j N + N \otimes \Gamma_j \tau = \Gamma_j \tau \otimes N + \Gamma_j N \otimes \tau$ (for all $\tau \perp N$). Applying this to

τ yields $\Gamma_j N = \tau(\Gamma_j N) \cdot \tau$. If $\Gamma_j N \neq 0$, then for all $\tau \perp N$, τ has the direction of $\Gamma_j N$, which is false for $n \geq 3$. Thus, $\Gamma_j N = 0$ so (since $Q_j = 0$) $\Gamma_j = 0$.

(3) So far, Ω (and the corresponding λ) have been kept fixed. Allowing Ω to vary near Ω_0 there exists $\lambda \in \Lambda$ so that $\operatorname{Im} D(A(\lambda)) = 0$ and $Q_1(N_\Omega) \otimes Q_2(N_\Omega) = 0$ on $\partial\Omega \cap V$. By the transversality theorem, whenever $\tau \perp N$, we also have

$$Q_1'(N)\tau \otimes Q_2(N) + Q_1(N) \otimes Q_2'(N)\tau = 0.$$

Assuming $n \geq 3$, if $Q_1(N) \neq 0$ at some point of $\partial\Omega \cap V$, then $Q_2(N) = 0$ and $Q_2'(N)\tau = 0$ for all $\tau \perp N$, hence $\Gamma_2 = 0$ at that point. Similarly $Q_2(N) \neq 0$ implies $\Gamma_1 = 0$. If $\Gamma_1 \otimes \Gamma_2 \neq 0$, then both Γ_j are nonzero so both $Q_j(N) = 0$. Thus we have, on $\partial\Omega \cap V$.

$$\operatorname{Im} D(A(\lambda)) = 0 \quad \text{and} \quad (\Gamma_1 \otimes \Gamma) \otimes (Q_1(N_\Omega), Q_2(N_\Omega)) = 0.$$

Applying the transversality theorem to this condition, we see also

$$(\Gamma_1 \otimes \Gamma_2) \otimes (Q_1'(N_\Omega)\tau, Q_2'(N_\Omega)\tau) = 0$$

for all $\tau \perp N_\Omega$. Suppose $\Gamma_1 \otimes \Gamma_2 \neq 0$ at some point of $\partial\Omega \cap V$; then for both $j = 1, 2$, $Q_j(N_\Omega) = 0$ and $Q_j'(N_\Omega)\tau = 0$ when $\tau \perp N_\Omega$, so $\Gamma_1 = \Gamma_2 = 0$, a contradiction. Hence at each point of $\partial\Omega \cap V$, $\operatorname{Im} D(A(\lambda)) = 0$ and $\Gamma_1 \otimes \Gamma_2 = 0$ (or $|\Gamma_1||\Gamma_2| = 0$). By the lemma of 6.8, we obtain the long-sought contradiction.

We summarize

Theorem. *Assume $n \geq 3$ and let $b_j(x,\lambda)$, $c(x,\lambda)$ be polynomials in λ with $x \mapsto b_j(x,\lambda) \in \mathbb{C}$ of class C^2, $x \mapsto c(x,\lambda) \in \mathbb{C}$ of class $C^1 (j = 1, 2, \cdot, n)$. Let $\Lambda \subset \mathbb{C}$ and $V \subset \mathbb{R}^n$ be open sets such that:*

> *for any $\lambda \in \Lambda$, if $(c - \frac{1}{2}\operatorname{div} b - \frac{1}{4}b^2)$ is real everywhere in a neighborhood of some point of ∇, then neither* $\operatorname{Re} b(\cdot,\lambda)$ *nor* $\operatorname{Im} b(\cdot,\lambda)$ *is a gradient in any neighborhood of that point.*

Then for most C^2 bounded regions $\Omega \subset \mathbb{R}^n$ such that $\partial\Omega$ meets V, all eigenvalues $\lambda \in \Lambda$ of

$$\Delta u + b \cdot \nabla u + cu = 0 \text{ in } \Omega, \qquad u = 0 \text{ on } \partial\Omega, \qquad u \not\equiv 0,$$

are simple and freely moveable.

(For $n = 2$, we have the same conclusion assuming $\operatorname{Im}(c - \frac{1}{2}\operatorname{div} b - \frac{1}{4}b^2) \neq 0$ on a dense set of V, for each $\lambda \in \Lambda$; see Example 6.8. Whether the above weaker hypothesis suffices for $n = 2$ is not known.)

Remark. If the b_j, c are real-valued for real λ, then for real λ, $c - \frac{1}{2}\operatorname{div} b - \frac{1}{4}b^2$ is real and $\operatorname{Im} b(\cdot,\lambda) \equiv 0$, so Λ cannot meet the real axis.

8.5 Generic Simplicity of Solutions of a System

We will study solutions of

$$\Delta u_j + f_j(x, u_1, u_2, \ldots, u_p) = 0 \text{ in } \Omega, \quad u_j = 0 \text{ on } \partial\Omega \, (j = 1, 2, \ldots, p)$$

for $p \geq 2$, $n \geq 2$, where $f : \mathbb{R}^n \times \mathbb{R}^p \to \mathbb{R}^p$ is a given smooth function, and the region Ω is assumed open, bounded and C^2 (or better). We assume Ω is connected during the argument, but this is not important – we may perturb each component separately. We write $f'(x, u)$ for the $p \times p$ matrix $[\partial f_j(x, u)/\partial u_k]_{j,k=1}^p$.

If zero is a regular value of $u \mapsto \Delta u + f(\cdot, u)$, from $W_q^2 \cap W_q^1(\Omega, \mathbb{R}^p)_0$ to $L_q(\Omega, \mathbb{R}^p)$, $n < q < \infty$, then all solutions u of $\Delta u + f(\cdot, u) = 0$ in Ω, $u = 0$ on $\partial\Omega$, are simple. This would hold for most Ω if zero were a regular value of $(h, u) \mapsto h^*(\Delta + f)h^{*-1}u$. We prove this by contradiction, under appropriate hypotheses on f. Effective hypotheses are far from obvious, so we start merely assuming reasonable smoothness and that, for every Ω near a certain Ω_0, there is a solution u which is not simple. This will imply a sequence of additional conditions, and solvability of ever more over-determined problems. Then it will be clear how to select our hypotheses for an argument by contradiction.

Roughly speaking, we treat the cases

(1) $f(x, 0) \not\equiv 0$;
(2) $f(x, 0) \equiv 0$ and $f'(x, 0) \not\equiv 0$, depending on x;
(3) $f(u)$ is independent of x, $f(0) = 0$, and $f'(0)$ has only simple eigenvalues (and p is not too large);
(4) $p = 2$, $f : \mathbb{R}^2 \to \mathbb{R}^2$ is independent of $x \in \mathbb{R}^n$, $f(0) = 0$, $f'(0) = \lambda I$ for some real λ, $f''(0) \neq 0$.

(The exact conditions are more complicated: see Theorem 8.5.1 for (1), (2); Thm. 8.5.3 for (3); Thm. 8.5.4 for (4).) In each case, if f is fairly smooth, we conclude all solutions of $\Delta u + f(x, u) = 0$ in Ω, $u = 0$ on $\partial\Omega$, are simple, for most choices of $\Omega \subset \mathbb{R}^n$.

As in the scalar case (6.5), if there is a non-simple solution u for every Ω near Ω_0, then there is a solution (u, ψ) of

$$(*)_\Omega \qquad \begin{cases} \Delta u + f(x, u) = 0 \text{ in } \Omega, & u = 0 \text{ on } \partial\Omega \\ \Delta\psi + {}^T f'(x, u)\psi = 0 \text{ in } \Omega, & \psi = 0 \text{ on } \partial\Omega, \quad \psi \not\equiv 0 \end{cases}$$

with $\frac{\partial\psi}{\partial N} \cdot \frac{\partial u}{\partial N} \equiv 0$ on $\partial\Omega$. Also, by uniqueness in the Cauchy problem, $\partial\psi/\partial N \neq 0$ on a dense set of $\partial\Omega$; but since $p \geq 2$, we cannot conclude that $\partial u/\partial N$ vanishes.

Applying the transversality theorem, we find: for every Ω near Ω_0 there is a solution (u, ψ) of $(*)_\Omega$ such that $\psi_N \cdot u_N = 0$ and also

$$\sigma \mapsto \frac{\partial}{\partial N}\{\mathcal{B}_{L^*} \cdot (\sigma\psi_N) - \mathcal{A}_{L^*} \cdot {}^T f''(\,\cdot\,, u)\psi \mathcal{B}_L(\sigma u_N)\} \cdot u_N$$
$$+ \psi_N \cdot \frac{\partial}{\partial N}\mathcal{B}_L(\sigma u_N) + \sigma\psi_N \cdot f_0$$

has finite rank, $L = \Delta + f'(\,\cdot\,, u)$, $L^* = \Delta + {}^T f'(\,\cdot\,, u)$, $f_0 = (\,\cdot\,, 0)$. (We have used the facts $\psi_{NN} + H\psi_N = 0$, $u_{NN} + Hu_n + f_0 = 0$ on $\partial\Omega$.) Taking $\sigma = \gamma e^{i\omega\theta}$, $|\nabla_{\partial\Omega}\theta| = 1$ in supp γ, as usual, we find

$$\dot\Psi_0 \cdot u_N + \psi_N \cdot \dot U_0 + \gamma\psi_N \cdot f_0 = 0, \quad \dot\Psi_k \cdot u_N + \psi_N \cdot \dot U_k = 0 \text{ for } k \geq 1$$

where

$$\Lambda U_k + LU_{k-1} = 0 \quad (U_{-1} = 0), \qquad U_k \mid \partial\Omega = \begin{cases} \gamma u_N & \text{if } k = 0 \\ 0 & \text{if } k > 0 \end{cases}$$

$$\Lambda\Psi_k + L^*\Psi_{k-1} = -{}^T f''(u)\psi U_{k-1}, \qquad \Psi_k \mid \partial\Omega = \begin{cases} \gamma\psi_N & \text{if } k = 0 \\ 0 & \text{if } k > 0 \end{cases}$$

and $\Lambda = \nabla S \cdot \nabla + \frac{1}{2}\Delta S$ is the same for L or L^*.

It is convenient to treat a more general case, so assume: for every Ω near Ω_0, there is a solution (u, ψ) of $(*)_\Omega$ such that

$$\psi_N \cdot \mathcal{A} = 0 \text{ on } \partial\Omega \qquad \text{for all } a \in \mathcal{A}$$

and

$$\psi_N \cdot Mu_N = 0 \text{ on } \partial\Omega \qquad \text{for all } M \in \mathcal{M}$$

where $\mathcal{A} \subset C(\mathbb{R}^n, \mathbb{R}^p)$, $\mathcal{M} \subset C(\mathbb{R}^n, \mathbb{R}^{p\times p})$ are given sets (or linear spaces). This certainly holds for $\mathcal{A} = \{0\}$, $\mathcal{M} = \{I\}$, the constant functions. The same argument as above yields:

$$\dot\Psi_0 \cdot a - \gamma\partial_N(\psi_N \cdot a) = 0 \;\; \dot\Psi_0 \cdot Mu_N + \psi_N \cdot M\dot U_0 - \sigma\,\partial_N(\psi_n \cdot Mu_k) = 0$$

and

$$\dot\Psi_k \cdot a = 0 \qquad \dot\Psi_k \cdot Mu_N + \psi_N \cdot M\dot U_k = 0, \quad \text{for } k \geq 1.$$

Step 1. Take $k = 0$: if a, M are C^1,

$$0 = i\gamma\partial_\theta\psi_N \cdot a + \gamma\partial_N(\psi_N \cdot a),$$
$$0 = i\gamma(\partial_\theta\psi_N \cdot Mu_N + \psi_N \cdot M\partial_\theta u_N) + \gamma\partial_N(\psi_N \cdot Mu_N).$$

The first of these (using $\psi_{NN} + H\psi_N = A\psi \mid \partial\Omega = 0$) yields

$$\psi_N \cdot \partial a/\partial x_j = 0 \text{ on } \partial\Omega \qquad (1 \le j \le n), \quad a \in \mathcal{A} \cap C^1.$$

The second yields (since $u_{NN} + Hu_N + f_0 = 0$ on $\partial\Omega$, $f_0 = f(\cdot, 0)$)

$$\psi_N \cdot \partial_{x_j} Mu_N - N_j\psi_N \cdot Mf_0 = 0 \text{ on } \partial\Omega \quad (1 \le j \le n).$$

This is not quite the form treated before, but a similar transversality argument yields (in the principal symbol, for a change)

$$0 = \psi_N \cdot \partial_{x_j} Mu_N - ie_j \cdot \nabla_{\partial\Omega}\theta\psi_N \cdot Mf_0$$

where e_j is the unit vector on the x_j-axis. This shows each term vanishes separately: for $M \in \mathcal{M} \cap C^1$, on $\partial\Omega$,

$$\psi_N \cdot \partial_{x_j} Mu_N = 0 \qquad (1 \le j \le n), \quad \psi_N \cdot Mf_0 = 0.$$

Step 2. Next consider $k = 1$, with a, M in C^2. Writing $f_0' = f'(\cdot, 0)$, $\dot\Psi_1 \cdot a = 0$ yields

$$0 = (\ddot\Psi_0 + H\dot\Psi_0 + A_{\partial\Omega}\gamma\psi_N + {}^T f_0'(\gamma\psi_N)) \cdot a.$$

We have $0 = \Delta_{\partial\Omega}(\gamma\psi_N \cdot a)$ so, by previous results, $0 = \Delta_{\partial\Omega}(\gamma\psi_N) \cdot a$, and clearly $0 = \dot\Psi_0 \cdot a$. Finally $\ddot\Psi_0 \cdot a = -i\partial_\theta\dot\Psi_0 \cdot a = i\dot\Psi_0 \cdot \partial_\theta a = 0$ so

$$\psi_N \cdot f_0' a = 0.$$

Similarly, $0 = \dot\Psi_1 \cdot Mu_N + \psi_N \cdot M\dot U_1$ yields

$$0 = \psi_N \cdot (f_0' M + Mf_0')u_N + \Delta_{\partial\Omega}\psi_N \cdot Mu_N + \psi_N \cdot M\Delta_{\partial\Omega}u_N$$
$$- \partial_\theta^2\psi_N \cdot Mu_n - \psi_N \cdot M\partial_\theta^2 u_N.$$

If $n = 2$, θ is arclength in the curve $\partial\Omega$ and $\Delta_{\partial\Omega} = \partial^2/\partial\theta^2$ so

$$0 = \psi_N \cdot (f_M' + Mf_0')u_N.$$

We will show this also holds for $n \ge 3$, by more calculation.

Step 3. From $\partial_\theta^2(\psi_N \cdot Mu_N) = 0$, we see

$$-(\partial_\theta^2\psi_N \cdot Mu_N + \psi_N \cdot M\partial_\theta^2 u_N) = 2\partial_\theta\psi_N \cdot M\partial_\theta u_N,$$

so $\partial_r \psi_N \cdot M \partial_r u_N = A = \text{constant}$ for all directions τ with $\tau \perp N$, $|r| = 1$, so (in component form)

$$\begin{aligned} A &= \sum_{\alpha,\beta=1}^{p} M_{\alpha\beta} \arg\{\tau \cdot \nabla_{\partial\Omega} \psi_N^\alpha \tau \cdot \nabla_{\partial\Omega} u_n^\beta\} \\ &= \frac{1}{n-1} \sum_{\alpha,\beta} M_{\alpha\beta} \nabla_{\partial\Omega} \psi_N^\alpha \cdot \nabla_{\partial\Omega} u_N^\beta \\ &= -\frac{1}{2(n-1)} (\Delta_{\partial\Omega} \psi_N \cdot M u_N + \psi_N \cdot M \Delta_{\partial\Omega} u_N). \end{aligned}$$

Then for $\tau \perp N$,

$$2(n-2)\partial_\tau \psi_N \cdot M \partial_\tau u_N = |\tau|^2 \psi_N \cdot (f_0' M + M f_0') u_N.$$

Further, previous results imply $\partial_\tau \psi_N \cdot M u_{NN} + \psi_{NN} \cdot M \partial_\tau u_N = 0$, $\psi_{NN} \cdot M u_{NN} = 0$ so, for any $z \in \mathbb{R}^n$, with $\partial_z = z \cdot \nabla$,

$$2(n-2)\partial_z \psi_N \cdot M \partial_z u_N = (|z|^2 - (z \cdot N)^2) \psi_N \cdot (f_0' M + M f_0') u_N.$$

We apply transversality again ($M \in \mathcal{M} \cap C^2$) with this condition to conclude

$$\begin{aligned} \sigma \mapsto\ & 2(n-2)\{\partial_z(\partial_N(\delta\psi) - \dot{N} \cdot \nabla\psi) \cdot M \partial_z u_N \\ & + \partial_z \psi_N \cdot M \partial_z(\partial_N(\delta u) - \dot{N} \cdot \nabla u)\} \\ & - 2z \cdot N z \cdot \dot{N} \psi_N \cdot M_1 u_N \\ & - |z_{\tan}|^2((\delta\psi)_N \cdot M_1 u_N + \psi_N \cdot M_1(\delta u)_N) \\ & - \sigma \partial_N(2(n-2)\partial_z \psi_N \cdot M \partial_z u_N - |z_{\tan}|^2 \psi_N \cdot M_1 u_N) \end{aligned}$$

has finite rank, where $M_1 = f_0' M + M f_0'$ and

$$\begin{aligned} \Delta\delta u + f'(u)\delta u &= \text{(finite)}, & \delta u|_{\partial\Omega} &= \sigma u_N \\ \Delta\delta\psi + {}^T\! f'(u)\delta\psi + {}^T\! f''(u,\psi)\delta u &= \text{(finite)}, & \delta\psi|_{\partial\Omega} &= \sigma\psi_N. \end{aligned}$$

Since $\psi = 0$, on $\partial\Omega$, also, $\dot{N} \cdot \nabla\psi = 0$ on $\partial\Omega$ and we find $\partial_z(\dot{N} \cdot \nabla\psi) = z \cdot N \dot{N} \cdot \nabla\psi_N = -z \cdot N \nabla_{\partial\Omega}\sigma \cdot \nabla_{\partial\Omega}\psi_N$ (assuming $\partial N/\partial N = 0$ on $\partial\Omega$); and similarly for u. We take $\sigma = \gamma e^{i\omega\theta}$ as usual, and as usual, the principal part (coefficient of ω^2) says nothing new. But the coefficient of ω is

$$\begin{aligned} 0 = 2(n-2)\gamma\{&(-z \cdot (N + i\theta')i\partial_\theta \psi_N + z_{\tan} \cdot \nabla_{\partial\Omega\psi_N}) \cdot M \partial_z u_N \\ &+ \partial_z \psi_N \cdot M(-z \cdot (N + i\theta')i\partial_\theta u_N + z_{\tan} \cdot \nabla_{\partial\Omega} u_N)\} \\ &+ (2z \cdot N z \cdot \theta' i\gamma - |z_{\tan}|^2 \gamma)\psi_N \cdot M_1 u_N. \end{aligned}$$

The part odd in θ must vanish, but this tells us nothing new. The even part also vanishes, so taking $z = \tau, \theta' = \sigma$, unit tangent vectors ($\perp N$), we obtain $0 = (n-2)\gamma\tau \cdot \sigma(\partial_\tau \psi_N \cdot M\partial_\sigma u_N + \partial_\sigma \psi_N \cdot M\partial_\tau u_N)$. Choosing $\tau = \sigma$ we see

$$(n-2)\partial_\tau \psi_N \cdot M\partial_\tau u_N = 0 = \psi_N \cdot M_1 u_N$$

so $\psi_N \cdot (Mf_0' + f_0' M)u_n = 0$ on $\partial\Omega$ for $n \geq 2$.

Thus the hypothesis.

> $(\mathcal{A}, \mathcal{M})$: For all Ω near Ω_0, there is a solution (u, ψ) of $(*)_\Omega$ such that on $\partial\Omega$, $\psi_N \cdot a = 0$ for all $a \in \mathcal{A}$, $\psi_N \cdot Mu_N = 0$ for all $M \in \mathcal{M}$,

implies this still holds with $\mathcal{A}_1, \mathcal{M}_1$ in place of $\mathcal{A}, \mathcal{M}$,

$$\begin{aligned}
\mathcal{A}_1 = \mathcal{A} &\cup \left\{ \frac{\partial a}{\partial x_j} (j = 1, \ldots, n) \text{ for } a \in \mathcal{A} \cap C^1 \right\} \\
&\cup \{\mathcal{M} f_0, M \in \mathcal{M} \cap C^1\} \\
&\cup \{f_0' a, a \in \mathcal{A} \cap C^2\} \\
\mathcal{M}_1 = \mathcal{M} &\cup \left\{ \frac{\partial M}{\partial x_j} (j = 1, \ldots, n) \text{ for } M \in \mathcal{M} \cap C^1 \right\} \\
&\cup \{f_0' M + M f_0', M \in \mathcal{M} \cap C^2\}.
\end{aligned}$$

We apply this, starting from $\mathcal{A} = \{0\}, \mathcal{M} = \{I\}$, to obtain first $\mathcal{A} = \{f_0\}, \mathcal{M} = \{I, 2f_0'\}$; then

$$\mathcal{A} = \left\{ f_0, \frac{\partial}{\partial x_j} f_0, f_0' f_0 \right\}, \qquad \mathcal{M} = \left\{ I, 2f_0', 2\partial_{x_j} f_0', 4f_0'^2 \right\};$$

etc.

Theorem 8.5.1. *Suppose $n \geq 2$, $p \geq 2$, and let $f : \mathbb{R}^n \times \mathbb{R}^p \to \mathbb{R}^p$, $f' : \mathbb{R}^n \times \mathbb{R}^p \to \mathbb{R}^{p \times p}$ be at least C^1 and $f_0 = f(\cdot, 0)$, $f_0' = f'(\cdot, 0)$ at least C^2. Define linear spaces $\mathcal{A} \subset C(\mathbb{R}^n, \mathbb{R}^p)$, $\mathcal{M} \subset C(\mathbb{R}^n, \mathbb{R}^{p \times p})$ by*

$$\mathcal{A} = \bigcup_{j \geq 0} \mathcal{A}_j \qquad \mathcal{M} = \bigcup_{j \geq 0} \mathcal{M}_j$$
$$\mathcal{A}_0 = \{0\}, \qquad \mathcal{M}_0 = \{I\},$$

and for $j \geq 0$,

$$\mathcal{A}_{j+1} = \text{span}\left\{ \mathcal{A}_j; \frac{\partial a}{\partial x_k} (1 \leq k \leq n, a \in \mathcal{A}_j \cap C^1); f_0' a \; (a \in \mathcal{A}_j \cap C^2); \right.$$
$$\left. \mathcal{M} f_0 \, (M \in \mathcal{M}_j \cap C^1) \right\}$$

$$\mathcal{M}_{j+1} = \text{span}\left\{\mathcal{M}_j; \frac{\partial M}{\partial x_k} \ (1 \le k \le n,\ M \in \mathcal{M}_j \cap C^1); \right.$$
$$\left. f_0' M + M f_0' \ (M \in \mathcal{M}_j \cap C^2)\right\}.$$

$\mathcal{M}$ contains (at least) all polynomials in f_0' and $\mathcal{A}$ contains (polynomial in f_0') $\times f_0$.

(1) Assume, for a dense set of $x \in \mathbb{R}^n$, that

$$\{a(x), \quad \textit{all } a \in \mathcal{A}\} = \mathbb{R}^p.$$

Then for most C^2 bounded regions $\Omega \subset \mathbb{R}^n$, all solutions of $\Delta u + f(x, u) = 0$ in Ω, $u = 0$ on $\partial\Omega$, are simple.

(2) In case $f(x, 0) \equiv 0$, we have $\mathcal{A} = \{0\}$, but assume $\mathcal{M}$ is large so that: for a dense set of $x \in \mathbb{R}^n$, whenever $\beta \neq 0$ in $\mathbb{R}^p$, $\{M(x)\beta$, all $M \in \mathcal{M}\} = \mathbb{R}^p$. Then we have generic simplicity of solutions, as in case (1).

Remark. An important case is when $f(u)$ is independent of x and $f(0) = 0$; then $\mathcal{A} = \{0\}$ and $\mathcal{M} = \{\sum_{k=0}^{p-1} c_k f'(0)^k (c_k \in \mathbb{R})\}$, and hypothesis (2) ordinarily fails. (The only exception is when $p = 2$ and $f'(0)$ has complex eigenvalues.) We will treat this case below.

Proof. The difficult point, as usual, is density. We suppose this fails: for every Ω near some Ω_0 (connected, bounded, smooth), suppose there is a non-simple solution. Then there is a solution (u, ψ) of $(*)_\Omega$ such that $\psi_N \cdot a = 0$, $\psi_N \cdot M u_N = 0$ on $\partial\Omega$ for all $a \in \mathcal{A}$, $M \in \mathcal{M}$.

In case (1), we may suppose $\{a(x)$ all $a \in A\} = \mathbb{R}^p$ for a dense (open) set in $\partial\Omega$ and conclude $\psi_N = 0$ on $\partial\Omega$, hence $\psi = 0$ in Ω, a contradiction. (It suffices to consider a finite subset of $\mathcal{A}$, depending on Ω_0.)

In case (2), we may suppose the dense (open) set of the hypothesis is also dense in $\partial\Omega$. Since $\psi_N \neq 0$ on a dense set of $\partial\Omega$, we conclude $u_N \equiv 0$ on $\partial\Omega$. By uniqueness in the Cauchy problem, $u \equiv 0$. Thus for all Ω near Ω_0, there is a nontrivial solution ϕ of $\Delta\phi + f_0'\phi = 0$ in Ω, $\phi = 0$ on $\partial\Omega$, and it is not simple. This is a particular (linear) case of our initial hypothesis, and $\mathcal{M}$ is unchanged when we replace $f(x, u)$ by $f'(x, 0)u$, so we obtain the same conclusion: $\phi = 0$, $\phi_N = 0$ on $\partial\Omega$, so $\phi \equiv 0$ in Ω, contrary to hypothesis.

Now suppose $f(u)$ is independent of x and $f(0) = 0$. Our hypothesis for contradiction is that, for any Ω near Ω_0, there is a solution (u, ψ) of

$$(*)_\Omega \qquad \begin{cases} \Delta u + f(u) = 0 \text{ in } \Omega, & u = 0 \text{ on } \partial\Omega \\ \Delta\psi + {}^T f'(u)\psi = 0 \text{ in } \Omega, & \psi = 0 \text{ on } \partial\Omega, \quad \psi \not\equiv 0 \end{cases}$$

such that $\psi_N \cdot f'(0)^j u_N = 0$ on $\partial\Omega$ for all $j \geq 0$ (or $0 \leq j \leq p-1$). For any Ω, $u \equiv 0$ solves the first equation, but the zero solution is usually simple. In fact, it fails to be simple if and only if some eigenvalue λ of the matrix $f'(0)$ is also an eigenvalue of the scalar Dirichlet problem

$$\Delta\phi + \lambda\phi = 0 \text{ in } \Omega, \qquad \phi = 0 \text{ on } \partial\Omega, \qquad \phi \not\equiv 0.$$

For a dense open set of Ω, we avoid the eigenvalues of $f'(0)$ so assume henceforth: the zero solution is simple. Then our hypothesis above includes the condition $u \not\equiv 0$.

Consider the case where all eigenvalues of $f'(0)$ are simple (distinct). Say $f'(0)p_j = \lambda_j p_j$, ${}^T f'(0)q_j = \lambda_j q_j$, with $q_j \cdot p_j = 1$, $q_j \cdot p_k = 0$ for $j \neq k$ $(1 \leq j, k \leq p)$. Then the condition $\psi_N \cdot f'(0)^j u_N = 0$ for all $j \geq 0$ becomes

$$\sum_k \lambda_k^j \psi_N \cdot p_k q_k \cdot u_N = 0 \qquad \text{for all } j \geq 0,$$

hence $\psi_N \cdot p_k q_k \cdot u_N = 0$ for $k = 1, 2, \ldots, p$. Suppose we can prove $q_k \cdot u_N \neq 0$ on a dense set of $\partial\Omega$, for every k; then $\psi_N \cdot p_k = 0$ for every k, $\psi_N = 0$ on $\partial\Omega$, $\psi = 0$ in Ω, contrary to hypothesis.

The argument is a bit more general: if each eigenvalue of $f'(0)$ has only one eigenvector (as in the case $f'(0) = \left(\begin{smallmatrix} 0 & 1 \\ 0 & 0 \end{smallmatrix}\right)$) and $q_k \cdot u_N \neq 0$ densely $\partial\Omega$ for every eigenvector q_k of ${}^T f'(0)$, we again conclude $\psi \equiv 0$.

In the following lemma we make no hypothesis about $f'(0)$.

Lemma 8.5.2. *Assume $f : \mathbb{R}^p \to \mathbb{R}^p$ is C^4 with $f(0) = 0$. Let $Q \subset C^p$ be any subspace $\neq \{0\}$ and let $V \subset \mathbb{R}^n$ be open and assume: for every Ω near Ω_0, there is a solution u of*

$$\Delta u + f(u) = 0 \text{ in } \Omega, \qquad u = 0 \text{ on } \partial\Omega, \qquad u \not\equiv 0,$$

such that $q \cdot \partial u/\partial N = 0$ on $\partial\Omega \cap V$ for every $q \in Q$. Then, for every Ω near Ω_0, there is a solution $u \not\equiv 0$ such that for all $q \in \operatorname{span}\{{}^T f'(0)^j Q : j \geq 0\}$,

$$q \cdot u_N = 0, \qquad q \cdot f''(0)u_N^2 = 0, \quad q \cdot f'''(0)u_N^3 = 0,$$

$$q \cdot f''(0)(u_N, f'(0)u_N) = 0, \quad \text{and } q \cdot f^{iv}(0)u_N^4 = 0$$

on $\partial\Omega \cap V$.

Proof. The transversality theorem says there is a solution $u \not\equiv 0$ such that, for $q \in Q$, $q \cdot u_N = 0$ on $\partial\Omega \cap V$ and also

$$\sigma \mapsto q \cdot \partial_N \mathcal{B}_L(\sigma u_N)|_{\partial\Omega\cap V}, \qquad L = \Delta + f'(u)$$

has finite rank. (It suffices to use a finite set of q in Q.) Taking $\sigma = \gamma^{e^{i\theta}}$, $|\nabla_{\partial\Omega}\theta| = 1$ on supp γ, we conclude

$$q \cdot \dot{U}_k = 0 \text{ on } \partial\Omega \cap V \text{ for } k \geq 0 \qquad (\text{at least}, 0 \leq k \leq 4.)$$

Now $U_0 \mid \partial\Omega = \gamma u_N$ and $U_k \mid \partial\Omega = 0$ for $k \geq 0$, so also $q \cdot U_k = 0$ on $\partial\Omega \cap V$, or $q \cdot U_k(y + tN(y)) = O(t^2)$ as $t \to 0$ with $y \in \partial\Omega \cap V, k \geq 0$.

Since $\Lambda = \nabla S \cdot \nabla + {}^1\!/\!_2 \Delta S$ is a scalar operator, acting separately on each component, and q is constant, we have $\Lambda_q \cdot U_0 = q \cdot \Lambda U_0 = 0$ with $q \cdot U_0|\partial\Omega = 0$, and so $q \cdot U_0(y + tN(y)) = o(|t|^N)$ as $t \to 0$, $y \in \partial\Omega$, for arbitrary N – assuming $\partial\Omega, \theta, \gamma$ are smooth, as we may.

For $k \geq 1$

$$\Lambda q \cdot U_k + \Delta(q \cdot U_{k-1}) + q \cdot f'(u)U_{k-1} \cong 0, \qquad q \cdot U_k = O(t^2)$$

Define c_k by q $\cdot f'(u)U_0(y + tN(y)) = c_0 + tc_1 + t^2c_2 + t^3c_3 + o(t^3)$, recalling f' is C^3, and then

$$\Lambda q \cdot U_1 = -(c_0 + c_1 t + c_2 t^2 + c_3 t^3) + o(t^3) \text{ as } t \to 0, \text{ in } \partial\Omega \cap V.$$

At $t = 0$, $q \cdot \dot{U}_1 = -c_0 = 0$ or $q \cdot f'(0)\gamma u_N = 0$ for $q \in Q$.Thus our hypothesis holds also with span $\{Q, {}^T\!f'(0)Q\}$ in place of Q. We may, and shall, assume ${}^T\!f'(0)Q \subset Q$.

Since $c_0 = 0, q \cdot U_1 = -c_1 t^2/2 + O(t^3)$ so $\Delta(q \cdot U_1)|_{\partial\Omega} = -c_2$ so $0 = \Lambda q \cdot U_2 + \Delta q \cdot U_1 + q \cdot f'(u)U_1|_{t=0} = q \cdot \dot{U}_2 - c_1$, so $c_1 = 0$. (We compute c_1, c_2, c_3 below.)

Since $c_1 = 0$, $q \cdot U_1 = -c_2 t^3/3 + O(t^4)$ so

$$0 = \Lambda q \cdot U_2 + (-2c_2 t + O(t^2)) + q \cdot f'(0)t\dot{U}_1 + O(t^2),$$

so $q \cdot U_2 = c_2 t^2 + O(|t|^3)$ (recalling $q \cdot f'(0)\dot{U}_1 = 0$). Then $0 = \Lambda q \cdot U_3 + \Delta q \cdot U_2 + q \cdot f'(u)U_2|_{\partial\Omega} = q \cdot \dot{U}_3 + 2c_2$ so also $c_2 = 0$.

Now $q \cdot U_1 = -c_3 t^4/4 + o(t^4)$, and this step is more delicate. We have $q \cdot f'(u)(t\dot{U}_1 + \frac{t^2}{2}\ddot{U}_1 + \cdots) = tq \cdot f'(0)\dot{U}_1 + t^2(q \cdot f''(0)u_N\dot{U}_1 + {}^1\!/\!_2 q \cdot f'(0)\ddot{U}_1) + o(t^2)$, and the t-coefficient vanishes, so $q \cdot U_2 = at^3 + o(t^3)$ where $a = c_3 - {}^1\!/\!_3(q \cdot f''(0)u_N\dot{U}_1 + {}^1\!/\!_2 q \cdot f'(0)\ddot{U}_1)$. Then $0 = \Lambda q \cdot U_3 + 6at + tq \cdot f'(0)\dot{U}_2 + o(t)$ so $q \cdot U_3 = -3at^2 + o(t^2)$, and $0 = q \cdot \dot{U}_4 - 6a$, or $a = 0$. In fact $q \cdot U_1 = O(t^4)$ for any $q \in Q$ so also $q \cdot f'(0)U_1 = O(t^4)$ and $q \cdot f'(0)\ddot{U}_1 = 0$ in $\partial\Omega \cap \nabla$. Thus we have

$$\begin{aligned} q \cdot f'(u)U_0 &= c_0 + c_1 t + c_2 t^2 + c_3 t^3 + o(t^3) \\ &= \frac{t^3}{3}(q \cdot f''(0)\dot{U}_1) + o(t^3) \\ &= q \cdot (f'(u) - f'(0))U_0 + o(t^3) \end{aligned}$$

$$
\begin{aligned}
&= q \cdot f''(0)\left(tu_N + \frac{t^2}{2}u_{NN} + \frac{t^3}{6}u_{NNN} + \cdots\right)\\
&\quad \cdot \left(\gamma U_N + t\dot{U}_0 + \frac{t^2}{2}\ddot{U}_0 + \cdots\right)\\
&\quad + \tfrac{1}{2} q \cdot f'''(0)\left(tu_N + \frac{t^2}{2}u_{NN} + \cdots\right)^2\\
&\quad \cdot (\gamma U_N + t\dot{U}_0 + \cdots)\\
&\quad + \tfrac{1}{6} q \cdot f^{iv}(0)(tu_N + \cdots)^3(\gamma U_N + \cdots) + o(t^3).
\end{aligned}
$$

The coefficient of t vanishes, so $q \cdot f''(0)u_N^2 = 0$ in $\partial\Omega \cap V$, which implies $0 = \partial_\tau(q \cdot f''(0)u_N^2) = 2q \cdot f''(0)u_N \partial_\tau u_N$ for any $\tau \perp N$, so $q \cdot f''(0)u_N \dot{U}_0 = 0$. Thus the coefficient of t^2 is

$$0 = \frac{\gamma}{2} q \cdot f''(0)u_N u_{NN} + \frac{\gamma}{2} q \cdot f'''(0)u_N^3$$

since $u_{NN} + Hu_N = 0 (= \Delta u + f(u)|\partial\Omega)$ we have $q \cdot f'''(0)u_N^3 = 0$ on $\partial\Omega \cap V$. Omitting some terms clearly zero, we find from the coefficient of t^3,

$$
\begin{aligned}
&\gamma(q \cdot f^{iv}(0)u_N^4 + q \cdot f''(0)u_N u_{NNN})\\
&\quad = 2q \cdot f''(0)u_N \dot{U}_1 - 3q \cdot f''(0)u_N \ddot{U}_0\\
&\quad = -5q \cdot f''(0)u_N \ddot{U}_0 - 2q \cdot f''(0)u_N(H\dot{U}_0 + \Delta_{\partial\Omega}\gamma u_N + f'(0)\gamma u_N)\\
&\quad = 5\gamma q \cdot f''(0)u_N \partial_\theta^2 u_N - 2\gamma q \cdot f''(0)u_N(\Delta_{\partial\Omega}u_N + f'(0)u_N).
\end{aligned}
$$

Since $0 = \partial_N(\Delta u + f(u))|\partial\Omega = u_{NNN} - (H^2 + H_2)u_N + \Delta_{\partial\Omega}u_N + f'(0)u_N$, we have, in supp $\gamma \subset \partial\Omega \cap V$,

$$5q \cdot f''(0)u_N \partial_\theta^2 u_N = q \cdot f^{iv}(0)u_N^4 + q \cdot f''(0)u_N(\Delta_{\partial\Omega}u_N + f'(0)u_N).$$

Since $\partial_\theta^2(q \cdot f''(0)u_N^2) = 0$, the left-side is $-5q \cdot f''(0)(\partial_\tau u_N)^2$ when $\nabla_{\partial\Omega}\theta = \tau \perp N, |\tau| = 1$, and this is independent of the choice of τ. Thus

$$
\begin{aligned}
q \cdot f''(0)(\partial_\tau u_N)^2 &= \underset{\tau \perp N, |\tau|=1}{\text{average}}\ q \cdot f''(0)(\tau \cdot \nabla_{\partial\Omega}u_N)^2\\
&= \frac{1}{n-1} q \cdot f''(0)(\nabla_{\partial\Omega}u_N)^2\\
&= -\frac{1}{n-1} q \cdot f''(0)u_N \Delta_{\partial\Omega}u_N
\end{aligned}
$$

(The penultimate term is ambiguous, but clear in component form.) Then

$$(n-6)q \cdot f''(0)(\partial_\tau u_N)^2 = Q(u_N) \qquad \text{for } |\tau| = 1, \quad \tau \perp N,$$

where $Q(u_N) = q \cdot f^{iv}(0)u_N^4 + q \cdot f''(0)(u_N, f'(0)u_N)$.

A similar condition occurred above, just before Theorem 5.1, and this is treated the same way. Since $q \cdot f''(0)\partial_\tau u_N u_{NN} = 0 = q \cdot f''(0)u_{NN}^2(u_{NN} = -Hu_N)$ for $z \in \mathbb{R}^n$,

$$(n-6)q \cdot f''(0)(\partial_z u_N)^2 - |z|_{\tan}^2 Q(u_N) = 0 \text{ on } \partial\Omega \cap V.$$

We apply the transversality theorem, and as usual the principal part says nothing new, but the sub-principal part gives (with $z = \tau \perp N$, $\nabla_{\partial\Omega}\theta = \sigma \perp N$, $|\sigma| = |\tau| = 1$)

$$\begin{aligned} 2(n-6)q \cdot f''(0)(\partial_\tau u_N)(\partial_\tau u_N + \tau \cdot \sigma \partial_\sigma u_N) &= Q'(u_N)u_N \\ &= 2\{Q(u_N) + (\tau \cdot \sigma)^2 Q(u_N)\}. \end{aligned}$$

Choose $\tau = \sigma$: then $Q'(u_N)u_N = 4Q(u_N)$, so the quadratic part vanishes, $q \cdot f''(0)(u_N, f'(0)u_N) = 0$. if $n \geq 3$, we may choose $\tau \perp \sigma$, so also $q \cdot f^{iv}(0)u_N^4 = 0$. This completes the proof of the lemma.

Theorem 8.5.3. *Let $f : \mathbb{R}^p \to \mathbb{R}^p$ be C^4 with $f(0) = 0$. Assume $f'(0)$ has only simple eigenvalues, or more generally, each eigenvalue of $f'(0)$ has only one eigenvector. Further assume, whenever $\beta \neq 0$ in $\mathbb{R}^p$, that*

$$\mathbb{R}^p = \operatorname*{span}_{j \geq o} f'(0)^j \{\beta, f''(0)\beta^2, f'''(0)\beta^3, f''(0)(\beta', f(0)\beta), f^{iv}(0)\beta^4\}.$$

Then for most bounded C^2 regions $\Omega \subset \mathbb{R}^n$, all solutions of

$$\Delta u + f(u) = 0 \; in \; \Omega \subset \mathbb{R}^n, \qquad u = 0 \; on \; \partial\Omega,$$

are simple.

Proof. Let $\{q_1, \ldots, q_m\}$ be a basic for the eigenfunctions of ${}^T f'(0)$ and let $\{V_1, V_2, \ldots\}$ be a countable basis for the topology of $\mathbb{R}^n$ – for example, all open balls of rational radius whose center has rational coordinates. By the lemma, for each $j \in \{1, 2, \ldots, m\}$ and each $k \in \{1, 2, \ldots\}$, there is an ample (= category II) set of regions Ω such that, for any solution $u \not\equiv 0$, we have $q_j \cdot \partial u/\partial N \neq 0$ at some point of $\partial\Omega \cap V_k$ (unless the intersection is empty). In the countable intersection, which is still ample, any solution $u \not\equiv 0$, satisfies $q_j \cdot \partial u/\partial N \neq 0$ on a dense set of $\partial\Omega$, for all $j \in \{1, \ldots, m\}$. As note above (before the lemma), this implies denseness of the property "only simple solutions."

Example. $(p = 2)$ $f : \mathbb{R}^2 \to \mathbb{R}^2$ is C^4 with

$$f(u) = f'(0)u + \begin{pmatrix} Q_1(u^2) + K_1(u^3) + L_1(u^4) + o(|u|^4) \\ Q_2(u^2) + K_2(u^3) + L_2(u^4) + o(|u|^4) \end{pmatrix}$$

where Q_1, Q_2 are (real homogeneous) quadratic polynomials; K_1, K_2 are cubics; L_1, L_2 are quartics; and we consider the cases

(i) $f'(0)$ has two complex eigenvalues

(ii) $f'(0) = \left(\begin{smallmatrix}\lambda & 0\\ 0 & \mu\end{smallmatrix}\right)$ has real distinct eigenvalues

(iii) $f'(0) = \left(\begin{smallmatrix}\lambda & 1\\ 0 & \lambda\end{smallmatrix}\right)$ has one eigenvalue and only one eigenvector

(iv) $f'(0) = \left(\begin{smallmatrix}\lambda & 0\\ 0 & \lambda\end{smallmatrix}\right)$ has one real eigenvalue with two eigenvactors.

(We may always change the u-variable to put $f'(0)$ in real Jordan canonical form.)

Case (i): Theorem 8.5.1, part (2), applies. Say $f'(0)(p+iq) = (\alpha + i\beta)(p+iq)$ where $\alpha, \beta \in \mathbb{R}$ and $p, q \in \mathbb{R}^2$, $\beta \neq 0$ and $(p, q) \neq (0, 0)$. It is easy to see p, q are independent when $\beta \neq 0$, and verify $\{\gamma, f'(0)\gamma\}$ are independent for any $\gamma \neq 0$ in $\mathbb{R}^2$.

Case (ii): Let $e_1 = \text{col}(1, 0)$, $e_2 = \text{col}(0, 1)$: then assuming

$$0 \neq |Q_1(e_2^2)| + |K_1(e_2^3)| + |L_1(e_2^4)|$$

and

$$0 \neq |Q_1(e_1^2)| + |K_2(e_1^3)| + |L_2(e_1^4)|,$$

the hypotheses of Theorem 8.5.3 hold.

Case (iii): If $\alpha \cdot f'(0)^j \beta = 0$ for $j = 0, 1$, and $\alpha \neq 0$, $\beta \neq 0$ in $\mathbb{R}^2$, then $\alpha_1 = 0$, $\beta_2 = 0$, and we may suppose without loss of generality $\alpha = e_2$, $\beta = e_1$. The hypotheses of Theorem 8.5.3 hold if

$$0 \neq |Q_2(e_1^2)| + |K_2(e_1^3)| + |L_2(e_1^4)|.$$

Case (iv): This case is not included above, but will be permitted in the next theorem provided for some $\gamma \in \mathbb{R}^2$

$$\det \begin{bmatrix} \gamma_1 & Q_1(\gamma^2) \\ \gamma_2 & Q_2(\gamma^2) \end{bmatrix} \neq 0$$

and for all $\gamma \neq 0$ in $\mathbb{R}^2$, the 2×4 matrix

$$[\gamma, Q(\gamma^2), K(\gamma^3), L(\gamma^4)]$$

has rank 2.

Remark. Many cases remain open – for example Case (iv) when f is odd, $f(-u) = -f(u)$, or $f'(0) = \lambda I$ with $n \geq 3$.

Theorem 8.5.4. *Let $f : \mathbb{R}^2 \to \mathbb{R}^2$ be C^4, $f(0) = 0$, $f'(0) = \left(\begin{smallmatrix}\lambda & 1\\ 0 & \lambda\end{smallmatrix}\right)$ for some real λ, and assume*

(i) for some $\gamma \neq 0$ in $\mathbb{R}^2$, $f''(0)\gamma^2$ is not a real multiple of γ, or

$$\det[\gamma, f''(0)\gamma^2] \not\equiv 0;$$

(ii) for every $\gamma \neq 0$ in $\mathbb{R}^2$, the 2×4 matrix

$$[\gamma, f''(0)\gamma^2, f'''(0)\gamma^3, f^{iv}(0)\gamma^4]$$

has rank 2.

Then for most C^2 bounded regions $\Omega \subset \mathbb{R}^2$ $(n \geq 2)$, all solutions u of $\Delta u + f(u) = 0$ in Ω, $u = 0$ on $\partial\Omega$ are simple.

Proof. As noted earlier, we may assume the zero solution is simple. If there is a non-simple solution for every Ω near some Ω_0, then there is a solution (u, ψ) of $(*)_\Omega$ with $u \not\equiv 0$, $\psi \not\equiv 0$ and $\psi_N \cdot u_N = 0$ on $\partial\Omega$. We apply the transversality theorem to conclude that also (in usual notation)

$$\dot{\Psi}_k \cdot u_N + \psi_N \cdot \dot{U}_k = 0 \text{ on } \partial\Omega, \quad \text{for } k \geq 0.$$

The equation for $k = 0$ says nothing new; consider $k = 1$.

$$\begin{aligned} 0 &= -\dot{\Psi}_1 \cdot u_N - \psi_N \cdot \dot{U}_1 \\ &= (\ddot{\Psi}_0 + H\dot{\Psi}_0 + \Delta_{\partial\Omega}\gamma\psi_N + f'(0)\gamma\psi_N) \cdot u_N \\ &\quad + \psi_N \cdot (\ddot{U}_0 + H\dot{U}_0 + \Delta_{\partial\Omega}\gamma u_N + f'(0)\gamma u_N), \end{aligned}$$

which (with $f'(0) = \lambda I$) simplifies to

$$\gamma(\Delta_{\partial\Omega}\psi_N \cdot u_N + \psi_N \cdot \Delta_{\partial\Omega}u_N) = \gamma\left(\partial_\theta^2\psi_N \cdot u_N + \psi_N \cdot \partial_\theta^2 u_N\right).$$

If $n = 2$, then $\Delta_{\partial\Omega} = \partial_\theta^2$ so we have nothing new; but assuming $n \geq 3$, it follows (as above) that, where $\gamma \neq 0$,

$$\begin{aligned} \partial_\theta^2\psi_N \cdot u_N + \psi_N \cdot \partial_\theta^2 u_N &= -2\partial_\theta\psi_N \cdot \partial_\theta u_N \\ &= \frac{1}{n-1}(\psi_N \cdot \Delta_{\partial\Omega}u_N + \Delta_{\partial\Omega}\psi_N \cdot u_N) \end{aligned}$$

so $\partial_\tau\psi_N \cdot \partial_\tau u_N = 0$ for any $\tau \perp N$. For any tangent vector field σ, $D_\sigma =$ (scalar) $\sigma \cdot \nabla_{\partial\Omega}$ + (scalar) is a general first-order operator, and a second-order (tangential) operator is a sum of terms $D_{\sigma_1}D_{\sigma_2} + D_{\sigma_3}$, and we have

$$\psi_N \cdot u_N = 0, \quad D_\sigma \psi_N \cdot u_N + \psi_N \cdot D_\sigma u_N = 0,$$
$$D_{\sigma_1}\psi_N \cdot D_{\sigma_2} u_N + D_{\sigma_2}\psi_N \cdot D_{\sigma_1} u_N = 0$$

and, for any second-order tangential operator D^2,

$$D^2\psi_N \cdot u_N + \psi_N \cdot D^2 u_N = 0.$$

(For example, $\Delta_{\partial\Omega}(\gamma\psi_N) \cdot u_N + \psi_N \cdot \Delta_{\partial\Omega}(\gamma u_N) = 0$.)

This greatly simplifies the calculation for $k = 2$ – in fact, almost everything vanishes and $0 = \psi_N \cdot \dot{U}_2 + \dot{\Psi}_2 \cdot u_N$ reduces to $0 = 2\gamma\psi_N \cdot f''(0)u_N^2$.

In case $n = 2$, the equations for $k = 0, k = 1$, tell us nothing. We must go directly to $\psi_N \cdot \dot{U}_2 + \dot{\Psi}_2 \cdot u_N = 0$, knowing only $\psi_N \cdot u_N = 0$ and $n = 2$. Fortunately, the calculation simplifies for $n = 2$ since $q = H = \kappa$ (the scalar curvature of $\partial\Omega$), θ is arc-length along $\partial\Omega$, $\Delta_{\partial\Omega} = \partial^2/\partial\theta^2$, and $\sigma_0 = -q + H + i\Delta_{\partial\Omega}\theta = 0$. Again everything cancels but one term, though the calculations are more difficult and we conclude, for $n \geq 2$, $\psi_N \cdot f''(0)u_N^2 = 0$ on $\partial\Omega$.

Now let $\{\beta_1 \ldots, \beta_m\}$ be the finite set of $\beta \in \mathbb{R}^2$ with $|\beta| = 1$ and $\det[\beta, f''(0)\beta^2] = 0$. Choose $\gamma_k = \beta_k^\perp \in \mathbb{R}^2$, β_k rotated 90°. By Lemma 8.5.2 and hypothesis (ii), we may suppose any non-zero solution u satisfies $\gamma_k \cdot u_N \neq 0$ on a dense set of $\partial\Omega$ for each $k = 1, \ldots, m$, so $\{u_N, f''(0)u_N^2\}$ is a basis for $\mathbb{R}^n$. But $\psi_N \cdot u_N = 0$ and $\psi_N \cdot f''(0)u_N^2 = 0$ on $\partial\Omega$, so $\psi_N \equiv 0$ on $\partial\Omega$, $\psi \equiv 0$ in Ω, contrary to hypothesis.

Summary of 8.2

(1) $|\nabla_{\partial\Omega}\theta| \equiv 1$, $\nabla S \cdot \nabla S = 0$, $S|\partial\Omega = i\theta$, $\operatorname{Re}\frac{\partial s}{\partial N} > 0$

$$\begin{aligned} S(y + tN(y)) &= \sum_{k\geq 0} \frac{t^k S_k(y)}{k!} \\ &= i\theta(y) + t - \frac{qt^2}{2!} + \frac{t^3}{3!}(-q^2 + 3r + i\partial_\theta q) \\ &\quad + \frac{t^4}{4!}(-i\partial_\theta S_3 + 3_q S_3 - 12s - 6i\nabla_{\partial\Omega q} \cdot K\nabla_{\partial\Omega}\theta) + O(t^5) \end{aligned}$$

where $q = \theta' \cdot K\theta', r = \theta' \cdot K^2\theta', S = \theta' \cdot K^3\theta', \theta' = \nabla_{\partial\Omega}\theta, \partial_\theta = \theta' \cdot \nabla_{\partial\Omega}$.

(2) $\Lambda = \nabla S \cdot \nabla + \frac{1}{2}\sigma, \sigma = \Delta S + b \cdot \nabla S, b(y + tN(y)) = b(y) + t\dot{b}(y) + \frac{t^2}{2!}\ddot{b}(y) + \cdots$

$$\begin{aligned} \sigma(y + tN(y)) &= \sigma_0(y) + t\sigma_1(y) + \frac{t^2}{2!}\sigma_2(y) + \cdots \\ &= -q + H + i\Delta_{\partial\Omega}\theta + b \cdot (N + i\theta') \\ &\quad + t\{S_3 - qH - H_2 + i\partial_\theta H - 2i\operatorname{div}_{\partial\Omega}(K\theta') \end{aligned}$$

$$+\dot{b}\cdot(N+i\theta')+b\cdot(-qN-iK\theta')\}$$
$$+\frac{t^2}{2!}\{S_4+qHS_3+2qH_2+2H_3-i\partial_\theta H_2$$
$$-4i(K\theta')\cdot H'+\ddot{b}\cdot(N+i\theta')+2\dot{b}\cdot(-qN-iK\theta')$$
$$+6i\,\mathrm{div}_{\partial\Omega}(K\theta')-\Delta_{\partial\Omega}q+b\cdot(S_3N-q'+2iK^2\theta')\}$$
$$+O(t^3)$$

$H_m = \text{trace } K^m$, $H = H_1$, $K = DN = \text{curvature}$.

$$\nabla S\cdot\nabla=\partial_t+i\partial_\theta+t(-q\partial_t-2iK\theta'\cdot\partial_y)$$
$$+\frac{t^2}{2}(S_3\partial_t+6iK^2\theta'\cdot\partial_y-q'\cdot\partial_y)+\cdots$$

(3) $\overbrace{(\Delta+b\cdot\nabla+c)}^{=L}e^{\omega S}(U_0+\frac{1}{2\omega}U_1+\frac{1}{(2\omega)^2}U_2+\cdots)=2\omega e^{\omega S}(F_0+\frac{1}{2\omega}F_1+\cdots)$

$$U_j|\partial\Omega=G_j,\qquad \Lambda U_j+LU_{j-1}=F_j\ (U_{-1}=0)$$

$$\Lambda U_0=F_0,\quad U_0(y+tN(y))=G_0(y)+t\dot{U}_0(y)+\frac{t^2}{2!}\ddot{U}_0(y)+\cdots$$

At $t=0$:

$$F_0=\dot{U}_0+i\partial_\theta G_0+{}^1\!/\!_2\sigma_0G_0$$
$$\dot{F}_0=\ddot{U}_0+(i\partial_\theta+{}^1\!/\!_2\sigma_0-q)\dot{U}_0+(-2iK\theta'\cdot\partial_y+{}^1\!/\!_2\sigma_1)G_0$$
$$\ddot{F}_0=\dddot{U}_0+(i\partial_\theta+{}^1\!/\!_2\sigma_0-2q)\ddot{U}_0+(-4iK\theta'\cdot\partial_y+\sigma_1+S_3)\dot{U}_0$$
$$+(6iK^2\theta'\cdot\partial_y-q'\cdot\partial_y+{}^1\!/\!_2\sigma_2)G_0$$

$$\Lambda U_1+LU_0=F_1,\qquad U_1|\partial\Omega=G_1,\qquad U_1=G_1+t\dot{U}_1+\frac{t^2}{2}\ddot{U}_1+\cdots$$

At $t=0$:

$$F_1=\dot{U}_1+(i\partial_\theta+{}^1\!/\!_2\sigma_0)G_1+\ddot{U}_0+(H+b\cdot N)\dot{U}_0$$
$$+\Delta_{\partial\Omega}G_0+b\cdot\nabla_{\partial\Omega}G_0+cG_0$$
$$\dot{F}_1=\ddot{U}_1+(i\partial_\theta+{}^1\!/\!_2\sigma_0-q)\dot{U}_1+(-2iK\theta'\cdot\partial_y+{}^1\!/\!_2\sigma_1)G_1$$
$$+\dddot{U}_0+(H+b\cdot N)\ddot{U}_0+\Delta_{\partial\Omega}\dot{U}_0+b\cdot\nabla_{\partial\Omega}\dot{U}_0+c\dot{U}_0$$

$$+ (\dot{b} \cdot N - H_2)\dot{U}_0 - 2\,\mathrm{div}_{\partial\Omega}(K G_0')$$
$$+ (\dot{b} + H') \cdot G_0' - b \cdot K G_0' + \dot{c} G_0$$

$$\Lambda U_2 + L U_1 = F_1 \qquad U_2|\partial\Omega = G_2$$

At $t = 0$:

$$F_2 = \dot{U}_2 + \left(i\partial_\theta + {}^1\!/_2\sigma_0\right)G_2 + \ddot{U}_1 + (H + b \cdot N)\dot{U}_1 + \Delta_{\partial\Omega} G_1$$
$$+ b \cdot \nabla_{\partial\Omega} G_1 + c G_1$$

Appendix 1

Eigenvalues of the Laplacian in the Presence of Symmetry

A. L. Pereira, in his doctoral thesis investigated more general symmetry conditions than the one discussed in 6.2. We describe some of his results (and sketch some of the arguments) that appear with more details in [28], which is the general reference for this section. We proved (Example 6.1) that, for most C^2-regular bounded regions $\Omega \subset \mathbb{R}^n$, all eigenvalues of

$$\Delta u + \lambda u = 0 \text{ in } \Omega, \qquad u = 0 \text{ in } \Omega$$

are simple. Given this result, it may be surprising to find higher multiplicity in many specific examples. This is well-known in the case of the disc or the rectangle in $\mathbb{R}^2$. Another example is the equilateral triangle, where eigenvalues with arbitrarily high multiplicity exist (see Pinsky [29]). It seems that higher multiplicity appears whenever a detailed analysis can be done. Now, a common characteristic in these cases is the symmetry of the regions. In fact, it is the symmetry that makes a detailed analysis possible. To be more precise, we start with some definitions. Let G be a subgroup of the orthogonal group $O(n)$ on $\mathbb{R}^n$. We say that Ω is *G-symmetric* (or *G-invariant*) if $g(\Omega) = \Omega$ for every $g \in G$. A map $h : \Omega \to \mathbb{R}^n$ is (G-) *equivariant* if $h \circ g = g \circ h$ for all $g \in G$, so $h(\Omega)$ is G-symmetric when Ω is G-symmetric.

A function ϕ on $\mathbb{R}^n$ is *even* or *G-invariant* if $\phi \circ g = \phi$ for all $g \in G$. If Ω is G-symmetric and u is an eigenfunction then $u \circ g^{-1}$ is also an eigenfunction for every $g \in G$, so the multiplicity of λ is at least the dimension of span $\{u \circ g^{-1} \mid g \in G\}$.

In fact, suppose there exists a point $x \in \Omega$ such that $\{g \in G : gx = x\} = \{Id\}$ (such a point is called a free point under the natural action of G in Ω; there is always a free point in Ω if G is, for example, finite or commutative). Then, if some irreducible representation of G has real dimension m, there are infinitely many eigenvalues of the Laplacian with multiplicity $\geq m$ in any G-symmetric bounded open set.

If we want only simple eigenvalues, the irreducible representations of G must have real dimension one, i.e., G is isomorphic to $Z_2 \oplus Z_1 \oplus \cdots \oplus Z_2$ (m times) or G is generated by m commuting reflections. In any other case the symmetry forces higher multiplicity.

Suppose Ω is G-symmetric, λ is an eigenvalue of the Laplacian in Ω with an eigenfunction u; if span $\{u \circ g^{-1} : g \in G\} = \mathcal{N}(\lambda + \Delta)$; i.e., the orbit of some eigenfunction generates all eigenfunctions corresponding to λ, we say that λ is a G-simple eigenvalue. [If we use complex-valued functions, we should also include the G-orbit of the conjugate $\bar{u}$.] Equivalently, λ is G-simple if the natural action of $G(u \mapsto u \circ g^{-1})$ on $\mathcal{N}(\Delta + \lambda)$ is irreducible.

Suppose Ω is a G-symmetric bounded open set in $\mathbb{R}^n$ which is C^2 regular, and λ is a G-simple eigenvalue of multiplicity m. Then, for every Ω, C^2, G-equivariant region $\tilde{\Omega}$ near Ω there is a unique eigenvalue $\lambda(\tilde{\Omega})$ for the Laplacian in $\tilde{\Omega}$ which is close to $\lambda = \lambda(\Omega)$, $\lambda(\tilde{\Omega})$ is also G-simple of multiplicity m, and $\tilde{\Omega} \mapsto \lambda(\tilde{\Omega})$ is analytic.

Thus a G-simple eigenvalue in a G-symmetric region behaves like a simple eigenvalue when no symmetry is imposed. The most we could hope for then is: in generic G-symmetric C^2-domains, all eigenvalues of the Laplacian are G-simple. This is proved in certain cases – for finite commutative groups (argument sketched below) and for finite subgroups of $O(2) \subset O(n)$ (such as the non-commutative dihedral groups) along with partial results for some other cases (see [28]).

The commutative case is comparatively simple, since the irreducible representations all have real or complex dimension one it suffices to use characters. A *character* of G is a map $\chi : G \to \mathbb{C}$ with $|\chi(g)| = 1$, $\chi(Id) = 1$, $\chi(g_1 \bullet g_2) = \chi(g_1) \bullet \chi(g_2)$ for $g, g_1, g_2 \in G$. The character is real if it only has values in $\{-1, 1\}$. If χ is a character, define

$$M\chi = \{u \in L^2(\Omega, \mathbb{C}) \mid u \circ g^{-1} = \chi(g)u, g \in G\}$$

If χ_1, χ_2 are distinct characters, then $M\chi_1 \perp M\chi_2$. In fact, if $u \in M_{\chi_1}$, $v \in M_{\chi_2}$, then

$$\int_\Omega u\bar{v} = \int_\Omega u \circ g^{-1} \bar{v} \circ g^{-1} = \chi_1(g)\overline{\chi_2(g)} \int_\Omega u\bar{v},$$

for all $g \in G$, so if $\int_\Omega u\bar{v} \neq 0$, we must have $\chi_1(g) = \chi_2(g)$ for all $g \in G$, or $\chi_1 = \chi_2$. Since G is commutative $L_2(\Omega, \mathbb{C}) = \bigoplus_\chi M_\chi$ (see [20] p. 51). Further $\Delta(H^2 \cap H_0^1(\Omega, \mathbb{C}) \cap M_\chi) \subset M_\chi$ for every character χ and G-symmetric Ω. It suffices then to prove the following:

(A) For every character χ and integer k, there is a dense open set of G-invariant C^3 regions Ω such that the restriction

$$\Delta \mid H^2 \cap H_0^1(\Omega, \mathbb{C}) \cap M_\chi \to M_\chi$$

has only simple eigenvalues λ with $|\lambda| \leq k$.

(B) For an integer k and every pair of characters with $\chi_1 \neq \chi_2$ and $\chi_1 \neq \bar{\chi}_2$, there is a dense open set of G-invariant C^3 regions Ω such that there is no eigenvalue λ with $|\lambda| \leq k$ with both an eigenfunction in M_{χ_1} and one in M_{χ_2}.

For real characters the argument is like in Example 6.1. In any case, openness is easy to obtain; to obtain density we use transversality arguments.

For (A) with a complex χ, given a C^3, G-invariant $\Omega_0 \subset \mathbb{R}^n$ consider the map

$$(u, \lambda, h) \xrightarrow{F} h^*(\Delta + \lambda)h^{*-1}u \in M_\chi$$

with $u \in H^2 \cap H_0^1(\Omega, \mathbb{C}) \cap M_\chi \backslash \{0\}$, $\lambda \in \mathbb{C}$, $|\lambda| \leq k$, $h \in E^3$ where $E^3 = \{C^3$ G-equivariant imbeddings of $\Omega_0 \subset \mathbb{R}^n\}$. It is enough to show that 0 is a regular value of F. If this is not true, we find that there exists $\lambda \in \mathbb{C}$, $|\lambda| \leq k$ and $u, \psi \in H^2 \cap H_0^1(\Omega, \mathbb{C}) \cap M_\chi \backslash \{0\}$ such that

$$(*) \qquad \begin{cases} \Delta u + \lambda u = 0 \text{ in } \Omega, & \Delta\psi + \lambda\psi = 0 \text{ in } \Omega \\ \int_\Omega \bar{\psi} u = 0, & \operatorname{Re}\left(\frac{\partial\bar{\psi}}{\partial N}\frac{\partial u}{\partial N}\right) = 0 \text{ on } \partial\Omega. \end{cases}$$

(Note that $\psi = iku$, $k \in \mathbb{R}\backslash\{0\}$ satisfies all conditions except that $\int_\Omega \bar{\psi} u \neq 0$).

To see this is not possible for most Ω (i.e., except in a closed set with empty interior), we consider the map

$$(u, \psi, \lambda, h) \xrightarrow{G} \Big(h^*(\Delta + \lambda)h^{*-1}u,\ h^*(\Delta + \lambda)h^{*-1}\psi,$$
$$\int_\Omega \bar{\psi} u,\ h^* B h^{*-1}(u, \psi)|_{\partial\Omega}\Big)$$

where $B(u, v) = \operatorname{Re}\nabla u \bullet \nabla\bar{v}$, $u, \psi \in H^2 \cap H_0^1(\Omega, \mathbb{C}) \cap M_\chi \backslash \{0\}$, $\lambda \in \mathbb{C}$, $|\lambda| \leq k$, $h \in E^3$ and the values are in $M_\chi \times M_\chi \times \mathbb{C} \times L^1(\Omega, \mathbb{C})$. Again, it is easy to see that G^{-1} (0, 0, 0, 0) is closed. If it is (0, 0, 0, 0) for some choice of (u, ψ, λ) for every h in an open set, we may suppose i_{Ω_0} is in its interior and then, condition (2.β) of Theorem 5.4 must fail, so the quotient

$$\frac{\mathcal{R}(DG(u, \psi, \lambda, h))}{\mathcal{R}\left(\frac{\partial G}{\partial(u, \psi, \lambda)}\right)}$$

must be finite dimensional. This implies that for every Ω, $\mathcal{C}^3$-close to Ω_0, there is a solution (u, ψ, λ) of $(*)$ such that

$$\sigma \overset{\Gamma}{\longmapsto} \mathrm{Re}\{u_N \partial_N \mathcal{B}_{\Delta+\lambda}(\sigma \bar{\psi}_N) + \psi_N \partial_N \mathcal{B}_{\Delta+\lambda}(\sigma u_N)\}$$

with σ even, has finite rank. Here $\mathcal{B}_{\Delta+\lambda}$ is given by $v = \mathcal{B}_{\Delta+\lambda} g$, where $\mathcal{B}_{\Delta+\lambda}$ is defined by

$$\begin{cases} (\Delta+\lambda)v \in \mathcal{N}(\Delta+\lambda) \\ v|_{\partial\Omega} = g, \qquad v \perp \mathcal{N}(\Delta+\lambda) \end{cases}$$

The method of rapidly oscillating solutions shows, as $\omega \to \infty$

$$\Gamma(\gamma \cos\omega\theta) = \omega^{-1}\gamma \cos\omega\theta \,\mathrm{Re}(\nabla_{\partial\Omega} u_N \bullet \nabla_{\partial\Omega}\bar{\psi}_N - \partial_\theta u_N \partial_\theta \bar{\psi}_N) + O(\omega^{-2})$$

where γ, θ are even $|\nabla_{\partial\Omega}\theta| = 1$ in supp γ and γ is supported in a small neighborhood of $\{Gx\}$ for some $x \in \partial\Omega$ – recall G is finite. Then $\mathrm{Re}(\partial_\tau u_N \partial_\tau \psi_N) = \mathrm{Re}(\nabla_{\partial\Omega} u_N \bullet \nabla_{\partial\Omega}\bar{\psi}_N)$ for all $\tau \perp N$, $|\tau| = 1$ which implies

$$\mathrm{Re}(\partial_\tau u_N \partial_\tau \bar{\psi}_N) = \frac{1}{n-1}\,\mathrm{Re}(\nabla_{\partial\Omega} u_N \bullet \nabla_{\partial\Omega}\bar{\psi}_N).$$

If $n \geq 3$ we conclude $\mathrm{Re}(\partial_\tau u_N \partial_\tau \bar{u}_N) = 0$ for all $\tau \perp N$. If $n = 2$, more calculation is needed – we only quote the result:

$$\Gamma(\gamma\cos\omega\theta) = -\omega^{-3}\lambda\gamma \mathrm{Re}(\partial_\theta \bar{\psi}_N \partial_\theta u_N) + O(\omega^{-4}).$$

Thus in every case, on $\partial\Omega$

$$\mathrm{Re}(u_N, \bar{\psi}_N) = 0 \quad \text{and} \quad \mathrm{Re}(u_n', \bar{\psi}_N') = 0$$

where $'$ indicates any tangential derivative. Consider a point where $u_N \neq 0$, $\psi_N \neq 0$

$$u_N = q_1 + iq_2 \quad \psi_N = r_1 + ir_2 \quad (q_j, r_j \text{ real}),$$

then $q_1r_1 + q_2r_2 = 0$, $q_1'r_1' + q_2'r_2' = 0$ so, for some α, $r_1 = \alpha q_2$, $r_2 = -\alpha q_1$ and $0 = \alpha'(q_2q_1' - q_1'q_2')$. If $\nabla_{\partial\Omega}\alpha \neq 0$, almost all tangential directions have $\alpha^1 \neq 0$ so q_1/q_2 (or q_2/q_1) is locally constant. It follows, for some $z \in \mathbb{C} - \{0\}$, that u/z is real valued in Ω. But this is impossible in $M_\chi \backslash \{0\}$, when χ is a complex character. If α is locally constant, for some $\alpha \in \mathbb{R}$, $\psi_N = i\alpha u_N$ on a nonempty open set of $\partial\Omega$, so $\psi = -i\alpha u$ in Ω, $0 = \int_\Omega \bar{\psi} \bullet u = i\alpha \int_\Omega |u|^2$. Thus $\alpha = 0$ so $\psi = 0$ which is false.

This contradiction shows, for an open dense set of E^3 $(*)$ is not solvable; restricting F to this set, we obtain (A).

(B) Suppose, for every G, C^3-near Ω_0, there exists u, v, λ with

$$(**)\qquad \begin{cases} u \in M_{\chi_1}, v \in M_{\chi_2}, & u, v \text{ in } H^2 \cap H_0^1(\Omega, \mathbb{C})\backslash\{0\} \quad |\lambda| \leq k \\ \Delta u + \lambda u = 0 \text{ in } \Omega, & \Delta v + \lambda v = 0 \text{ in } \Omega \end{cases}$$

We may assume λ is a simple eigenvalue for the restriction of $-\Delta$ in M_{χ_1} and in M_{χ_2}. For $\tilde{\Omega}$ G-symmetric near Ω there is a unique eigenvalue $\lambda_j(\tilde{\Omega})$ near λ for the restriction of Δ to M_{χ_j} and $\lambda_1(\tilde{\Omega}) = \lambda_2(\tilde{\Omega})$. Thus the derivative of $\lambda_1(\tilde{\Omega}) - \lambda_2(\tilde{\Omega})$ vanishes at $\tilde{\Omega} = \Omega$ so (assuming $\int_\Omega |u|^2 = \int_\Omega |v|^2 = 1$) we also have $|\frac{\partial u}{\partial N}|^2 = |\frac{\partial v}{\partial N}|^2$ on $\partial\Omega$. Thus, for every $h \in E^3$ near i_{Ω_0}, there exists $u, v \in H^2 \cap H_0^1(\Omega, \mathbb{C})\backslash\{0\}$ with $u \in M_{\chi_1}$, $v \in M_{\chi_2}$, and $\lambda \in \mathbb{C}$, with $|\lambda| \leq k$ and

$$(h^*(\Delta + \lambda)h^{*-1}u, h^*(\Delta + \lambda)h^{*-1}v, h^*Ch^{*-1}(u, v)|_{\partial\Omega}) = (0, 0, 0)$$

where $C(u, v) = |\nabla u|^2 - |\nabla v|^2$; the values are in $M_{\chi_1} \times M_{\chi_2} \times L_1(\partial\Omega)$. Theorem 5.4 (2.$\beta$) then implies for every G-symmetric region Ω, C^3-near Ω_0, there is a solution (u, v, λ) of $(**)$ which also satisfies $\int_\Omega |u|^2 = 1 = \int_\Omega |v|^2$, $|u_N|^2 = |v_N|^2$ on $\partial\Omega$ and

$$\sigma \overset{\Theta}{\longmapsto} \mathrm{Re}(\bar{u}_N \partial_N(\mathcal{B}_{\Delta+\lambda}(\sigma u_N)) + \bar{v}_N \partial_N \mathcal{B}_{\Delta+\lambda}(\sigma v_N))$$

has finite rank. This implies (with γ, θ even and $|\nabla_{\partial\Omega}\theta| = 1$ is supp γ)

$$\begin{aligned} \Theta(\gamma \cos\omega\theta) = 2\omega^{-1}\gamma\cos\omega\theta(|\nabla_{\partial\Omega}u_N|^2 - |\nabla_{\partial\Omega}v_N|^2 \\ - |\partial_\theta \dot{u}_N|^2 + |\partial_\theta v_N|^2) + O(\omega^{-2}) \end{aligned}$$

If $n = 2$, we have

$$\Theta(\gamma\cos\omega\theta) = -\omega^{-3}\frac{\lambda}{2}\cos\omega\theta(|\partial_\theta u_N|^2 - |\partial_\theta v_N|^2) + O(\omega^{-4})$$

Then – since G is finite – we find $|\partial_\tau u_N| = |\partial_\tau v_N|$ on $\partial\Omega$ for all $\tau \perp N$.

Let $u_N = \mathrm{Re}^{i\rho}$, $v_N = \mathrm{Re}^{i\sigma}$, then

$$|\partial_\tau u_N| = |R_\tau + iR\rho_\tau| = |R_\tau + iR\sigma_\tau|$$

so $\rho_\tau^2 = \sigma_\tau^2$, $\partial_\tau(\rho - \sigma)\partial_\tau(\rho + \sigma) = 0$ when $\tau \perp N$. Either $\rho - \sigma$ (and u_N/v_N) is locally constant or $\rho + \sigma$ (and $u_N/\bar{v}_N$) is locally constant, thus either u, v are linearly dependent (so $\chi_1 = \chi_2$) or u, $\bar{v}$ are linearly dependent (so $\chi_1 = \bar{\chi}_2$). Therefore we have a contradiction and (B) follows.

With (A) and (B) we conclude: for some G-symmetric C^2 regions $\Omega \subset \mathbb{R}^n (n \geq 2)$, all eigenvalues are G-simple: simple eigenvalues when the eigenfunction corresponds to a real character, double when it corresponds to a complex character with independent eigenfunctions u, $\bar{u}$ (or Re, u, Im u).

Appendix 2

On Micheletti's Metric Space

Anna Maria Micheletti [22] proved the set $\mathcal{M}_m\,(\Omega)(1 \leq m < \infty)$ of regions in $\mathbb{R}^n$ which are C^m-diffeomorphic to a given set $\Omega \subset \mathbb{R}^n$ is metrizable using the "Courant distance" d_m: for $\Omega_1, \Omega_2 \in \mathcal{M}_m(\Omega)$,

$$d_m(\Omega_1, \Omega_2) = \inf\left\{\sum_{k=1}^{N} ||f_k - Id||_{C^m} + \left\|f_k^{-1} - Id\right\|_{C^m} \mid f_1 \circ f_2 \circ \cdots \circ f_N(\Omega_1) = \Omega_2\right\}$$

where we consider all $N \geq 1$ and all C^m diffeomorphisms $f_k : \mathbb{R}^n \to \mathbb{R}^n$ such that $D^j(f_k(x) - x) \to 0$ as $|x| \to \infty$ for $0 \leq j \leq m$. Further, the metric space $(\mathcal{M}_m(\Omega), d_m)$ is complete and separable. (In fact, we only consider regions diffeomorphic to Ω by diffeomorphisms which are near the identity at infinity. Diffeomorphic images of Ω could be "knotted" in different ways; we ignore such complications, so $\mathcal{M}_m(\Omega)$ does not usually include *all* diffeomorphs of Ω.) In the metric topology, a neighborhood of $\tilde{\Omega} \in \mathcal{M}_m(\Omega)$ contains a neighborhood of the form $\{F(\tilde{\Omega}) \mid F : \mathbb{R}^n \to \mathbb{R}^n$ a C^m map with $||F - Id||_{C^m} < \epsilon$ and $D^j(F(x) - x) \to 0$ as $|x| \to \infty, 0 \leq j \leq m\}$ for small $\epsilon > 0$; and such a neighborhood contains a ball $\{\Omega' : d_m(\Omega', \tilde{\Omega}) < \delta\}$ for some $\delta > 0$. These results are proved below (A.9), in the manner of Micheletti (who considered $m = 3$).

If Ω is bounded and C^m-regular, $\mathcal{M}_m(\Omega)$ may be considered a C^0 Banach manifold, modeled on the Banach space $C^m(\partial\Omega, \mathbb{R})$ (see (A.10)). Given a C^m vector field $V : \mathbb{R}^n \to \mathbb{R}^n$ transverse to $\partial\Omega$ and (small) $\sigma \in C^m(\partial\Omega, \mathbb{R})$ we define $\Omega(\sigma)$ near Ω by

$$\partial\Omega(\sigma) = \{x + \sigma(x)V(x) : x \in \partial\Omega\},$$

which defines a homeomorphism $\sigma \mapsto \Omega(\sigma)$ from $C^m(\partial\Omega, \mathbb{R})$ [near zero] onto a neighborhood of Ω in $\mathcal{M}_m(\Omega)$; various such choices of coordinates are C^0-compatible.

This coordinate system has the advantage of uniqueness – given V – but is otherwise difficult to use. We will usually employ the set of C^m imbeddings $h : \Omega \to \mathbb{R}^n$ with $h(\Omega) \in \mathcal{M}_m(\Omega)$, so the restriction $h : \Omega \to h(\Omega) \subset \mathbb{R}^n$ is a diffeomorphism. This is an open set $\text{Diff}^m(\Omega)$ in the Banach space $C^m(\Omega, \mathbb{R}^n)$ and naturally inherits the structure of an analytic Banach manifold. The map $h \mapsto h(\Omega) : \text{Diff}^m(\Omega) \to \mathcal{M}_m(\Omega)$ is continuous, surjective and open, with a local right-inverse near any point of $\mathcal{M}_m(\Omega)$. [We use the above (C^0) coordinate system to define the right inverse; it is assumed Ω is bounded and C^m-regular.] We call "σ-closed" a countable union of closed sets, also termed an "F_σ".

In our genericity problems, we consider sets $F \subset \text{Diff}^m(\Omega)$ defined by properties of the image $F(\Omega) = \{h(\Omega) : h \in F\} \subset \mathcal{M}_m(\Omega)$, so $F = \{h : h(\Omega) \in F(\Omega)\}$. Assuming this, F is closed [or σ-closed, or meager] in $\text{Diff}^m(\Omega)$ if and only if $F(\Omega)$ is closed [or σ-closed, or meager] in $\mathcal{M}_m(\Omega)$ (see (A.12)). More precisely, when F or $F(\Omega)$ is closed, the codimension of F in $\text{Diff}^m(\Omega)$ equals the codimension of $F(\Omega)$ in $\mathcal{M}_m(\Omega)$. ("Codimension" is defined in 5.16; positive codimension is equivalent to meagerness.)

We begin with the formula of Faá de Bruno: for $k \geq 2$, if f, g are C^k,

$$
\begin{aligned}
(f \circ g)^{(k)}(x)y^k &= \sum_{\substack{1 \leq j = |\alpha| \leq k \\ |\alpha'| = k}} \frac{k!}{\alpha!} f^{(j)}(g(x))(g'(x)y)^{\alpha_1} \\
&\quad \left(\frac{1}{2!}g''(x)y^2\right)^{\alpha_2} \cdots \left(\frac{1}{k!}g^{(k)}(x)y^k\right)^{\alpha_k} \qquad \text{(A.1)} \\
&= f^{(k)}(g(x))(g'(x)y)^k + f'(g(x))g^{(k)}(x)y^k \\
&\quad + \sum_{\substack{2 \leq j = |\alpha| \leq k-1 \\ |\alpha'| = k \\ \alpha_i = 0 \text{ for } i \geq k}} \cdots
\end{aligned}
$$

where $\alpha = (\alpha_1, \alpha_2, \alpha_3, \ldots)$, $\alpha' = (\alpha_1,\ 2\alpha_2,\ 3\alpha_3, \ldots, j\alpha_j, \ldots)$, the α_j are integers ≥ 0. $|\alpha| = \alpha_1 + \alpha_2 + \alpha_3 + \ldots$, $\alpha! = \alpha_1!\alpha_2!\alpha_3!\ldots$, and the second equality of (A.1) isolates the terms with a highest-order (k^{th}) derivative.

It is enough to prove this when f, g are polynomials, and then it becomes the multinomial formula:

$$
g(x+y) = g(x) + \gamma_1 + \gamma_2 + \ldots, \qquad \gamma_j = \frac{g^{(j)}(x)y^j}{j!}
$$

and

$$
\begin{aligned}
f(g(x+y)) &= \sum_{k \geq 0} \frac{1}{k!}(f \circ g)^{(k)}(x)y^k = f(g(x) + \gamma_1 + \gamma_2 + \ldots) \\
&= \sum \frac{1}{j!} f^{(j)}(g(x))(\gamma_1 + \gamma_2 + \ldots)^j \\
&= \sum_{|\alpha| = j \geq 0} \frac{1}{\alpha!} f^{(j)}(g(x))\gamma_1^{\alpha_1}\gamma_2^{\alpha_2} \ldots .
\end{aligned}
$$

Counting the powers of y in $\gamma_1^{\alpha_1}\gamma_2^{\alpha_2}, \ldots$ gives $\alpha_1 + 2\alpha_2 + 3\alpha_3 + \ldots = |\alpha^j|$.

Define $|\psi^{(j)}(x)| = \sup_{|y|\leq 1}|\psi^{(j)}(x)y^j|$, $||\psi^{(j)}|| = \sup_x|\psi^{(j)}(x)|$, and $[\psi]_{k,t} = \sum_{j=2}^{k}\frac{t^j}{j!}||\psi^{(j)}||$ for $k \geq 2$, $[\psi]_{k,t} = 0$ for $k < 2$.

Lemma A.2. *For f, g of class C^m, we have*

$$||f \circ g|| \leq ||f||, \qquad ||(f \circ g)'|| \leq ||f'||||g'||$$

and if $2 \leq k \leq m, t > 0$,

$$[f \circ g]_{k,t} \leq [f]_{k,t||g'||} + ||f'||[g]_{k,t} + [f]_{k-1,eT_{k-1}(g)}$$

where $T_{k-1}(g) = t||g'|| + [g]_{k-1,t}$.

For $2 \leq k \leq m$, from (A.1)

$$[f \circ g]_{k,t} = \sum_{j=2}^{k}\frac{t^j}{j!}\left\|(f \circ g)^{(j)}\right\| \leq [f]_{k,t||g'||} + ||f'||[g]_{k,t}$$
$$+ \sum_{\substack{2\leq|\alpha|=j\leq k-1 \\ 2\leq|\alpha'|\leq k}} \frac{e^j}{\alpha!}\left\|f^{(j)}\right\|\prod_{i=1}^{k-1}\left(\frac{t^i}{i!}\left\|g^{(i)}\right\|\right)^{\alpha_i}$$

and the final sum is $\sum_{j=2}^{k-1}\frac{e^j}{j!}||f^{(j)}||(\Sigma_1^{k-1}\frac{t^i}{i!}||g^{(i)}||)^j$. We used the fact about symmetric j-linear maps, such as $f^{(j)}(x)$, that

$$\frac{\left|f^{(j)}(x)h_1h_2\ldots h_j\right|}{(|h_1||h_2|\ldots|h_j|)} \leq e^j\left|f^{(j)}(x)\right|, \qquad h_i \neq 0,$$

proved in a stronger form (with $j^j/j!$ in place of e^j) by the polarization identity in, for example, H. Cartan, *Differential Calculus*.

Now we generalize the fundamental lemma of Micheletti [22].

Lemma A.3. *Let each $\phi_j : \mathbb{R}^n \to \mathbb{R}^n$ be C^m-bounded and write $||\phi||_{C^m} = \Sigma_0^m\frac{1}{k!}||\phi^{(k)}||$, $||\phi_1^{(k)}|| + ||\phi_2^{(k)}|| + \ldots + ||\phi_N^{(k)}|| \leq \sigma_k$ $(0 \leq k \leq m)$; then*

$$||\,(Id + \phi_N) \circ (Id + \phi_{N-1}) \circ \ldots \circ (Id + \phi_1) - Id||_{C^m}$$
$$\leq \sum_{k=0}^{m}\frac{\sigma_k}{k!}P_m(e^{\sigma_1}, \sigma_2, \ldots, \sigma_{m-1})$$

for a certain polynomial P_m with positive coefficients, dependent only on m. Observe that the number of factors N is not mentioned in the estimate.

Proof. Let $\Phi_0 = Id$, $\Phi_j = (Id + \phi_j) \circ \ldots \circ (Id + \phi_1) = (Id + \phi_j) \circ \Phi_{j-1}$ for $j \geq 1$, so

$$\begin{aligned}||\Phi_j - Id|| &= ||\Phi_{j-1} - Id + \phi_j \circ \Phi_{j-1}|| \\ &\leq ||\Phi_{j-1} - Id|| + ||\phi_j|| \leq \sum_1^j ||\phi_i|| \leq \sigma_0.\end{aligned}$$

Also

$$||\Phi'_j|| \leq (1 + ||\phi'_j||)||\Phi'_{j-1}|| \leq \prod_1^j (1 + ||\phi'_i||) \leq e^{\sigma_1}$$

$$||\Phi'_j - Id|| \leq -1 + \prod_1^j (1 + ||\phi'_j||) \leq e^{\sigma_1} - 1 \leq \sigma_1 e^{\sigma_1}.$$

From $m \geq 2, t > 0$

$$\begin{aligned}[\Phi_j]_{m,t} &= [\Phi_j - Id]_{m,t} \leq [\Phi_{j-1}]_{m,t} + [\phi_j \circ \Phi_{j-1}]_{m,t} \\ &\leq (1 + ||\phi'_j||)[\Phi_{j-1}]_{m,t} + [\phi_j]_{m,te^{\sigma_1}} + [\phi_j]_{m-1,eT_{m-1}(t,\Phi_{j-1})}\end{aligned}$$

so

$$[\Phi_k]_{m,t} \leq \sum_{j=1}^k \prod_{j+1}^k (1 + ||\phi'_i||)([\phi_j]_{m,te^{\sigma_1}} + [\phi_j]_{m-1,eT_{m-1}(\Phi_{j-1})}$$

Define the polynomials $Q_k : \mathbb{R}^{k-1} \to \mathbb{R}$ by: $Q_1 \equiv 1$,

$$Q_k(\tau_2, \ldots, \tau_k) = 1 + \sum_2^k \frac{\tau_i}{i!} + \sum_2^{k-1} \left(\frac{e^i \tau_i}{i!} \right) (Q_{k-1}(\tau_2 \ldots \tau_{k-1}))^i.$$

Each $Q_k(0) = 1$ and Q_k is a polynomial with coefficients ≥ 0. By induction on $m \geq 2$, we have with $z = te^{\sigma_1}$, $j \geq 1$.

$$\begin{aligned}T_{m-1}(t, \Phi_{j-1}) &= \sum_{i=1}^{m-1} \frac{t^i}{i!} \| \Phi_{j-1}^{(i)} \| \\ &\leq z Q_{m-1}(z e^{\sigma_1} \sigma_2, z^2 e^{\sigma_1} \sigma_3, \ldots, z^{m-2} e^{\sigma_1} \sigma_{m-1}).\end{aligned}$$

(For the induction, let $\tau_k = e^{\sigma_1}(te^{\sigma_1})^{k-1}\sigma_k$ in the expression for Q_m.) Then taking $t = 1, k \geq 0$,

$$\| \Phi_k - Id \|_{C^m} \leq \sigma_0 + \sigma_1 e^{\sigma_1} + e^{\sigma_1}(Q_m(\tau_2, \ldots, \tau_m) - 1)) \quad \text{with } \tau_j = e^{j\sigma_1}\sigma_j.$$

We have $Q_m(\tau_2, \ldots, \tau_m) - 1 = \sum_{i=2}^m \tau_i Q_{m,i}(\tau_2, \ldots, \tau_m)$ for polynomials $Q_{m,i}$ with positive coefficients and thus obtain the form claimed.

Let $C_0^m = \{f : \mathbb{R}^n \to \mathbb{R}^n$ of class C^m with $f^{(k)}(x) \to 0$ as $|x| \to \infty, 0 \leq k \leq m\}$ which is a Banach space in the norm $\|f\|_{C^m}$. Let $\mathrm{Diff}^m(\mathbb{R}^n)$ be the translate of an open set in C_0^m, with the induced topology,

$$\mathrm{Diff}^m(\mathbb{R}^n) = \left\{F : \mathbb{R}^n \to \mathbb{R}^n \text{ diffeomorphism with } F - Id \in C_0^m\right\}. \quad \text{(A.4)}$$

It is easy to show that $F, G \in \mathrm{Diff}^m(\mathbb{R}^n)$ imply that $F \circ G$ and F^{-1} are in $\mathrm{Diff}^m(\mathbb{R}^n)$. Further, since each $F \in \mathrm{Diff}^m(\mathbb{R}^n)$ has $x \to D^j F(x) (0 \leq j \leq m)$ uniformly continuous, $(F, G) \mapsto F \circ G$ and $F \mapsto F^{-1}$ are continuous (in the induced topology): $\mathrm{Diff}^m(\mathbb{R}^n)$ is a topological group.

Define $\delta_m : \mathrm{Diff}^m(\mathbb{R}^n) \to \mathbb{R}$ by

$$\begin{aligned}\delta_m(F) = \inf \Bigg\{ &\sum_{k=1}^{N} \|f_k - Id\|_{C^m} + \left\|f_k^{-1} - Id\right\|_{C^m} : N \geq 1, \\ &f_k \in \mathrm{Diff}^m \text{ and } F = f_N \circ f_{N-1} \circ \ldots \circ f_1 \Bigg\}\end{aligned} \quad \text{(A.5)}$$

Lemma A.6. *$\delta_m(F) = \delta_m(F^{-1}) \geq 0$, $\delta_m(F \circ G) \leq \delta_m(F) + \delta_m(G)$, and for $F, G \in \mathrm{Diff}^m(\mathbb{R}^n)$ and there exist $r_1 > 0$, $r_2 > 0$, $K > 1$ so that $\|F - Id\|_{C^m} \leq r_1$ implies $\delta_m(F) \leq r_2$ and $\delta_m(F) \leq r_2$ implies*

$$K^{-1} \leq \frac{\delta_m(F)}{\|F - Id\|_{C^m}} \leq K.$$

In particular, $\delta_m(F) = 0$ only when $F = Id$.

Proof. Let $F \in \mathrm{Diff}^m (m \geq 1)$ and define $G = F^{-1}$; if $y = F(x)$, $G'(y) = F'(x)^{-1}$ and for $2 \leq k \leq m$, the kth derivative of $G(F(x) = x$ gives

$$\begin{aligned}0 = &G^{(k)}(y)h^k + F'^{-1}(x)F^{(k)}(x)(F'^{-1}(x)h)^k \\ &+ \sum_{\substack{2 \leq j = |\alpha| \leq k-1 \\ |\alpha'| = k}} \frac{k!}{\alpha!} G^{(j)}(y) \prod_{i=1}^{k-1} \left(\frac{1}{i!} F^{(i)}(x)(F'^{-1}(x)h)^i\right)^{\alpha_i}\end{aligned}$$

(This shows $G - Id \in C_0^m$ so $F^{-1} = G \in \mathrm{Diff}^m$ and also shows $F \mapsto F^{-1}$ is continuous.) It follows, for an increasing polynomial $P_m(a, b)$, that

$$\|F^{-1} - Id\|_{C^m} \leq \|F - Id\|_{C^m} \cdot P_m(\|(F')^{-1}\|, \|F - Id\|_{C^m})$$

so $\delta_m(F) \le \|F - Id\|_{C^m} + \|F^{-1} - Id\|_{C^m} \le K\|F - Id\|_{C^m}$ for $\|F - Id\|_{C^m}$ small (so, for example, $|F'(x) - I| \le 1/2$, $|F'^{-1}(x)| \le 2$) with $K > 1 + P_m(2, 0)$.

If $F = f_N \circ f_{N-1} \circ \ldots \circ f_1$ with all $f_j \in \mathrm{Dif}^m(\mathbb{R}^n)$ and $S = \sum_{k=1}^N \|f_k - Id\|_{C^m}$ we have, by Lemma A.3, $\|F - Id\|_{C^m} \le S \cdot Q_m(S)$ for an increasing polynomial Q_m. We may choose the N and f_j so $S \le 2\delta_m(F)$ and then

$$\|F - Id\|_{C^m} \le 2\delta_m(F) \cdot Q_m(2\delta_m(F))$$

to obtain the inequality $K^{-1} \le \frac{\delta_m(F)}{\|F - Id\|_{C^m}} \le K$ for F near Id.

For $F, G \in \mathrm{Diff}^m(\mathbb{R}^n)$ define

$$d_m(F, G) = \delta_m(F \circ G^{-1}) \tag{A.7}$$

with δ_m defined in (A.5) above.

Lemma. *d_m is a distance (metric) on* $\mathrm{Diff}^m(\mathbb{R}^n)$*, giving the induced topology from $Id + C_0^m$ and in this metric,* $(\mathrm{Diff}^m(\mathbb{R}^n), d_m)$ *is separable and complete. (This is also a topological group under composition $(F, G) \mapsto F \circ G$.)*

Proof. From (A.6) it is clear we have a metric which is topologically equivalent to the (translated-norm) metric of $C_0^m + Id$. Since C_0^m is separable, so is $\mathrm{Diff}^m(\mathbb{R}^n)$. Let $\{F_k\}_{k\ge1}$ be a Cauchy sequence in $(\mathrm{Diff}^m(\mathbb{R}^n), d_m)$; it is enough to show some subsequence converges, or equally, to treat the case

$$d_m(F_{k+1}, F_k) = \delta_m\left(F_{k+1} \circ F_k^{-1}\right) \le r \cdot 2^{-k} \quad (k = 1, 2, \ldots)$$

for some fixed $r > 0$. We choose $r = \frac{1}{2}\min\{r_2, 1/4k\}$ where r_2, k are the constants of (A.6). Since $\delta_m(F_k \circ F_j^{-1}) \le 2r \cdot 2^{-j} < r_2$ when $k \ge j \ge 1$, we have $\|F_k \circ F_J^{-1} - Id\|_{C^m} \le 2Kr \cdot 2^{-j}$.

Now $\|F_{k+1} - F_k\|_{C^m} = \|F_{k+1} \circ F_k^{-1} - Id) \circ (F_k \circ F_1^{-1}) \circ F_1\|_{C^m} \le O(2^{-k})$ for all $k \ge 1$, so $\{F_k - Id\}_{k\ge1}$ is a Cauchy sequence in C_0^m, with limit $f = \lim_{k\to\infty} F_k - Id$. Similarly, $\|F_k^{-1} - Id\|_{C^m}$ is bounded and $\|F_k \circ F_{k+1}^{-1} - Id\|_{C^m} < Kr \cdot 2^{-k}$ so $\|F_k \circ F_j^{-1} - Id\|_{C^m} \le 2kr \cdot 2^{-k}$ for $j \ge k \ge 1$

$$\left\|F_{k+1}^{-1} - F_k^{-1}\right\|_{C^0} = \left\|F_k^{-1} \circ \left(F_k \circ F_{k+1}^{-1}\right) - F_k^{-1} \circ Id\right\|_{C^0} \le O(2^{-k})$$

since $m \ge 1$, so $F_k^{-1} - Id \to g$ uniformly (in C^0) and $(Id + f) \circ (Id + g) = Id = (Id + g) \circ (Id + f)$. It follows that $Id + f$ is a bijection as well as locally C^m diffeomorphism (implicit function theorem) so $Id + f \in \mathrm{Diff}^m(\mathbb{R}^n)$. Since $F_k \to Id + f$ as $k \to \infty$, also $F_k^{-1} \to (Id + f)^{-1}$ in Diff^m as $k \to \infty$.

Now define

$$\mathcal{M}_m(\Omega) = \{F(\Omega) : F \in \mathrm{Diff}^m(\mathbb{R}^n)\}$$

for any fixed open set $\Omega \subset \mathbb{R}^n$. For Ω_1, Ω_2 in $\mathcal{M}_m(\Omega)$, define

$$\begin{aligned} d_m(\Omega_1, \Omega_2) &= \inf\{\delta_m(F) : F \in \mathrm{Diff}^m(\mathbb{R}^n),\ F(\Omega_1) = \Omega_2\} \\ &= \inf\Bigg\{ \sum_{j=1}^{N} \|f_j - Id\|_{C^m} + \left\|f_j^{-1} - Id\right\|_{C^m} : N \geq 1, \\ &\qquad f_j \in \mathrm{Diff}^m(\mathbb{R}^n) \quad \text{and } f_n \circ \ldots \circ f_1(\Omega_1) = \Omega_2 \Bigg\} \end{aligned} \tag{A.8}$$

Theorem A.9. *$(\mathcal{M}_m(\Omega), d_m)$ is a complete separable metric space.*

Proof. Since $\mathrm{Diff}^m(\mathbb{R}^n)$ is separable, so is $\mathcal{M}_m(\Omega)$. Let $\{\Omega_j\}_{j\geq 1}$ be a sequence in $\mathcal{M}_m(\Omega)$ with $d_m(\Omega_{j+1}, \Omega_j) < 2^{-j}$. There exists $F_j \in \mathrm{Diff}^m(\mathbb{R}^n)$ with $F_j(\Omega_j) = \Omega_{j+1}; \delta_m(F_j) < 2^{-j}$. Let $G_j = F_j \circ F_{j-1} \circ \ldots \circ F_1$ so $G_j \in \mathrm{Diff}^m(\mathbb{R}^n)$, $G_j(\Omega_1) = \Omega_j$ and for $k \geq j$,

$$d_m(G_k, G_j) = \delta_m\left(G_k \circ G_j^{-1}\right) \leq \delta_m(F_k) + \ldots + \delta_m(F_{j+1}) < 2^{-j}.$$

Since $\{G_k\}_{k\geq 1}$ is a Cauchy sequence in $\mathrm{Diff}^m(\mathbb{R}^n)$, there exists $G \in \mathrm{Diff}^m(\mathbb{R}^n)$ with $d_m(G_j, G) \to 0$. Therefore $G(\Omega_1) \in \mathcal{M}_m(\Omega)$ and $d_m(\Omega_j, G(\Omega_1)) = d_m(G_j(\Omega_1), G(\Omega_1)) \leq \delta_m(G_j \circ G_1^{-1}) \to 0$ as $j \to +\infty$, or $\Omega_j \to G(\Omega_1)$ in $\mathcal{M}_m(\Omega)$.

Remark. Micheletti defines the closed subgroup $Z_m(\Omega) \subset \mathrm{Diff}^m(\mathbb{R}^n)$ by $Z_m(\Omega) = \{F \in \mathrm{Diff}^m(\mathbb{R}^m) : F(\Omega) = \Omega\}$ and obtains $\mathcal{M}_m(\Omega)$ as the quotient space $\mathrm{Diff}^m(\Omega)/Z_m(\Omega)$ with the quotient norm

$$d_m(G_1(\Omega), G_2(\Omega)) = \inf\{d_m(G_1 \circ H_1, G_2 \circ H_2) : H_j \in Z_m(\Omega)\}.$$

This is isometric with our definition.

Theorem A.10. *Suppose Ω is a bounded C^m-regular region in $\mathbb{R}^m (1 \leq m < \infty)$; then $\mathcal{M}_m(\Omega)$ is a C^0 Banach manifold modeled on $C^m(\partial\Omega, \mathbb{R})$. (A coordinate system was mentioned in the introduction; here we give details).*

Proof. First we define local coordinates on a neighborhood of Ω. (If $\tilde{\Omega} \in \mathcal{M}_m(\Omega)$, then $\mathcal{M}_m(\Omega) = \mathcal{M}_m(\tilde{\Omega})$; so coordinates are similarly defined near any point of $\mathcal{M}_m(\Omega)$.)

Let $V : \mathbb{R}^n \to \mathbb{R}^n$ be a C^m vector field transverse to $\partial\Omega$, i.e., $V \cdot N \neq 0$ on $\partial\Omega$ where N is the outward unit normal. (V need only be defined near $\partial\Omega$.) By the inverse function theorem, for small $r > 0$,

$$(x, t) \mapsto x + tV(x) : \partial\Omega \times (-r, r) \to \mathbb{R}^n$$

is a C^m imbedding of $\partial\Omega \times (-r, r)$ onto a neighborhood U of $\partial\Omega$, and there is a C^m inverse $y \mapsto (\pi(y), \tau(y)) : U \to \partial\Omega \times (-r, r)$, $y = x + tV(x)$ if and only if $x = \pi(y)$, $t = \tau(y)$ for $y \in U$, $x \in \partial\Omega$, $-r < t < r$.

For any $\tilde{\Omega}$ in a small C^1-neighborhood of Ω $[d_1(\tilde{\Omega}, \Omega) < \epsilon]$ we have $\partial\tilde{\Omega} \subset U$ and V transverse to $\partial\tilde{\Omega}$ so $\pi \mid \partial\tilde{\Omega} : \partial\tilde{\Omega} \to \partial\Omega$ is a diffeomorphism and $\partial\tilde{\Omega} = \{\pi(y) + \tau(y)V(\pi(y)) : y \in \partial\tilde{\Omega}\} = \{x + \sigma(x, \tilde{\Omega})V(x) : x \in \partial\Omega\}$ where $\sigma(\cdot, \tilde{\Omega}) = \tau \circ (\pi|\partial\tilde{\Omega})^{-1}$. We prove the continuity of $\tilde{\Omega} \to \sigma(x, \tilde{\Omega})$ later.

Choose C^∞ $\theta : \mathbb{R}^n \to [0, 1]$ with $\theta \equiv 1$ near $\partial\Omega$, $\theta \equiv 0$ outside U and define $H(\sigma) : \mathbb{R}^n \to \mathbb{R}^n$ by

$$H(\sigma)(x) = \begin{cases} x + \theta(x)\sigma(\pi(x))V(x), & x \in U \\ x, & x \notin U \end{cases}$$

for $\sigma \in C^m(\partial\Omega, \mathbb{R})$. $H(0) = Id$ and $H(\sigma) \in \text{Diff}^m(\mathbb{R}^n)$ for small $\|\sigma\|_{C^m}$, with $\sigma \mapsto H(\sigma) : C^m(\partial\Omega, \mathbb{R})$ [near zero] $\to \text{Diff}^m(\mathbb{R}^n)$ continuous. Since $(\sigma, x) \mapsto H(\sigma)(x)$ is C^m, by smoothness of the evaluation map. We define

$$\Omega(\sigma) = H(\sigma)(\Omega) \in \mathcal{M}_m(\Omega),$$

and $\sigma \mapsto \Omega(\sigma) \in \mathcal{M}_m(\Omega)$ is continuous; we show it is a homeomorphism of $C^m(\partial\Omega, \mathbb{R})$ [near zero] onto a neighborhood of Ω.

The inverse map is $\tilde{\Omega} \mapsto \sigma(\cdot, \tilde{\Omega})$ mentioned above:

$$\Omega(\sigma(\cdot, \tilde{\Omega})) = \tilde{\Omega} \quad \text{and} \quad \sigma(\cdot, \Omega(\tau)) = \tau.$$

For $\tilde{\Omega}$ near Ω, $\tilde{\Omega} = F(\Omega)$ for $F \in \text{Diff}^m(\mathbb{R}^n)$ near Id, say $\delta_n(F) \le 2d_m(\tilde{\Omega}, \Omega)$. Now $\pi \circ F|\partial\Omega : \partial\Omega \to \partial\Omega$ is a C^m diffeomorphism, depending continuously on F (near the identity) and

$$\sigma(\cdot, F(\Omega)) = \tau \circ F \circ (\pi \circ F \mid \partial\Omega)^{-1} \in C^m(\partial\Omega, \mathbb{R})$$

also depends continuously on F. If $\tilde{\Omega}_\nu \to \tilde{\Omega}$ in $\mathcal{M}_m(\Omega)$, we may choose $F_\nu \in \text{Diff}^m(\mathbb{R}^n)$ with $F_\nu \to F$ in $\text{Diff}^m(\mathbb{R}^n)$, $\tilde{\Omega}_\nu = F_\nu(\Omega)$, $\tilde{\Omega} = F(\Omega)$, so $\sigma(\cdot, \tilde{\Omega}_n) \to \sigma(\cdot, \tilde{\Omega})$ in $C^m(\partial\Omega, \mathbb{R})$.

We note, for later use, that $h \mapsto h(\Omega)$ $[h \in C^m(\Omega, \mathbb{R}^n)$ near the inclusion $i_\Omega : \Omega \subset \mathbb{R}^n]$ has a continuous local right-inverse p. In fact, let $p(\Omega(\sigma)) = H(\sigma)|\Omega$; then $p(\Omega(\sigma))(\Omega) = H(\sigma)(\Omega) = \Omega(\sigma)$ for small $\sigma \in C^m(\partial\Omega, \mathbb{R})$. Such a local right-inverse exists near each point of $\mathcal{M}_m(\Omega)$, and shows $h \mapsto h(\Omega)$ is an open map (restricted to C^m imbeddings h).

$$\begin{aligned} \text{Diff}^m(\Omega) &= \{h : \Omega \to \mathbb{R}^n \ C^m \quad \text{imbedding with } h(\Omega) \in \mathcal{M}_m(\Omega)\} \\ &= \{H_{|\Omega} : H \in \text{Diff}^m(\mathbb{R}^n)\}, \end{aligned} \tag{A.11}$$

which is an open set in $C^m(\Omega, \mathbb{R}^n)$ if Ω is bounded, with the induced metric and Banach manifold structures. As noted above

$$h \mapsto h(\Omega) : \mathrm{Diff}^m(\Omega) \to \mathcal{M}_m(\Omega)$$

is continuous, surjective and an open map with a local right-inverse near each point provided Ω is bounded and C^m-regular.

Theorem A.12. *Let $\Omega \subset \mathbb{R}^n$ be a bounded open set with $\partial\Omega$ C^m-regular, and let $F \subset \mathrm{Diff}^m(\Omega)$ be a set defined by its image $F(\Omega) = \{h(\Omega) : h \in F\}$, i.e., $F = \{h \in \mathrm{Diff}^m(\Omega) : h(\Omega) \in F(\Omega)\}$. Then F is closed [or σ-closed, or meager] in $\mathrm{Diff}^m(\Omega)$ if and only if $F(\Omega)$ is closed [or σ-closed, or meager] in $\mathcal{M}_m(\Omega)$. More precisely, if F and $F(\Omega)$ are σ-closed,*

$$\operatorname{codim} F \ (\textit{in } \mathrm{Diff}^m(\Omega)) = \operatorname{codim} F(\Omega) \ (\textit{in } \mathcal{M}_m(\Omega)).$$

Remark. "Codimension" is defined in 5.15, and positive codimension is equivalent to meagerness.

Proof. According to 5.16 (3),

$$\operatorname{codim}\{F_1 \cup F_2 \cup \ldots\} = \min\{\operatorname{codim} F_j : j \geq 1\}$$

so it suffices to treat closed sets F. For the same reason, we may suppose $F(\Omega)$ has small diameter. In fact, for each $\tilde{\Omega} \in \mathcal{M}_m(\Omega)$, there exists $r > 0$ and a local right-inverse $p : B_r(\tilde{\Omega}) \to \mathrm{Diff}^m(\Omega)[p(\Omega')(\Omega) = \Omega'$ for $d_m(\Omega', \tilde{\Omega}) < r]$. The sets $\{B_{r/2}(\tilde{\Omega}) : \tilde{\Omega} \in \mathcal{M}_m(\Omega)\}$ cover $\mathcal{M}_m(\Omega)$ which is separable, so a countable subcover suffices, and it is enough to treat each $F(\Omega) \cap \bar{B}_{r/2}(\tilde{\Omega})$. We thus suppose F is closed, $F(\Omega) \subset \bar{B}_{r/2}(\tilde{\Omega})$ and there is a right-inverse p defined on $B_r(\tilde{\Omega})$. (Since $h \mapsto h(\Omega)$ is continuous and open, F is closed if and only if $F(\Omega)$ is closed.)

Suppose codim $F > j$, so on an open dense set of $V \in C(I^j, \mathrm{Diff}^m(\Omega))$ we have $\gamma(I^j) \cap F = \emptyset$. Given continuous $\psi : I^j \to \mathcal{M}_m(\Omega)$ we show – by arbitrarily small changes in ψ – we can achieve $\psi(I^j) \cap F(\Omega) = \emptyset$. This will prove codim $F(\Omega) \geq$ codim F.

If $\psi(I^j) \subset B_r(\tilde{\Omega})$, this is easy: $p \circ \psi \in C(I^j, \mathrm{Diff}^m(\Omega))$ and there exists γ near $p \circ \psi$ with $\gamma(I^j) \cap F = \emptyset$, so $\gamma(t)(\Omega)$ is near $p(\psi(t))(\Omega) = \psi(t)$ for $t \in I^j$ and $\gamma(I^j)(\Omega) \cap F(\Omega) = \emptyset$.

In general, divide I^j in small congruent cubes $\{C_i\}$ with disjoint interiors, with diameter so small that diam $\psi(C_i) < r/2$. For each i, either $\psi(C_i)$ is outside $\bar{B}_{r/2}(\tilde{\Omega})$ (hence, disjoint from $F(\Omega)$), or inside $B_r(\tilde{\Omega})$ when the previous argument applies. In modifying $\psi \mid C_i$ to avoid $F(\Omega)$, we make sufficiently small changes so we don't upset previous changes in other sub-cubes – which is possible since $F(\Omega)$ is closed. Thus codim $F(\Omega) \geq$ codim F in every case.

If codim $F = \infty$, codim $F(\Omega) = \infty$. If codim $F = 0$, then F has interior so also $F(\Omega)$ has interior, codim $F(\Omega) = 0$. Suppose k is a positive integer and codim $F = k$; we show codim $F(\Omega) \leq k$, i.e., there exists $\psi_0 \in C(I^k, F(\Omega))$ so that for any ψ near ψ_0, $\psi(t) \in F(\Omega)$ for some $t \in I^k$.

We know codim $F = k$, so there exists $\gamma_0 \in C(I^k, F)$ so that any nearby γ has $\gamma(I^k) \cap F \neq \emptyset$, and we may suppose $\gamma_0(I^k) \cap F = \emptyset$ [dim $\partial I^k < k$].

If $\gamma_0(I^k)(\Omega) \subset B_{2r/3}(\tilde{\Omega})$, this is again simple. Define $\tilde{\gamma}_0(t) = p(\gamma_0(t)(\Omega))$, $t \in I^k$; we have $\tilde{\gamma}_0(t)(\Omega) = \gamma_0(t)(\Omega)$ so there is a (unique) map $H(t): \Omega \to \Omega$ with $\tilde{\gamma}_0(t) \circ H(t) = \gamma_0(t)$, and (by the implicit function theorem) it is a C^m diffeomorphism with $t \mapsto H(t) \in C^m$ continuous. Then γ near $\tilde{\gamma}_0$ in $C(I^k, \mathrm{Diff}^m(\Omega))$ implies $\gamma \circ H$ is near $\tilde{\gamma}_0 \circ H = \gamma_0$ so $\gamma(t) \circ H(t) \in F$ and $\gamma(t) \in F$ for some $t \in I^k$. Choose $\psi \in C(I^k, \mathcal{M}_m(\Omega))$ near $\psi_0 = \{t \mapsto \tilde{\gamma}_0(t)(\Omega)\}$; then $\psi(I^k) \subset B_r(\tilde{\Omega})$ so $p \circ \psi(t)$ is well-defined and near $p(\tilde{\gamma}_0(t)(\Omega)) = \tilde{\gamma}_0(t)$ for all $t \in I^k$, so $p \circ \psi(t) \in F$ for some t, hence $\psi(t) \in F(\Omega)$.

In general, we may choose a small k-dimensional cube $C \subset I^k$ so that $\gamma_0(C)(\Omega) \subset B_{2r/3}(\tilde{\Omega})$, and any $\gamma \mid C$ near $\gamma_0 \mid C$ has $\gamma(C) \cap F \neq \emptyset$, so the above argument applies and the proof is complete.

Estimates of $\prod_{n=1}^{\infty}(1 + z/2^n)$ and Extension Operator

Working with the extension operator E (for a half-space) of Thm. 1.9, we sometimes want estimates of $\sum_{k\geq 1} |a_k||b_k|^p$ or $\sum_{k\geq 1} |a_k| 2^{kp}$ ($b_k = 1 - 2^k$) for large p. Let $f(z) = \prod_1^{\infty}(1 + z/2^n)$ so $A(z) = zf(-z)/f(-i) = \sum_1^{\infty} a_k z^k$ and $\max_{|z|=r} |A(z)| = rf(r)/f(-1) = \sum_1^{\infty} |a_k| r^k$.

(1) $f(z) = \sum_{1\leq n_1<n_2<\ldots<n_k} \frac{z^k}{2^{n_1+n_2+\ldots+n_k}} = 1 + z + \frac{z^2}{3} + \frac{z^3}{3\cdot 7} + \ldots$
$+ \frac{z^k}{(3\cdot 7\cdot (2^k-1)} + \ldots$ so $a_1 = -a_2 = 1/f(-1)$,
$a_k = \frac{(-1)^{k-1}}{(f(-1)\cdot 3\cdot 7\cdot 15\cdot\ldots\cdot(2^{k-1}-1))}$ for $k \geq 3$ and $\frac{1}{f(-1)} \approx 3.462746619$.

(2) Let $\phi(z) = f(z)f(1/z)\exp(\frac{-(\log z)^2}{2\log 2})\cdot(\sqrt{z} + 1/\sqrt{z})$ for $z > 0$, we have $\phi(1/z) = \phi(z) = \phi(2z)$ for all $z > 0$ so $\phi(z)$ is bounded from zero and infinity on $0 < z < \infty$; it is enough to know $\phi(z)$ in $1 \leq z \leq \sqrt{2}$. Numerically, it appears that ϕ is increasing in $(1, \sqrt{2})$ and is constant

to twelve figures:

$$\max \phi = \phi(\sqrt{2}) = 11.369\ 11519\ 96115$$
$$\min \phi = \phi(1) = 11.369\ 11519\ 95920.$$

This gives us the behavior of $f(z)$ as $z \to +\infty$. Then for $p \geq 0$

$$\sum_{k \geq 1} |a_k| 2^{kp} \leq 40 \cdot 2^{(P^2+p)/2}.$$

References

1. R. Abraham and J. Robbin, *Transversal Mappings and Flows*, W. A. Benjamin (1967).
2. S. Agmon, *Unicité et Convexité dans les Problèmes Différentielles*, Univ. De Montreal (19??)
3. I. Babuška, On the stability of domains with respect to the fundamental problems of the theory of PDEs, especially in relation to those of the theory of elasticity, I *Czech. Math. J.* **11** (1961), 76–105 and 165–203 (in Russian).
4. A. P. Calderón, *Lecture Notes on Pseudo-Differential Operators and Elliptic Boundary Value Problems, I*, Instituto Argentino de Matematica, Buenos Aires (1976).
5. R. Courant and D. Hilbert, *Methods of Mathematical Physics*, Vol. 1 (1953), vol. 2 (1962), Interscience.
6. P. Garabedian, *Partial Differential Equations*, J. Wiley (1964).
7. P. Garabedian and M. Schiffer, Convexity of domain functionals, *J. Analyse Math.* **2** (1952–53), 281–368.
8. D. Gilbarg and N. Trudinger, *Elliptic PDEs of Second Order*, Springer Verlag (Grundlehren 244) (1977).
9. J. Hadamard (1908), Mémoire sur le problème d'analyse relatif à l'équilibre des plaques élastiques encastrées, *Ouvres de J. Hadamard* **2** ed. C.N.R.S. Paris (1968).
10. J. Hale and J. Vegas, A nonlinear parabolic equation with varying domain, *Arch. Rat. Mech. Anal.* **86** (1984), 99–124.
11. D. B. Henry, *Topics in Nonlinear Analysis*, Trabalho de Matemática 192, Março, 1982, Univ. Brasilia.
12. L. Hörmander, *Linear Partial Differential Operators*, Springer-Verlag (Grundlehren 116) (1964).
13. D. Joseph, Parameter and domain dependence, *Arch. Rat. Mech. Anal.* **24** (1967).
14. T. Kato, *Perturbation Theory of Linear Operators*, Springer-Verlag (Grundlehren 132) (1966).
15. O. D. Kellog, *Foundations of Potential Theory*, Dover (1953).
16. I. Kupka, Counterexample to the Morse-Sard theorem in the case of infinite dimensional manifolds, *Proc. Am. Math. Soc.* **16** (1965), p. 954.
17. S. Lang, *Introduction to Differentiable Manifolds*, Wiley (1962).

18. P. Levy, Sur l'allure des fonctions de Green et de Neumann dans le voisinage du contour, *Acta Math.* **42** (1920), 207–267.
19. J. Lindenstrauss and L. Tzafiri, On the complemented subspaces problem, *Israel J. Math.* **9** (1971), 263–269.
20. G. MacKey, *Unitary Representations in Physics, Probability and Number Theory*, The Benjamin/Cummings Publishing Company, Inc. (1978).
21. A. M. Micheletti, Perturbazione dello spettro dell'operatore di Laplace in relazione ad una variazione del campo, *Annali della Scuola Norm. Sup. Pisa* Ser. II V. 26 (1972), 151–169.
22. A. M. Micheletti, Metrica per famiglie di domini limitati e proprietà generiche degli autovalori, *Annali della Scuola Norm. Sup. Pisa* Ser. II, v. 26 (19??), 683–694.
23. A. M. Micheletti, Perturbazione dello spetro di un operatore ellitico di tipo variazionale, in relazione ad una variazione del campo, *Annali di Matem. Pura ed Applic.* Ser. IV, v. 97 (1973), 267–281.
24. F. Mignon, F. Murat et J. Puel, Variation d'un point de retournement par rapport au domaine, in Springer Lecture Notes 782 (1980): *Bifurcation and Nonlinear Eigenvalue Problems*, 222–254.
25. E. H. Moore, On certain crinkly curves, *Trans. Amer. Math. Soc.* **1** (1900), 72–90 (esp. p. 77).
26. J. Nečas, *Les Méthodes Directes en Théorie des Équations Élliptiques*, Masson, et Cie (1967).
27. J. Peetre, On Hadamard's variational formula, *J. Diff. Eq.* **36** (1980), 335–346.
28. A. Pereira, Eigenvalues of the Laplacian on symmetric regions, *NoDEA – Nonlinear Differential Equations and Applications* **2** (1995) 63–109.
29. M. A. Pinsky, The eigenvalues of an equilateral triangle, *SIAM J. Math. Anal.* **11** (1980), 819–327.
30. G. Polya, Torsional rigidity, principal frequency, electrostatic capacity and symmetrization, *Q. Appl. Math.* **6** (1948), 267–277.
31. G. Polya and G. Szëgo, *Isoperimetric Inequalities in Mathematical Physics*, Ann. Math. Studies 27, Princeton U. Press (1951).
32. F. Quinn, Transversal approximation on Banach manifolds, *Proc. Symp. Pure Math.* **15**, Amer. Math. Soc. (1970).
33. J. W. S. Rayleigh, *Theory of Sound*, Dover (1945) (second edition of 1894).
34. J. C. Saut and R. Temam, Generic properties of nonlinear boundary-value problems, *Commun. in PDEs* (1979), 293–319.
35. R. Seeley, Extensions of C^∞ functions defined in a halfspace, *Proc. Am. Math. Soc.* **15** (1964), 625–626.
36. J. Serrin, A symmetry problem of potential theory, *Arch. Rat. Mech. Anal.* **43** (1971), 304–318.
37. S. Smale, An infinite-dimensional version of Sard's theorem, *Amer. J. Math.* **87** (1969), 861–866.
38. E. Stein, *Singular Integrals and Differentiability Properties of Functions*, Princeton U. Press (1970).
39. S. Sternberg, *Lectures on Differential Geometry*, Prentice-Hall (1964).
40. F. Treves, *Introduction to Pseudo-Differential Operators and Fourier Integral Operators*, v. 1 and 2, Plenum Press (1980).

41. K. Uhlenbeck, Eigenfunctions of Laplace operators, *Bull. Am. Math. Soc.* **78** (1972), 1073–1076.
42. K. Uhlenbeck, Generic properties of eigenfunctions, *Amer. J. Math.* **98** (1976), 1059–1078.
43. H. Whitney, A function not constant on a connected set of critical points, *Duke Math. J.* **1** (1935), 514–517.

Index

For EU product safety concerns, contact us at Calle de José Abascal, 56–1°, 28003 Madrid, Spain or eugpsr@cambridge.org.

www.ingramcontent.com/pod-product-compliance
Lightning Source LLC
LaVergne TN
LVHW090358210926
848480LV00002B/379
* 9 7 8 0 5 2 1 5 7 4 9 1 4 *